Biological Spectroscopy

Two wc
loэ

D1420307

Biological Spectroscopy

Iain D. Campbell
and
Raymond A. Dwek

University of Oxford

The Benjamin/Cummings Publishing Company, Inc.
Advanced Book Program

Menlo Park, California • Reading, Massachusetts
London • Amsterdam • Don Mills, Ontario • Sydney

Sponsoring Editor   Paul Elias
Production   Jan deProsse, Pat Burner
Copy Edit   Barbara Liguori
Cover Design   Robin Gold
Book Design   Richard Kharibian

Library of Congress Cataloging in Publication Data

Campbell, Iain D.
    Biological spectroscopy.

    (Biophysical techniques series)
    Bibliography: p.
    Includes index.
    1. Spectrum analysis.   2. Biology—Technique.
I. Dwek, Raymond A.   II. Title.   III. Series.
QH324.9.S6C35   1984   574'.028   84-6232
ISBN 0-8053-1847-X
ISBN 0-8053-1849-6

CDEFGHIJ-HA-8987

The Benjamin/Cummings Publishing Company, Inc.

2727 Sand Hill Road
Menlo Park, California 94025

Preface

The molecules that make up the living cell continue to excite great interest. The methods available for investigating these molecules have improved dramatically in the last decade. Some of the most powerful new techniques involve the application of electromagnetic radiation. These techniques include nuclear magnetic resonance and visible spectra, which detect transitions between energy levels, and others such as microscopy, scattering, and diffraction. In this book the term *spectroscopy* is used to cover all these methods. The close relationships among these topics are rarely emphasized in textbooks; the reading level is often either too superficial or too detailed for most readers. This book tries to bring out the recurring concepts in this field and to steer a course between the specialist text and the introductory.

Our overall aim is to lead undergraduates, graduate students, and other readers to an appreciation of the current literature. Many people find the concepts involved in spectroscopy rather difficult, so our approach here is to use worked examples and problems to emphasize important points and to illustrate biological applications rather than complex mathematical derivations. Most chapters in this text deal with one particular technique and can thus be read independently of other chapters, but we have also included a central chapter on concepts and definitions and various appendices designed to give more information on difficult points.

We owe thanks to many people who have given advice and read chapters of the manuscript. Among these we should mention D. Ashford, P. Atkins, J. Boyd, S. Easterbrook-Smith, Z. Luz, M. Moody, R. Parekh, J. Paton, S. Perkins, R. Perutz, T. W. Rademacher, W. G. Richards, J. Seelig, R. J. Simpson, B. Sutton, and A. Watts. We also thank D. Kozlow, S. Hoeft, and F. Caddick for their skillful work on the diagrams.

IAIN D. CAMPBELL
RAYMOND A. DWEK

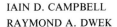

Contents

3 Infrared Spectroscopy   37

5 Fluorescence

6 Nuclear Magnetic Resonance   127

8 Scattering   217

9 Raman Scattering

10 Optical Activity

11 Microscopy   279

12 Diffraction   299

13 Other Spectroscopic Methods   333

Appendixes

Biological Spectroscopy

CHAPTER 1

Introduction

Most objects we see in everyday life are visible only because they reemit part of the light that falls on them from some source, such as the sun. Interpretation of this reflected or transmitted light can yield a wealth of information not only about the color and shape of an object but also about the atomic and molecular mechanisms involved when light interacts with the object.

Light is, in fact, a form of **electromagnetic radiation** and this book is concerned with the study of the interaction of electromagnetic waves with matter and how this can be used to extract information about biological molecules and cells.

Electromagnetic radiation covers an enormous range of wavelengths (energies, frequencies). The two extremes in this range are usually taken to be radio waves, with wavelengths around 10^{-1} m, and gamma rays, with wavelengths around 10^{-11} m. Visible light covers a very small range, $4-7 \times 10^{-7}$ m.

When light interacts with an object, we can normally see only reflected or transmitted radiation. Three phenomena that occur when electromagnetic radiation interacts with matter can be defined more precisely as **scattering**, (e.g., the sky is blue because fluctuating particles in the atmosphere scatter blue light more than red light); **absorption**, (e.g., red light absorbed by a piece of glass causes the transmitted light to appear blue); and **emission**, (e.g., a fluorescent dye may emit green light after absorbing blue light). Another result of the interaction of electromagnetic radiation with matter is **photochemistry.** This is obviously extremely important in biology (such as in vision and photosynthesis) but this aspect is not dealt with here.

We thus define **spectroscopy** as the study of the interaction of electromagnetic radiation with matter, excluding chemical effects. (For the purposes of this book, neutrons and electrons are considered to give rise to electromagnetic radiation, although this is not strictly correct; see Chapter 2.)

All the techniques described in this book involve (1) irradiation of a sample with some form of electromagnetic radiation; (2) measurement of the scattering, absorption, or emission in terms of some measured parameters (e.g., scattering intensity at some angle θ, extinction coefficient at a particular wavelength, or

1

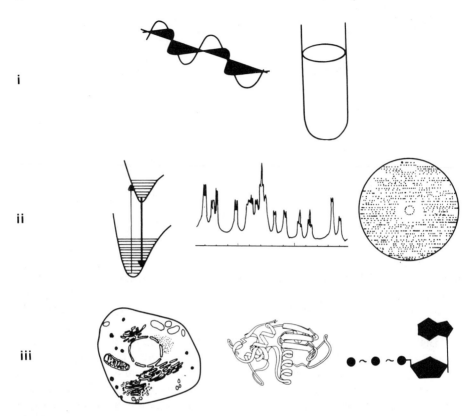

i

ii

iii

Figure 1.1   Spectroscopy involves (i) irradiation of the sample with some form of electromagnetic radiation. This results in scattering, absorption, and emission. In (ii) are shown the basic process of fluorescence (*emission*), a section of an NMR (*absorption*) spectrum of a protein, and a diffraction pattern from a crystal of a macromolecule, which arises from *scattering*. (iii) The interpretation of measurements of absorption, emission, and scattering leads to biological information on a wide range of systems, from cells to small molecules.

fluorescent lifetime); and (3) the interpretation of these measured parameters to give useful biological information. This last stage requires some understanding of the physical basis of the interaction, whether it is scattering by electrons or nuclei, absorption by excitation to a higher vibrational level, or emission from a triplet state. Aspects of these three stages of spectroscopy are illustrated in Figure 1.1.

The book is arranged as a series of chapters on different techniques, which measure different phenomena, except for Chapter 2, which deals with concepts and definitions. Each chapter can be read almost independently, but it will sometimes be necessary and helpful to use Chapter 2 and the Appendixes.

THE INFORMATION AVAILABLE
FROM SPECTROSCOPY

Detailed study of scattering, absorption, and emission yields biological information of various kinds. This information can be broadly classified as structural, dynamic, energetic, and analytical. The information available depends on the instrument used to make the measurements. While the eye is exceptionally powerful and versatile, instruments such as the microscope or the spectrometer can enhance and quantify the information discernible. Each technique has different advantages and disadvantages, both experimentally and in the interpretation of the measurements. Some spectroscopic techniques are rated (using asterisks) in Table 1.1 on their ability to give the different kinds of information.

The best techniques for determining structure or the coordinates of a biological system are microscopy and diffraction. Light microscopy is a technique that can give structural information directly and noninvasively about living systems, but the resolution that can be achieved (~ 1 μm) is not sufficient to study individual molecules. Electron microscopy can achieve higher resolution (~ 2 nm), but the sample must be studied in a vacuum and is normally covered with a metallic stain, which causes a problem with regard to the integrity of the structure determined. Diffraction studies of crystals of pure macromolecules can give structural information to the atomic level (~ 0.15 nm), but this technique requires crystals, and the structure is no longer obtained directly but must be interpreted from an observed diffraction pattern.

Some techniques that can be used to study macromolecular structure in solution include optical activity measurements, which give information on secondary structure of proteins; fluorescence and nuclear magnetic resonance (NMR), which can give information about interactions between pairs of centers; electron paramagnetic resonance (EPR), and resonance Raman, which can "fingerprint" certain types of structure around a metal ion or chromophore; and solution scattering, which can give information about the overall shape of a molecule in solution. The structural information available from these methods is nearly always equivocal, although it is often very useful and important because it can be obtained in solution and can be combined with other information on energetics and dynamics.

Dynamic information about biological systems can be obtained in a variety of ways. The best methods are those that give information in solution. Examples are dynamic light scattering studies of chemotaxis by bacteria, fluorescence depolarization studies of the rotational diffusion of macromolecules, EPR spin-label studies of lipid fluidity, and studies of the movement of fluorescent labels with the fluorescence microscope.

Information about energetics can be obtained by studying the influence of environment, such as temperature, ligand concentration, pH, and ionic strength of the system. The techniques of UV/visible spectroscopy, fluorescence, optical activity, and NMR are all good for distinguishing between bound and unbound forms of a ligand or a macromolecule, different ionization states, and different structural forms of a macromolecule.

Table 1.1 Information available from various spectroscopic techniques

Technique	Chapter	Structure		Dynamics		Energetics	
IR	3	Fingerprint	*	H-D exchange	*	Ionization states	*
UV/visible	4	Qualitative, DNA conformation	*	Follow reactions	*	Ligand binding	*
Fluorescence	5	Pairwise 2–5 nm (solution)	**	Molecular motion ~10^{-9} s	***	Environmental probe, pH	**
NMR	6	Pairwise 0.2–1 nm (solution)	***	10^{-3}–10^{-9} s	***	Ionization states, ligand binding	**
EPR	7	Fingerprint around one center	*	Diffusion of spin label	**	Environmental probe	*
Scattering	8	Overall shape (solution)	**	Net movement of cells and molecules	**	Association/dissociation	*
Raman	9	Fingerprint	*	H-D exchange	*	Ionization states	*
Optical Activity	10	Secondary structure (solution)	**	Conformational changes	*	Ligand binding	*
Microscopy	11	~2 nm (solid); 1 μm (solution)	****	Photobleaching recovery	*		
Diffraction	12	0.2 nm (crystals)	****	Temperature factors	*	Energy calculations from coordinates, "strain"	*

The best methods to obtain analytical information, by which we mean the identification of a particular compound and the determination of its concentration, are UV/visible spectroscopy, NMR, and atomic absorption spectroscopy (AAS). It should also be noted that different methods operate best in different concentration ranges. NMR is best for studying changes in the millimolar range, while flourescence is used to study much lower concentrations ($\sim 1 \ \mu M$).

This very brief description of the information available from spectroscopy is intended to indicate the wide variety of information that can be obtained. It is also intended to emphasize the complementarity of the information obtained from the different methods. A detailed study of a biological system will involve the application of several different techniques.

CHAPTER 2

Spectroscopy: The Study of the Interaction of Electromagnetic Radiation with Matter

WHAT IS ELECTROMAGNETIC RADIATION?

Electromagnetic radiation can be considered to behave as two **wave motions** (see Appendix XIII) at right angles to each other (Figure 2.1). One of these waves is magnetic (M) and the other is electric (E). Electromagnetic waves are generated by oscillating electric or magnetic dipoles and are propagated through a vacuum at the velocity of light (c). The energies associated with E and M are equal, but most optical effects are concerned with the electric wave.*

Polarization. Since the E- and the M-components are always perpendicular to each other, it is sufficient, in many cases, to consider only the E-component in de-

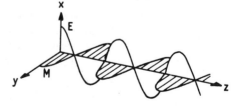

Figure 2.1 Electromagnetic radiation is made up of two wave motions perpendicular to each other. One is a magnetic (M) wave, the other an electric (E) wave. The waves are propagated along the z-direction.

---

*Orders of magnitude: A light beam of cross section 1 m^2, generated by a 100 W bulb, has an E-field of about 300 V•m^{-1} and a B-field of about 10^{-6} T. The earth's magnetic field is about 5×10^{-5} T. A coil of radius 10 cm and 100 turns of wire carrying 1.5 A generates a field of 10^{-3} T at its center.

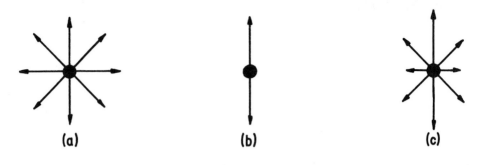

Figure 2.2   Directions of the electric vector in polarized and unpolarized light. In unpolarized light (a), or partly polarized light (c), the oscillations take place at all angles perpendicular to the direction of travel; in polarized light (b) they are restricted to one angle.

scribing the wave. Although the amplitude of the E-wave shown in Figure 2.1 oscillates in the zx-plane, it could oscillate equally well in *any* direction perpendicular to the direction of propagation (z). Unpolarized light contains oscillations of the E-components in *all* directions perpendicular to the direction of propagation. **Plane,** or **linearly, polarized** radiation has E-oscillations in only one plane.

It is convenient to introduce a parameter called the **degree of polarization** (P) to describe situations where the radiation is partially polarized (Figure 2.2).

$$P = \frac{(I_\parallel - I_\perp)}{(I_\parallel + I_\perp)}$$

where $I_\parallel$ and $I_\perp$ are the intensities of the radiation resolved in directions parallel and perpendicular to the direction of partial polarization.

Plane-polarized radiation of the sort pictured in Figure 2.2(b) can be considered to arise from a source that oscillates parallel to the x-axis. If the source also oscillates parallel to the y-axis in the same phase, then the two waves superimpose (see Appendix III) to produce another plane-polarized wave. For example, if the two waves have the same amplitude, the new wave will be oriented at 45° to the x-axis. When the oscillations are not in the same phase, then the superposition of the two oscillations does not lead to a fixed direction for E. For example, if the phase difference is $\pi/2$, as shown in Figure 2.3, then the path of the E-vector is helical. When the amplitudes of the two waves are equal and if the phase difference is $\pi/2$, then the resultant wave is said to be **circularly polarized.** If the two components of oscillation along x and y are unequal in amplitude, then the resultant wave is **elliptically polarized.**

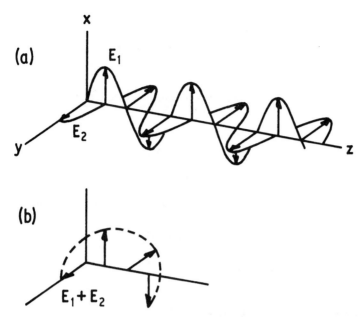

Figure 2.3 Illustration of (a) how two beams polarized along xz and yz, 90°phase shifted with respect to each other, generate a circularly polarized beam (b) when they are superimposed.

Frequency, Wavelength, Energy, and Wavenumber

The *frequency* (v) and *wavelength* (λ) of a wave are related by the equation

$$v = c/\lambda$$

where c is the velocity of propagation of the wave. For electromagnetic radiation in a vacuum, $c = 3 \times 10^8$ m·s^{-1}. Frequency can also be converted directly to units of energy using the relationship $E = hv$, where h is Planck's constant ($h = 6.63 \times 10^{-34}$ J·s). Units of energy (J·mol^{-1}), frequency (Hz), and wavelength (m), are all used in discussions of electromagnetic radiation.

Expression of the radiation as a frequency (Hz) gives results with very large numbers; therefore it is common to find, particularly for electromagnetic radiation in the microwave to X-ray range, the frequency expressed as a wavenumber (cm^{-1}). The **wavenumber** (v') is defined as the inverse of the wavelength in centimeters.

$$v' = \frac{1}{\lambda} = \frac{v}{c}$$

The wavenumber is thus the number of waves per centimeter.

Worked Example 2.1   Wavenumber to Energy

Calculate the conversion factor between wavenumber and joules per mole.

Solution

From $E = h\nu$ and $\nu' = \nu/c$, we have that $E = h\nu'c$.
Putting

$$\nu' = 1 \text{ cm}^{-1},$$
$$c = 3 \times 10^{10} \text{ cm} \cdot \text{s}^{-1}$$
$$h = 6.63 \times 10^{-34} \text{ J} \cdot \text{s}$$
$$E = 6.63 \times 10^{-34} \times 3 \times 10^{10} \text{ J}$$

This is the answer for one molecule. For one *mole,* we multiply by Avogadro's number, $\mathfrak{N} = 6.03 \times 10^{+23}$.

$$E = 6.03 \times 10^{23} \times 6.63 \times 10^{-34} \times 1 \times 3 \times 10^{10} \text{ J} \cdot \text{mol}^{-1}$$
$$= 12 \text{ J} \cdot \text{mol}^{-1}$$
$$1 \text{ cm}^{-1} = 12 \text{ J} \cdot \text{mol}^{-1}$$

Types of Electromagnetic Radiation

Table 2.1 summarizes some of the properties of electromagnetic radiation and Table 2.2 some of the associated energy scales. The names of the different types—X-rays, visible, radio frequency, and so forth—have largely arisen from the origin of the radiation and are somewhat arbitrary. There is no sharp transition between these names or between properties of the radiation.

The inclusion of neutrons and electrons in these tables is not strictly correct, since these particles have a finite rest mass (m), while electromagnetic radiation has zero rest mass. Particles of matter can, however, also be described by waves (see next section); hence the term **wave-particle duality.** The wavelength and the momentum (p) of a particle are related by the equations $\lambda = h/p$, where $p = (E^2 - mc^4)^{1/2}/c$ for a particle of mass m, and $\lambda = E/c$ for electromagnetic radiation. In the context of this book it is convenient to consider neutron waves and electron waves together with electromagnetic radiation.

WHAT IS MATTER?

The matter we are concerned with is the living cell, which consists of a wide variety of molecules in different environments. To develop some of the concepts important for this book, it is convenient to consider briefly some results of **wave mechanics,** which, in principle, is the most powerful model known for describing matter and is summarized by the Schrödinger wave equation (see Appendix IV). This equation can be solved to give the distribution of a particle in space and time and its energy. However, rigorous application of this equation to even an isolated molecule of reasonable size is a formidable task and the direct application to

Table 2.1 Properties of electromagnetic radiation

Name	Generated by	Dispersed by	Detected by	Particular Properties and Uses
(Electrons)	Acceleration of thermally produced electrons by a high voltage	(Electric and magnetic fields)	Photography; fluorescence	Deflected by electric and magnetic fields; used in electron microscope
(Neutrons)	Nuclear reactor		Photography (difficult); counting devices	Scattered by nuclei; diffracted by crystals
X-ray	Rapid deceleration of fast-moving electrons (e.g., by a tungsten target); changes in energy of innermost orbital electrons		Photography; ionization chamber	Can penetrate matter (e.g., radiography); reflected and diffracted by crystals; scattered by electrons; EXAFS
Ultraviolet	Electronic transitions of atoms and molecules (e.g., the sun and the mercury vapor lamp)	Quartz, fluorite	Photography; photoelectric cell; fluorescence	Absorbed by glass and conjugated molecules; can cause many chemical reactions (e.g. the tanning of the human skin); UV spectra
Visible light	Rearrangement of outer orbital electrons in atoms and molecules (e.g., gas discharge tube, laser)	Glass	Eye; photography; photocell	Can cause chemical action (e.g., photosynthesis); microscopy; visible spectra
Infrared	Change of molecular rotational and vibrational energies (e.g., incandescent matter)	Rock salt	Photography by special plate; special heating effect (e.g., radiometer, bolometer)	
Microwaves	Special electronic devices such as klystron tube; electron spin reorientation in a magnetic field	Paraffin wax	Valve circuit arranged as microwave receiver; point-contact diodes	Radar communication; ESR measurements
Radio waves	Oscillating electrons in special circuits coupled to radio aerials; nuclear-spin reorientation in a magnetic field		Tuned oscillatory electric circuit (e.g., radio receiver)	Radio communications; NMR measurements

Table 2.2   Electromagnetic radiation and the scales used

Name and (Approx.) Range of Radiation	Energy per Photon (J)	Frequency ν (Hz)	Wavelength λ (m)	Common Units of Length for Comparison
(Electrons)	10^{-11}	10^{22}	10^{-13}	
	10^{-12}	10^{21}	10^{-12}	picometer (pm)
γ-rays	10^{-13}	10^{20}	10^{-11}	
	10^{-14}	10^{19}	10^{-10}	
X-rays	10^{-15}	10^{18}	10^{-9}	nanometer (nm)
(Neutrons)	10^{-16}	10^{17}	10^{-8}	
	10^{-17}	10^{16}	10^{-7}	
u.v.	10^{-18}	10^{15}	10^{-6}	micrometer (μm)
Visible light	10^{-19}	10^{14}	10^{-5}	
i.r.	10^{-20}	10^{13}	10^{-4}	
	10^{-21}	10^{12}	10^{-3}	millimeter (mm)
	10^{-22}	10^{11}	10^{-2}	
Microwaves	10^{-23}	10^{10}	10^{-1}	
	10^{-24}	10^{9}	1	meter (m)
	10^{-25}	10^{8}	10^{1}	
	10^{-26}	10^{7}	10^{2}	
Radio frequencies	10^{-27}	10^{6}	10^{3}	kilometer (km)
	10^{-28}	10^{5}	10^{4}	
	10^{-29}	10^{4}	10^{5}	
	10^{-30}	10^{3}		

molecules of the cell is not yet feasible. There are, though, at least two concepts arising from wave mechanics that are very important to us here:

1. The distribution of a particle in space is given by the square of its wave-function. This leads to an understanding of orbitals (see Appendix V).

2. Energy states are **quantized;** thus any system has certain characteristic energy values or levels.

We shall see later that the distribution of a particle in space is important in interpreting and understanding scattering, and the concept of energy levels is the basis of absorption and emission spectroscopy.

An understanding of the properties of matter, whether they are describable by wave mechanics or not, depends on knowledge about its energy and its coordinates in space and time (that is, its shape and dynamic properties). It is important to realize that the three properties of energy, shape, and dynamics are closely

Figure 2.4 The angle θ in the water molecule and the folded shape of a polypeptide chain are determined by energetics.

interdependent. Thus, solution of the Schrödinger equation to give possible energy states can be obtained only if certain constraints are placed on the wave function. For example, as a model for an electron in a chemical bond, the equation can be solved for a particle of mass m if it is constrained to be in a one-dimensional box of width a. It turns out that the only energies available to the particle are $E_n = n^2h^2/8ma^2$ (where $n = 1, 2, 3$)—that is, the energies are related to shape (the width of the box, a). Two simple examples of the interrelationship between shape and energy are (1) the angle of 105° between the H atoms and the O atom of the H_2O molecule, which arises because this gives an energy minimum to the molecule; and (2) the folding of a polypeptide chain in solution to give a well-defined globular protein, which also depends on the energy of the system (Figure 2.4).

Molecules in solution have kinetic energy because they undergo **Brownian motion,** that is, they rotate and diffuse laterally (translate). In addition, the molecules vibrate because of their thermal energy. Thus, dynamic and energetic properties are also related.

In the remainder of this section we shall briefly describe some of the characteristics of the interactions of molecules with each other and with static magnetic fields. We shall also discuss the classification of different types of energies and some of the consequences of different energies and energy levels.

Interparticle Forces and Energies

The forces between various particles, such as a nucleus and an electron or between molecules, often result in a characteristic behavior for the energy of the system. This is represented by the diagram shown in Figure 2.5. If two particles are far apart, there is no interaction between them. As the particles approach, there may be direct attraction of positive and negative charges or there may be a change in charge distribution on the particle, resulting in a net attraction. The energy of the system decreases until an **equilibrium** value of the distance between the particles is reached. As the particles get closer than this, they begin to repel each other and the energy increases. This sort of behavior is typical of many situations. Examples include an electron orbiting a nucleus, the interaction between atoms in the formation of chemical bonds and crystals, the interaction between molecules in gases and liquids, and the specific binding of a substrate to an enzyme.

Figure 2.5   Variation of the energy of a pair of particles with their separation (r).

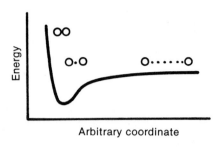

Energy Levels

The energy levels represent the **characteristic states** of the molecule. The properties of the characteristic states are related to the identity of the molecules, to the molecular structure, and to the energetics of any chemical processes that the molecules may undergo.

For convenience, the **ground state** is defined as the state of lowest energy. The ground state is the state that becomes progressively more occupied by the molecules as the matter is cooled toward absolute zero. States of higher energy are called **excited states.**

If two or more states of the molecule have the same numerical value of energy, they are said to be **degenerate.** This degeneracy may be removed (resolved or split) by the effect of some external influence, such as an electric or magnetic field (discussed later).

The Effect of an Applied Magnetic Field

All electrons and some nuclei possess a property known as **spin.** This property is a fundamental characteristic of a system and was introduced to explain spectroscopic observations in the presence of a magnetic field. New lines were observed and the multiplicity of these lines was given by the formula $2S + 1$, where S is the total spin of the atom or molecule. The spin can take values of $\frac{1}{2}$, $1, \frac{3}{2}$, etc. In fact, the spin orients in certain allowed (quantized) directions in the magnetic field, thus leading to new energy levels. The example of Mn^{2+} is illustrated in Figure 2.6.

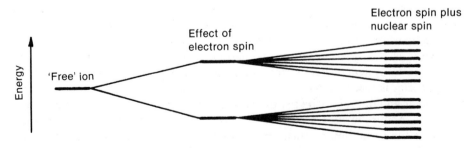

Figure 2.6   The energy levels of the Mn^{2+} ion in a magnetic field. This ion has electron spin $\frac{1}{2}$ and nuclear spin $\frac{5}{2}$. The multiplicity arising from the electron spin is 2 and that from the nuclear spin is 6, giving the 12 energy levels shown.

Classification of Energies

For most purposes, it is convenient to treat a molecule as if it possesses several distinct reservoirs of energy. The total energy is then given by

$$E_{total} = E_{translation} + E_{rotation} + E_{vibration} + E_{electronic}$$
$$+ E_{electron\ spin\ orientation} + E_{nuclear\ spin\ orientation}$$

Each E in the equation represents the appropriate energy as indicated by its subscript. Although a molecule has a large number of other types of energy, those listed are, for our purposes, the most important.

In solution, a molecule can obviously translate, rotate, and vibrate. The energies associated with each of these are quantized. Probably more familiar is the idea that the electrons have certain allowed energies, which are defined by orbitals (see Appendix V).

Probing Different Energies with Different Ranges of Electromagnetic Radiation

Figure 2.7 shows some examples of the arrangement of energy levels in a typical system. Although every type of energy is quantized, the separation between neighboring translational energy levels is so small that for practical purposes we can disregard the quantization of the translational energy. The separation between energy levels associated with the orientations that a nuclear or electron spin can take in a magnetic field are also very small (10^{-3} mol^{-1}), but they are very precise because their lifetimes are long (discussed below). Three or four orders of magnitude greater are the separations between the energy levels associated with the different rates of rotation of a molecule (10 J$\cdot$mol^{-1}). Two to three orders of magnitude greater still are vibrational energies of a molecule, which are determined by the mass of the atoms and the flexibility of the chemical bond joining them (1–40 kJ$\cdot$mol^{-1}). The separation between electronic energy levels is even greater (10^5–10^6 kJ$\cdot$mol^{-1}).

Using Worked Example 2.1, we can see that rotational energies occur in the microwave region of the electromagnetic spectrum. Similarly calculations will show that nuclear-spin orientation energies occur in the radio-frequency region. Vibrational energies occur in the infrared region. Electronic energies (when electrons are moved from one region of a molecule to another) occur in the visible and ultraviolet parts of the spectrum.

The range of electromagnetic radiation can thus be broken down into regions that are associated with a particular type of energy (see Table 2.1).

Population of Energy Levels

We have discussed the range of different types of energies available to one particular molecule. In any practical studies, we deal with very large numbers of

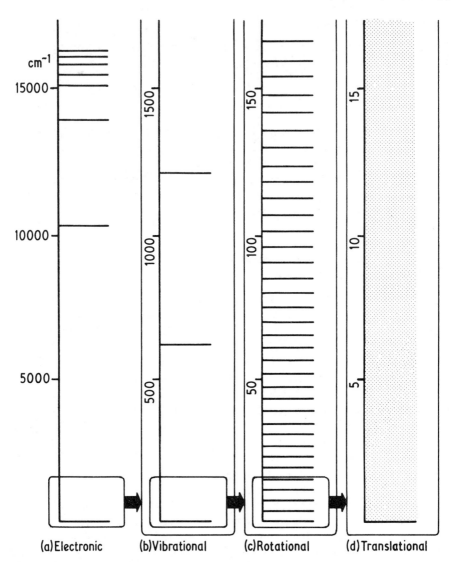

Figure 2.7 An illustration of the relative separations of the energy levels of atoms and molecules for the different types of energy. The translational energy levels are essentially continuous.

molecules. At any finite temperature, the molecules will be distributed among the energy levels available to them because of thermal agitation. The exact distribution will depend on the temperature (thermal energy) and on the separation between energy levels (ΔE) in the energy ladder. At a given temperature, the number of molecules in an upper state (n_{upper}) relative to the number in a lower state (n_{lower}) is given by Boltzmann distribution law (see Figure 2.8).

$$\frac{n_{upper}}{n_{lower}} = \exp\left(\frac{-\Delta E}{kT}\right)$$

where k is the Boltzmann constant (1.38×10^{-23} J·K^{-1}). When ΔE applies to 1 mol, the term on the right becomes $\exp(-\Delta E/RT)$, where R is the gas constant ($R = 8.31$ J·mol^{-1}).

It is clear that when $\Delta E \ll kT$, $\exp(-\Delta E/kT)$ approaches e^0, which is 1. The number of molecules in the upper and lower levels is then equal. Conversely, when $\Delta E \gg kT$, n_{upper} is negligible with respect to n_{lower}.

Worked Example 2.2   Populations of Energy Levels for Different Energy Separations

Calculate the ratio of molecules in the upper to those in the lower energy level when the separation between these is as follows. Assume the temperature is 300 K. The type of energy involved is shown.

a. 1.19×10^{-2} J·mol^{-1} (nuclear reorientation)

b. 11.9 J·mol^{-1} (rotational)

c. 11.9 kJ·mol^{-1} (vibrational)

d. 119 kJ·mol^{-1} (electronic)

Solution

We apply $n_{upper}/n_{lower} = \exp(-\Delta E/kT)$ in each case. Since E is given for 1 mol we must use the form

$$\frac{n_{upper}}{n_{lower}} = \exp\left(\frac{-\Delta E}{RT}\right)$$

where $R = 8.31$ J·mol^{-1}.

a. $\dfrac{n_{upper}}{n_{lower}} = \exp\left(\dfrac{-1.19 \times 10^{-2}}{8.31 \times 300}\right)$

$= 0.9999952$

that is, the two states are almost equally populated.

b. $\dfrac{n_{upper}}{n_{lower}} = 0.9952$

c. $\dfrac{n_{upper}}{n_{lower}} = 8.5 \times 10^{-3}$

d. $\dfrac{n_{upper}}{n_{lower}} = 1.86 \times 10^{-21}$

In contrast with (a), we see that for vibrational and electronic energies all the molecules are essentially in the lower state.

Figure 2.8 The dependence of the population of energy levels on temperature.

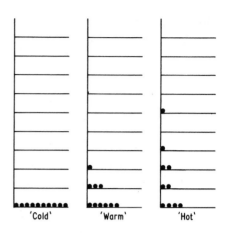

THE INTERACTION OF ELECTROMAGNETIC
RADIATION WITH MATTER

The consequences of the interaction of electromagnetic waves with matter are that the waves are *scattered, absorbed,* or *emitted.* With various modifications (in NMR, for example, the sample is placed in a static magnetic field, and in microscopy the scattered waves are refocused with a lens system), Figure 2.9 represents the basis of all the experiments discussed in this book.

Scattering

Scattering is usually detected by measuring the intensity of radiation at some angle θ to the incident wave, but it may also be detected by the reduction in the transmitted light with $\theta = 0$ (this is then often called **turbidity**). Scattering gives rise to refraction and diffraction phenomena. The frequency of the scattered waves is usually but not always the same as that of the incident wave. Electromagnetic waves incident on an isolated atom with an electron cloud cause

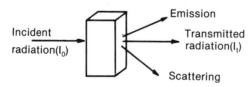

Figure 2.9 Electromagnetic radiation incident on a sample can give rise to absorption, emission, and scattering.

Figure 2.10   Effect of a static field E on an electron cloud. If E oscillates, the induced dipole will oscillate in sympathy with it.

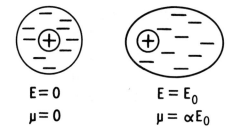

$$E = 0$$
$$\mu = 0$$

$$E = E_0$$
$$\mu = \alpha E_0$$

the electrons to oscillate about their equilibrium position, in sympathy with the applied wave (see Figure 2.10). The result is an electronic oscillator that emits radiation in all directions in a plane perpendicular to the oscillation. Some electron clouds are more susceptible to the applied wave than others, and their tendency to oscillate is defined by a parameter α, the polarizability. (Other parameters, including *cross sections* and *atomic scattering factors,* are sometimes used instead of α.) Electrons are the usual scatterers in molecules, while nuclei scatter neutrons.

Scattering and Interference.   Electromagnetic waves can superimpose (see Appendix III). Scattered waves, which usually all have the same frequency, are particularly susceptible to the phenomenon of **interference**, in which waves can add constructively or destructively. For example, two waves differing only in phase by 180° effectively cancel each other out. A simple example is the effect observed when light waves emerging from two slits interfere, as shown in Figure 2.11.

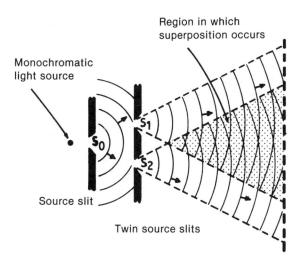

Figure 2.11   A simple illustration of interference effects. The light is diffracted at the slits and superposition of the two resulting waves gives rise to an interference pattern.

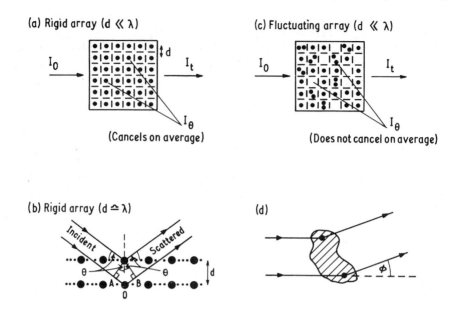

(a) Rigid array (d ≪ λ)

I_0 ⟶   ↕ d   ⟶ I_t

I_θ

(Cancels on average)

(c) Fluctuating array (d ≪ λ)

I_0 ⟶   ⟶ I_t

I_θ

(Does not cancel on average)

(b) Rigid array (d ≏ λ)

Incident   Scattered

θ   θ   ↕ d

A   B

0

(d)

φ

Figure 2.12   Situations where interference effects are important in scattering.

Several situations where interference is important can be identified and four cases will be illustrated here:

1. Large arrays of essentially rigid particles, e.g., crystals where the wavelength of the incident radiation is much greater than the dimensions of the array. In this case there is very little observed scattering at angles other than $\theta = 0$, since pairs of scatterers can always be found that cause destructive interference (see Figure 2.12(a)). Only forward scattering remains, and this leads to **refraction**.

2. Large arrays, as in (1), where the wavelength is much less than the dimension of the array. In this case, complex three-dimensional interference patterns, usually called **diffraction patterns**, may be generated. These can give information both about the array (or lattice) and the constituent molecules of the array. The simplest view that accounts for the observed patterns is that given by W. L. Bragg. In a regularly spaced array, such as a crystal, there are planes of atoms or molecules that will scatter in phase. Constructive interference will be observed when the phase difference between the scattering from these planes is a multiple of $360°$. As shown in Figure 2.12(b), the path difference between the two planes of scatterers for waves incident at an angle θ to the plane is $AO + OB = 2d \sin \theta$. If this path difference is an integral multiple of the wavelength, λ, then constructive interference will be observed. The relationship $2d \sin \theta = n\lambda$ is known as Bragg's law.

3. Fluctuating arrays, such as gases and liquids, where the dimensions of the scatterers are much less than λ. In this case, it is no longer possible to always find a pair that cancel each other, thus scattering is observed at angles other than $\theta = 0$ (see Figure 2.12(c)). This scattering is related to the concentration and size of the particle and can be analyzed to give molecular weight.

4. A particle in solution whose dimensions are greater than the wavelength. By analogy with the situation in (2), there will be certain angles at which destructive interference takes place (see Figure 2.12(d)). This occurs even if the particle tumbles in solution, causing an averaging over all angles. The net result is that the scattering decreases rapidly with increasing ϕ. This angular dependence can give information about the size and shape of the scattering particle. (Note that ϕ in Figure 2.12(d) $= 2\theta$.)

Scattering and Refraction. The phenomena of reflection and refraction can be explained using scattering theory. **Reflection** results when light is scattered in the direction *opposite* of that of the incident light. **Refraction** is the result of radiation scattered in the *same* direction as that of the incident wave (see Figure 2.13(a)). The phase of the scattered wave is different from that of the wave that passes straight through the sample. These two types of wave then recombine to give a wave that has apparently passed through the sample with a different velocity. The parameter used to describe this phenomenon is the **refractive index** (*n*), where

$$n = \frac{\text{velocity of wave in vacuum}}{\text{velocity of wave in sample}} = \frac{c_0}{c_s}$$

Figure 2.13 Illustration of refraction effects. (a) *Physical basis:* The transmitted and forward-scattered waves are 90° out of phase with respect to each other. This leads to an apparent change in the velocity of the superimposed waves. (b) *Macroscopic effect:* The result of the apparent change in velocity caused by the refractive index is that $\theta_1 \neq \theta_2$ in the ray diagram shown.

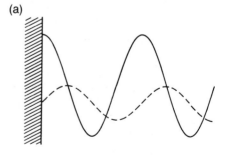

(a)

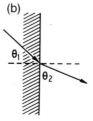

(b)

θ_1

θ_2

It can be shown (see Chapter 8, problem 3) that refraction results in the relationship $_1n_2 = \sin\theta_1/\sin\theta_2$, where $_1n_2$ is the refractive index in passing from medium 1 to medium 2 and θ_1 and θ_2 are the angles made between the directions of the propagated waves and the surface separating the two media (see Figure 2.13(b)). The consequence of this is familiar: *Lenses* can be made that focus some electromagnetic waves. The ability to manufacture lenses allows *microscopy* to be performed. Not all forms of radiation can be focused by a lens, e.g., X-rays cannot. (Note that electrons can be focused by electric and magnetic fields.)

The refractive index can be different for different directions of polarization—a phenomenon known as **birefringence**.

The polarizability, and therefore the scattering and the refractive index, are **frequency dependent** (see below). Frequency-dependent refraction is known as **dispersion.** This phenomenon gives rise to the familiar splitting of white light into colors by a prism. Dispersion occurs not only in the visible region but at all wavelengths. Dispersion arises from two main effects: (1) the ability of molecules to reorient in an applied field, and (2) absorption effects near **resonance**—a natural frequency of the scatterer.

1. If a molecule has a permanent dipole moment associated with it, as with water, then at low frequencies the molecules will reorient to follow the oscillating applied field. This will lead to a strong interaction (scattering) and a high refractive index. (This effect can also be observed by measuring the dielectric constant of the sample, which is related to n). If the applied frequency is high, then the molecule will not have time to follow the field and the observed refractive index will drop. The general tendency of the refractive index to decrease with increasing frequency is called **normal dispersion.**

2. When a dipole moment is induced in a molecule, there is the possibility that this induced charge separation may correspond to an absorption transition—for example, an electron might change its energy level (see the next section). The polarizability, and thus the scattered radiation, increases as the absorption transition is approached because of resonance, which gives rise to very large changes in the induced dipole moments. Moreover, the phase of the scattered radiation shifts by π on traversing the absorption. (This phase shift is discussed in Appendix II. The equation derived there for describing the behavior of a damped, driven harmonic oscillator describes the dispersion curves observed near resonance, that is, when $\nu \simeq \nu_0$). Since the refractive index is a result of forward-scattered radiation, the refractive index increases and changes sign as the frequency of radiation passes through an absorption band. This is known as **anomalous dispersion**. A plot of n against wavelength is called a **dispersion curve** (see Figure 2.14).

Figure 2.14 The variation of refractive index with wavelength. The dispersion curve observed near a natural resonance frequency of the system.

Refractive index

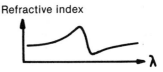

Worked Example 2.3   Normal and Anomalous Dispersion

Figure 2.15 is a schematic representation of the frequency dependence (dispersion) of the refractive index of a macromolecule in solution. Explain (a) the decreases in the refractive index in the radiowave and microwave regions and (b) the variations of the refractive index in the infrared, visible, and ultraviolet regions.

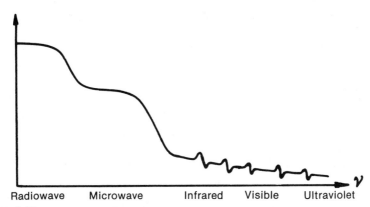

Radiowave   Microwave   Infrared   Visible   Ultraviolet

Figure 2.15

Solution

a. These are the normal dispersions (frequency dependences) that arise because the tumbling of the dipoles cannot keep pace with the alternating electric field. The radiowave dispersion arises from the (slow) tumbling of the macromolecule and the microwave dispersion from that of the water molecules in the solution.

b. The variations in refractive index are associated with absorption bands of the macromolecule in solution. Near an absorption band the refractive index behaves as shown in Figure 2.15, giving rise to anomalous dispersion.

Elastic and Inelastic Scattering. If the scattered radiation has the same frequency as the incident radiation (ν_0), the scattering is said to be **elastic** (energy is conserved). If the frequency changes, the scattering is **inelastic**. Changes in frequency can be detected if a very narrow bandwidth incident beam, such as from a laser (see page 32), is used to irradiate the sample. Figure 2.16 illustrates the kind of experimental setup used and the sort of plot that is obtained if the scattering intensity is plotted as a function of frequency.

Figure 2.16 A schematic representation of (a) the setup used to measure the frequency dependence of scattering and (b) a typical plot of intensity versus frequency.

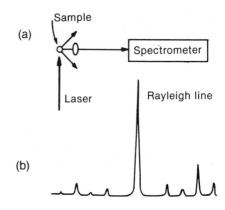

The line centered around ν_0 is called the **Rayleigh line**. It is much more intense than other lines in the spectrum (see the next section for definition of a spectrum) and has a finite linewidth. This linewidth arises because fluctuations and diffusion cause some molecules in the sample to move away from the detector and others to move toward it. This effect shifts the apparent frequency of the radiation (Doppler shifts) and causes broadening of the Rayleigh line. The linewidth can be analyzed to give information about the motion of the molecules in solution. (These movements can also be detected as "noise" in the scattered intensity; see Chapter 8.)

The changes in frequency caused by motion are relatively small. Much larger changes occur if a molecule changes its vibrational state while the scattering occurs. This results in the other lines shown in Figure 2.16(b). This phenomenon allows the vibrational states to be investigated and is known as **Raman scattering**.

Absorption

Absorption is usually measured by varying the frequency of the applied radiation and plotting the energy absorbed by the sample at each frequency. The resulting plot is called a **spectrum** (see Figure 2.17). A spectrum is a series of lines, bands, or resonances, and it contains information about position (on an energy scale), intensity (of absorption), and linewidth (usually at half-maximal intensity $\Delta\nu_{1/2}$). Note that lines a and b are resolved, while c, d, e, and f overlap to varying degrees.

The frequency dependence of absorption (a spectrum) arises because energy is absorbed by **transitions** induced between different energy states of the molecules in the sample. These transitions can occur only if there is a strong interaction between the incident radiation and the molecule. We have seen in discussing scattering that a molecule can oscillate in response to a wave without changing its energy (except in Raman scattering). We shall now discuss how changes between energy states can be induced. To do this we must, in general, use wave

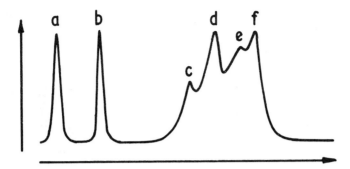

Figure 2.17 A spectrum showing six absorption lines, labeled a–f. The horizontal axis can be frequency, wavenumber, energy, or any other convenient scale.

mechanics, since classical physics cannot deal with quantization. There are several simple rules arising from wave mechanical treatments that help us to predict whether a certain interaction will result in a transition between energy levels.

Transition Probabilities and Selection Rules. A fundamental rule that helps us to understand whether a transition is probable or not is given by

$$\Delta E = h\nu$$

where E is the separation between the energy states of interest and ν is the frequency of the applied radiation. This equation means that absorption is most probable when the energy level separation matches (or resonates with) the energy of the incident radiation.

The next general rule that we must consider is one that applies to all electronic and vibrational spectra: *There must be a displacement of charge induced in going from one energy state to another.*

There are two relevant types of charge displacement during the transition:

1. One involving a linear displacement. This obviously constitutes an electric dipole because of the separation of charge; it is called an **electric transition dipole** μ_e. Note that it has a direction.

2. One involving a rotation of charge. This rotation will generate a magnetic moment perpendicular to the rotation (see Appendix VIII); it is called a **magnetic transition dipole**, μ_m. The applied electromagnetic radiation interacts with these dipoles to cause changes in energy states; the electric component interacts with μ_e and the magnetic component interacts with μ_m. The larger the transition dipole moment, the larger will be the transition probability. In fact, the transition probability is proportional to the square of the transition dipole moment. Separating charges, in general, creates a larger effect than rotating them, and since the radius of rotation is small (e.g., in an electronic transition the displacement is between molecular orbitals), the magnetic transition dipoles are much smaller than the electric ones (by $\sim 10^5$ times).

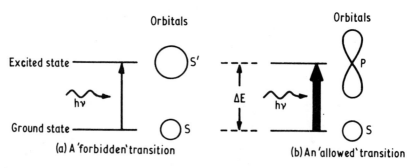

Figure 2.18   Illustration of selection rules. If the radiation applied has a frequency such that $E = h\nu$, then the transition probability between the levels will be a maximum. Another rule that applies says that there must be a charge displacement. In the figure there is no charge displacement in going between two *s*-type orbitals as shown in (a), thus the transition probability will be low—the transition is "forbidden". In (b) there is a charge displacement in going from an *s*-type to a *p*-type orbital; thus, the transition probability will be high—the transition is "allowed".

The rules that help us predict the probability of a transition are called **selection rules**. In fact, these rules are often not very strictly obeyed because molecules have many possible interactions and distortions. Some important absorption spectra arise from "forbidden" transitions, for example, in transition metal ion spectra (see Appendix VI).

In addition to the selection rules discussed so far and illustrated in Figure 2.18—the resonance condition, $E = h\nu$, and the charge-displacement rules for the existence of electric and magnetic transition dipole moments—others exist. The most important ones are related to the spin properties of the energy levels (see Chapters 6 and 7).

Worked Example 2.4   Vibrational Absorption Spectra Require a Charge Displacement

Which of the following molecules will show a vibrational spectrum in the infrared: CO_2, N_2, HCl, H_2O? Will any of these molecules scatter light?

Solution

We shall see later that a molecule can have many different vibrations. From the charge-displacement rule, any type of vibrations that result in a *change* in dipole moment will give rise to a spectral transition. We can imagine such vibrations for all the molecules except N_2, which is symmetrical. So CO_2, HCl, and H_2O will show a vibrational spectrum.

Scattering depends on inducing an electric dipole, so in theory all molecules will scatter light (How well depends on how tightly the electrons in the molecule are held, that is, on their polarizability).

Absorption Depends on the Populations of Energy Levels. We noted earlier that the distribution of molecules among energy levels depends on the energy difference between the levels. The relative populations of molecules in the different levels are given by the Boltzmann distribution law.

When electromagnetic radiation is applied to a sample, it is just as likely to cause transitions from an excited state to a ground state (emission) as it is to cause transitions from a ground state to an excited one. (This is because the transition dipole with which it interacts is associated with both energy states). Consequently, net absorption depends on the difference between the populations of the energy levels concerned to be significant. The perturbation of populations caused by the applied electromagnetic radiation can be important. For example, when the excess population in the ground state is small, the application of radiation can equalize the populations of the ground and excited state energy levels. This is known as **saturation** (see Figure 2.19). The mechanisms whereby molecules achieve a Boltzmann distribution after a perturbation are called **relaxation processes**. These will be discussed further under emission.

Worked Example 2.5   Populations and Intensity of Absorption

Typical energy-level separations corresponding to (a) nuclear-spin reorientation, (b) rotation, (c) vibration, and (d) electronic transition are given in Worked Example 2.2. The ratios of the populations between the ground and first excited states were found to be (a) 0.9999952, (b) 0.9952, (c) 8.5×10^{-3}, and (d) 1.86×10^{-21}. Which of the four transitions will give rise to the most intense absorption spectrum?

Solution

The relative populations show that in (d) *all* the molecules are essentially in the ground electronic energy level. On this basis the electronic transition will give rise to the most intense *net* absorption spectrum. In the vibrational case, only 8 molecules in every thousand are in the upper state, so there is still a very large excess of molecules for net absorption for transitions from the ground state. The net absorption for the rotational transition in this example depends on a smaller difference, about 5 molecules per 2 thousand.

For nuclear-spin reorientation transitions, however, net absorption will be observed for only 24 molecules in every 10 million. NMR, the technique that detects such transitions, is therefore inherently a very insensitive form of absorption spectroscopy. In general, the larger the energy separation, the more populated is the ground state and the more intense will be the absorption signal.

Absorption Spectra Depend on Concentration. The total absorption of a sample depends on the number of molecules in which transitions are induced. This means that absorption spectra can be used quantitatively. The effect of sample concentration on the spectral intensity is the basis of most analytical applications of absorption and emission spectra.

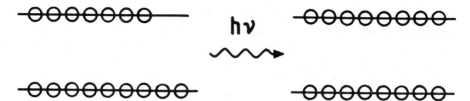

Figure 2.19   Absorption results in the excitation of molecules to excited states. If the number of molecules in the upper of two levels equals the number in the lower level no more absorption can take place. This is called **saturation**.

The Linewidth of an Absorption Line Depends on Lifetime.   Spectral lines are not infinitely sharp (i.e., truly monochromatic). Various factors contribute to the broadening of a line—the main ones being the *lifetime* of the excited state and *lack of resolution* caused by overlapping bands.

Quantum mechanics tell us that for a system that changes with time it is not possible to define the energy levels exactly.*

For a system that changes its energy state at a rate $1/\tau$ the energy levels become blurred and the corresponding uncertainty in energy δE is given by

$$\delta E \sim \hbar/\tau$$

where $\hbar = h/2\pi$.

If the lifetime of a molecule in an excited state is τ, then its energy is not exactly E but has a spread of energy δE around E. To specify the energy of a state exactly ($\delta E = 0$) requires that the molecule in that state have an infinitely long lifetime. Since no molecule in an excited state has an infinite lifetime, it follows that no excited state has a precisely defined energy and that spectral lines have a finite width. This is called **lifetime broadening**. Short-lived energy states give rise to broad spectral lines, while long-lived states give narrower spectral lines.

The actual mechanism affecting the lifetime of the excited state depends on various processes, such as collisions with other molecules, that cause molecules to relax back to their ground state.

Worked Example 2.6   Rotational Energies Are Not Resolved in Solution

Spectral lines from transitions between rotational energy levels are expected to be separated by about 2 cm^{-1}.

A molecule in a liquid undergoes 10^{13} collisions per second. Will the resulting lifetime broadening permit resolution of these spectral lines?

Solution

The lifetime of a small molecule in any given rotational state is about 10^{-13} s. From the Worked Example 2.1, this gives a value of ~ 50 cm^{-1} for the spectral linewidth. This is very much greater than the separation between the two lines, which will therefore *not* be resolved.

---

*This is a consequence of the Heisenberg uncertainty principle, which asserts that it is impossible to think of a measuring procedure in which the errors in the measurement of certain pairs of properties (e.g., position and momentum) can be made infinitely small at the same time.

Figure 2.20 The direction of the transition dipole of tyrosine at 270 nm.

Absorption Spectra Depend on the Direction of the Transition Dipole Moment. A transition dipole absorbs radiation by oscillating in sympathy with the applied radiation. This oscillation has a certain direction associated with it, so only those components of the electromagnetic radiation that are in the same direction as the transition dipole moment will cause transitions. In fact, if the angle between the direction of the applied wave and the direction of the transition moment is θ, then the effective value of the transition moment is proportional to $\cos \theta$ and the **transition probability** is proportional to $\cos^2\theta$.

To indicate the direction of oscillation with respect to the molecular axes, we shall use a two-headed arrow ($\leftrightarrow$). We shall sometimes find it convenient, however, especially when dealing with several interacting dipoles, to use a unidirectional arrow ($\rightarrow$). This will allow us to represent the relative phase of different dipoles and thus to consider the ways in which they combine and interact. For example, in tyrosine the transition dipole associated with a particular absorption band oscillates in the plane of the ring, as shown in Figure 2.20. For any one transition, the charge displacement may occur in either of the two possible directions.

Since the absorption depends on the angle θ between the transition dipole and the electromagnetic radiation, the absorption of radiation by a molecule also depends on θ. Absorption measurements are thus sometimes made using radiation polarized in two different directions. Two values of absorption are then obtained, say $A_{\parallel}$ and $A_{\perp}$. In solution, a molecule, and therefore its transition dipole moment, takes up all possible orientations. For plane-polarized radiation, $A_{\parallel}$ and $A_{\perp}$ are equal. In a system of oriented molecules, $A_{\parallel}$ and $A_{\perp}$ may be quite different for plane-polarized radiation.

The differential absorption of radiation polarized in two directions (e.g., $A_{\parallel}$ and $A_{\perp}$) as a function of frequency is called **dichroism**. For plane-polarized light, this is called **linear dichroism**; for circularly polarized light, circular dichroism (see next section).

Optically Active Molecules Differentially Absorb Left- and Right-Circularly Polarized Radiation. We have discussed transitions involving electric transition dipoles and have mentioned magnetic transition dipoles. The path of charge displacement during a transition that involves both electric and magnetic dipoles

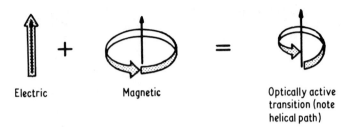

Electric   Magnetic   Optically active
                                              transition (note
                                              helical path)

Figure 2.21   Optical activity arises when both electric and magnetic transition dipole moments are present in an absorption process.

must have some helical character because a helix arises from a rotation plus a displacement (see Figure 2.21). Since a helix will have a handedness, we have the basis for optically active transitions. Optical activity can be detected using circularly polarized light, which has right- and left-handed circular components. This type of measurement is known as **circular dichroism**. A related technique is known as **optical rotatory dispersion** (see Chapter 10).

Absorption Spectra Can Arise from the Reorientation of Magnetic Moments in a Magnetic Field.   A rotating charge generates a magnetic field (see Appendix VIII). The two kinds of rotating charge that are most relevant here are (1) a system of electrons that has the net property of spin and (2) a system of nuclei with spin, S. Most electronic systems have paired electrons and thus no net spin, but sometimes $S \neq 0$, as in free radicals and some metal ions. Many nuclei possess spin, e.g., ^1H, ^{13}C, ^{31}P (see Chapter 6). The spinning electron or nucleus generates a magnetic field and thus has a magnetic moment associated with it.

Magnetic moments interact with applied fields. An analogy is a compass needle in the earth's magnetic field. There are three major differences, however, between the behavior of a compass needle and the atomic-level magnetic moments with which we are concerned. A brief discussion of these three effects and their consequences follows:

1. Quantization. The magnetic moments arising from spin are so small that the quantization of their energy states is very important. The number of energy levels that arise from their interaction with an applied magnetic field has already been discussed and is described by the term **multiplicity**. The splitting between the energy levels is directly proportional to both the applied field (B_0) and the magnitude of the magnetic moment (μ). The usual spectroscopic resonance condition ($\Delta E = h\nu$) is written as $2\pi\nu = \gamma B_0$, where γ is called the **magnetogyric ratio** and relates the observed energy level splitting to B_0. The magnetogyric ratio is a constant for a given nucleus.

Figure 2.22 In a magnetic field the magnetic moments of a $S = \frac{1}{2}$ system can take up two possible orientations in the applied magnetic field. These moments are not aligned perfectly along B_0 and they have angular momentum. The net result is that they precess around B_0.

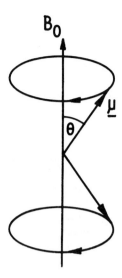

2. The magnetic moments do not completely align along B_0. Consider the case of a system with $S = \frac{1}{2}$. This will have multiplicity 2 and two possible orientations of the magnetic moments with respect to the applied field B_0. (see Figure 2.22). One has a component along B_0, the other has a component that opposes B_0. The lower energy state is the one along B_0. Note, however, that the moments make an angle θ with B_0. This is a result from quantum mechanics, which states that only certain *directions* of the magnetic moments in the applied field are allowed. The net result is that the moments always experience a torque in the magnetic field of magnitude $\mu B_0 \cos \theta$. Another important result is that there are always components of μ in a plane perpendicular to B_0.

3. The magnetic moments have angular momentum. The magnetic moments are generated by a rotation of charge. As a result of the rotation they have a property known as **angular momentum** associated with them. A system with angular momentum (e.g., a bicycle, a spinning ice skater, or a gyroscope) has some special properties. If a compass needle is perturbed, it oscillates linearly about its minimum energy position in a manner that depends on the friction of the bearing. If a system with angular momentum is perturbed, it behaves rather differently—it spirals *around* the minimum energy position rather than oscillating through it. In the situation described by Figure 2.22 this spiral is not possible because θ is fixed. The net result, then, of angular momentum plus torque is that μ **precesses** around B_0. The frequency of precession is the resonant frequency (remember that angular frequency $\omega = 2\pi\nu$).

Worked Example 2.7   Interaction of Electromagnetic Radiation with a Magnetic Moment

In discussions of electronic transitions, we described how applied electromagnetic radiation interacted with the transition dipole moment to produce oscillations and absorption. What do you predict the equivalent interaction between magnetic moments and electromagnetic radiation to be?

Solution

In electronic transitions the polarization of the electric vector of the electromagnetic radiation had to match up with the transition dipole. In the same way the magnetic vector of the electromagnetic radiation must rotate around B_0 in sympathy with the precessing magnetic moment. To produce the strongest interaction with μ (that is, to induce absorption), we need a magnetic field rotating at the resonant frequency in a plane perpendicular to B_0. This type of radiation is, in fact, easily applied with a tuned coil up to frequencies of about 1000 MHz.

Emission

Emission of radiation can occur when a molecule changes from an excited energy state to a lower energy state. As with some scattering measurements, emission is detected at some angle θ (usually 90°) to the incident beam; however there is an additional property of lifetime in an excited state that distinguishes emission from scattering.

We have said that absorption causes transitions to higher excited states and also that relaxation processes cause the distribution of energies to return to equilibrium and can be observed as an emission spectrum.

We shall consider three ways whereby a molecule can change its energy from an excited state to a lower one: (1) by stimulated emission, (2) by various intermolecular and intramolecular processes in the solution, and (3) by spontaneous emission.

1. Stimulated emission. We have already indicated that the interaction between the electromagnetic radiation and a molecule is equally likely to produce emission and absorption (see "Absorption Depends on the Population of Energy Levels" section). The probability that these transitions will occur involves the transition dipole moment connecting the two energy levels involved. Stimulated emission is very important in the laser (light amplification by stimulated emission of radiation). A laser operates in systems where a nonequilibrium distribution of energies has been set up by a pump that induces transitions to a higher excited state. In such a system, the emission of some radiation can be made to stimulate a cascade of emission. This emission will stop when the population of the excited state returns to equilibrium (see Figure 2.23).

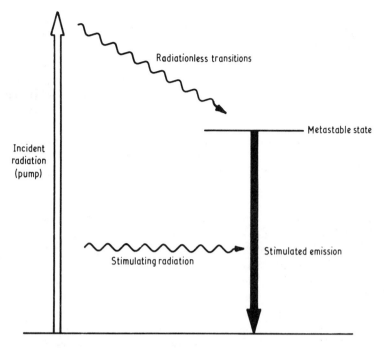

Figure 2.23   An illustration of the principle of the laser.

2. Intermolecular and intramolecular deexcitation processes. The most common way for a molecule to return to a lower energy state is by the generation of heat by collisions, vibrations, and general molecular motion. This type of decay is referred to as **thermal** or **radiationless** because there is usually no detectable emission.

     A special case of this type of deexcitation involves transfer to another molecule that has a similar energy level. This transfer is most efficient when the two energy levels have the same value—in other words, if there is a matching, or a resonance, between the levels. Resonant-energy transfer is important in photosynthesis, where sunlight absorbed by chlorophyll molecules is transferred via several molecules to the photoactive center (see Figure 2.24).

3. Spontaneous emission. In this case the molecule acts as an oscillator and radiates its energy $h\nu$ without any other interaction with its environment. This tendency for spontaneous emission to occur can be calculated using wave mechanics. The probability that spontaneous rather than stimulated emission will occur at a frequency ν is proportional to ν^3.

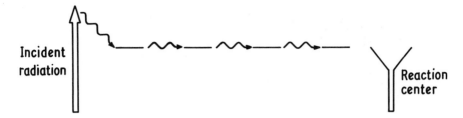

Figure 2.24   An illustration of the process of resonance-energy transfer in photosynthesis. The incident radiation is transferred to neighboring centers and relayed to the reaction center, where a chemical reaction takes place.

Fluorescence and Phosphorescence Are Two Particular Kinds of Spontaneous Emission.   **Fluorescence** is radiation emitted from a molecule in an excited energy level after absorption has taken place. The absorption and emission processes both involve transitions that are allowed. The lifetime of the fluorescent state is usually about 10^{-9} s. The absorption frequency is usually greater than the fluorescence frequency because the absorption process puts the molecule in an excited vibrational level of the excited electron state (see Figure 2.25). Rapid decay to the lowest vibrational level then occurs before emission. Usually only two electronic states are involved in fluorescence: the ground state (call it G) and the first excited state (call it F).

Phosphorescence involves emission from another energy level (call it P). The excitation energy in F can be transferred to P, but the probability of a P-G transition is relatively small—the transition is forbidden. The net result is that P may become populated, via F, but the lifetime of P may be very long because of the low probability of a P-G transition.

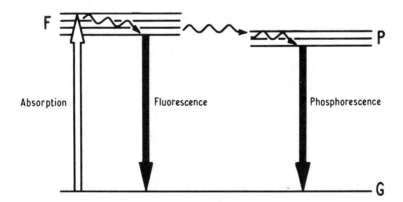

Figure 2.25   A representation of fluorescent and phosphorescent processes.

Worked Example 2.8 Lifetime of the Phosphorescent State

A sample fluoresces at 360 nm with a lifetime of 10^{-9} s. Phosphorescence is observed at 480 nm. The *G-P* transition probability is calculated to be 10^{-5} times the probability of a *G-F* transition because *G-P* involves only a magnetic transition dipole moment. What do you predict the lifetime of the phosphorescent state to be?

Solution

The relative probability that spontaneous emission will occur at the two frequencies is given by the ν^3 dependence, i.e., $(360/480)^3$. The emission also depends on the relative magnitude of the transition probabilities for stimulated emission (same as absorption), i.e., 10^{-5}. The lifetime of the phosphorescent state will thus be

$$10^{-9} \times (480/360)^3 \times 10^5 = 2.4 \times 10^{-4} \text{ s}$$

PROBLEMS

1. Show that plane-polarized light is equivalent to two circularly polarized waves rotating in opposite directions.

2. (a) Calculate the frequency and wavenumber corresponding to light of wavelength 600 nm. (b) What is the energy of this light?

3. Explain why the electronic energy levels of a large conjugated ring system, such as porphyrin, are closer together than those for a single aromatic ring, such as tyrosine.

4. The refractive index for quartz is 1.47 for $\lambda = 400$ nm (visible light), but is 0.9999 for $\lambda \sim 0.1$ nm (X-rays). Comment on the feasibility of making a lens for X-rays.

5. One of the $-C-H$ vibrations in a molecule has its electric transition dipole moment lying in the same direction as the $-C-H$ bond. What is the resultant electric transition dipole moment for a $-CH_2$ group with the vibrations illustrated in Figure 2.26?

Figure 2.26

6. Which of the following molecules will show a rotational microwave absorption spectrum: CO_2, N_2, HCl, H_2O?

7. Suggest a reason why most protein concentrations are usually measured at 280 nm, given the absorption spectra in Figure 2.27.

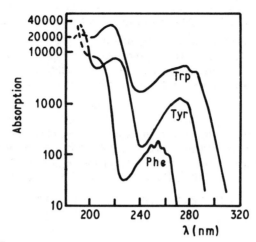

Figure 2.27

8. Calculate the absorption frequencies and the ratio of magnetic moments (a) for the spin resonance of a free electron, $\gamma_e = 1.760 \times 10^{11}\ T^{-1} \cdot s^{-1}$, and (b) for the nuclear-spin resonance of a H atom, $\gamma_H = 2.675 \times 10^8\ T^{-1} \cdot s^{-1}$, in an applied magnetic field of 1 T.

9. Calculate the ratio of the probability of spontaneous emission for an electronic transition at 300 nm (10^{15} Hz) and a rotational transition at 30 cm (10^9 Hz).

10. Calculate the lifetime broadening (in Hz and cm^{-1}) of a molecule that has a lifetime of (a) 10^{-13} s and (b) 10^{-1} s.

11. In magnetic resonance, spontaneous emission of radiation is negligible. Thermal decay by collision or vibration is also negligible and all transitions must be stimulated by magnetic fields with the correct resonant frequency. How then is a Boltzmann distribution set up?

12. Many molecules that fluoresce in solution are found to phosphoresce when trapped in a solid phase, such as a glass. Why is this?

13. The separation between energy levels in NMR is about $10^{-4}\ cm^{-1}$. This is four orders of magnitude closer spacing than between rotational energy levels. Transitions between rotational energy levels cannot be resolved in a solution (see Worked Example 2.6), yet well-resolved NMR transitions are commonly observed. Why is this?

CHAPTER 3

Infrared Spectroscopy

OVERVIEW

1. Infrared (IR) spectra give information on molecular vibrations.

2. The main experimental parameter is $\bar{\nu}_{max}$ in cm^{-1}, the position of the maximum of the absorption band.

3. The common infrared region of the spectrum is 1400–4000 cm^{-1}.

4. The physical basis for understanding molecular vibrations is the Morse (potential energy) curve. Quantum mechanics allows only discrete vibrational energy levels on the Morse curve. The lowest vibrational level corresponds to the zero-point energy.

5. The gross selection rule for a vibration to be infrared active is that the vibration must result in a change in the dipole moment.

6. The number of modes of vibration is $(3N - 6)$ for a molecule consisting of N atoms $(3N - 5$ if the molecule is linear). For a macromolecule there are thus very many vibrational transitions. Many of the vibrations can, however, be localized to particular bonds or groupings, such as the $-C=O$ group. This forms the basis of **characteristic group** frequencies.

7. The vibrational frequency ν_{vib} of a bond can be approximately described by the formula

$$\nu_{vib} = \frac{1}{2\pi} \sqrt{\frac{k}{\mu}}$$

where k is the force constant and μ is the reduced mass. The term μ can be changed by isotopic substitution.

8. Infrared absorption by solvent water is a problem because it leaves only selected "windows" in the range where studies can be done.

9. Infrared spectra of oriented samples give information on the direction of the transition dipole moment because the spectra are **dichroic**. These spectra can be

correlated with molecular conformation, but the spectral interpretation can be complicated because of interactions between neighboring transition dipoles.

10. The main biological applications involve studying ligand (including H^+) binding to macromolecules; probing hydrogen bonds; hydrogen–deuterium exchange; and, on oriented samples, molecular conformation.

INTRODUCTION

The energy of most molecular vibrations corresponds to that of the infrared region of the electromagnetic spectrum. These vibrations may be detected and measured either directly in an infrared spectrum or indirectly in a Raman scattering spectrum* (see Chapter 9). The main biochemical applications have been to monitor vibrations of selected groups, either on ligands or macromolecules, as the ligand binds or the pH changes. The technique has also been used to follow H–D exchange. With oriented samples, such as fibrous proteins and polysaccharides, dichroism measurements give information on the direction of the transition dipole moment, which can then be correlated with molecular conformation. Of the total infrared spectrum, only the range 1400–4000 cm^{-1} is useful in readily deriving any information. In this region the selected vibrations from many characteristic groups, e.g., $-C=O$, can be identified. However, the 600–1400 cm^{-1} region contains very many bands, including those from skeletal vibrations, so this region is sometimes helpful in "fingerprinting" molecules.

While the absorbance of a sample is determined by the Beer-Lambert Law, the extinction coefficient of a band is rarely used as a parameter of infrared spectra. The extinction coefficient reflects the transition probability, which is related to the magnitude of the transition dipole moment (see Chapter 2 for definition). This is so sensitive to environment that vibrations from characteristic groups in different molecules show enormous variation in their extinction coefficients. For this reason the intensities of infrared peaks are often indicated only by the (superscript) lettering, s(strong), m(medium), w(weak). Note that, in contrast, extinction coefficients in ultraviolet spectra are the basis for using intensities of ultraviolet bands to measure concentration.

EXPERIMENTAL PARAMETERS OF AN INFRARED SPECTRUM

An infrared spectrum usually consists of a plot of the absorption of radiation as a function of wavenumber and is characterized only in terms of the *positions* of the maxima of each of the absorption bands $\bar{\nu}_{max}$, expressed in cm^{-1}

---

*However, the Raman scattered light occurs in the visible and ultraviolet regions of the electromagnetic spectrum. The detection systems in this range are far superior to those in the infrared region.

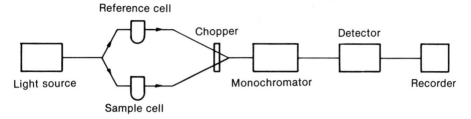

Figure 3.1   The double beam spectrometer.

MEASUREMENT OF INFRARED SPECTRA

Conventional Double Beam Spectrometer

The most common experimental arrangement for obtaining infrared spectra is the double beam spectrometer, which is shown schematically in Figure 3.1. The light source is usually a heated filament (or metal carbide rod) that emits a continuous infrared spectrum. The beam is split into two. One beam goes through the sample and the other throught a reference cell, which usually contains only the solvent. The beams then pass through a chopper, which allows alternate sampling of the two beams. These go through a monochromator, which allows a narrow wavelength band to be selected, and then to the detector, which measures the difference in energy between the two beams. This signal is then amplified and recorded. The cells are made of NaCl, KBr, CaF_2, or LiF, which are virtually transparent to infrared, unlike glass which has very high absorbance in the conventional infrared range. Studies of biological samples are usually done in water, so for these samples water-insoluble cells, such as CaF_2 and LiF, must be used. With water in the reference cell, the double-beam instrument, in principle, allows compensation   for the water absorbance. However, water has such a high absorbance throughout the infrared region that it is difficult to balance the beams sufficiently accurately to achieve complete compensation. The main problem is that the concentration of H_2O is $55M$, while most biologically interesting compounds can be obtained only in the millimolar or less concentration range. This poses a severe limitation on the use of infrared to study these systems. Measurements in water are thus practical only in selected regions or windows of the total infrared spectrum. This difficulty can be partially overcome by the use of D_2O, which has windows in those regions where H_2O has a strong absorption. However, for both D_2O and H_2O the background absorption from the solvent is still significant even in the windows (see Figure 3.6). We shall see later that the main advantage of Raman scattering spectroscopy, which gives essentially the same information as infrared spectroscopy, is that measurements can be carried out in aqueous solution.

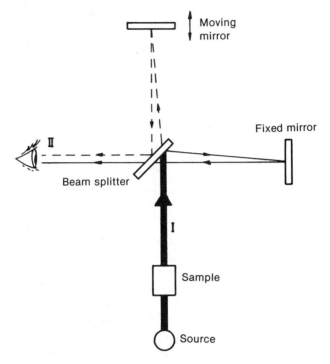

Figure 3.2   Methods of collecting an infrared spectra interferogram.

The Fourier Transform Method

It is often much more efficient to collect infrared spectra in a different way from that described above. Figure 3.2 shows a device invented in 1880 by Michelson. A half-reflecting mirror splits the beam into two equal parts. The two beams then traverse different paths, are reflected at different mirrors, and are recombined at a detector. The path length of one of the beams is varied by moving the appropriate reflecting mirror.

   If the two path lengths are the same, then there will be no phase difference between the beams and they will combine constructively for all frequencies present in the original beam. For different path lengths, the amplitude of the recombined signals will depend on the frequency and the distance the mirror moves. For example, low frequencies will interfere destructively (have a phase shift of 180°) for relatively large movements of the mirror, whereas high frequencies will require relatively small movements for this condition to occur. The result of this

Figure 3.3 An interferogram generated using Michelson's apparatus.

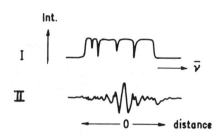

experiment is the production of an **interferogram** of the kind shown in Figure 3.3. This interferogram is the Fourier transform of the spectrum that is observed in conventional infrared methods and that would, in fact, be observed at position I in the figure.

Fourier Transform Infra Red (FTIR) is analogous to the methods used in NMR in which an interferogram or free induction signal is observed after a pulse and is Fourier transformed to produce the conventional spectrum. FTIR has several important advantages over conventional methods. The most important of which is the efficient and rapid collection of data. It is also much less susceptible to stray radiation. In addition, since a computer is already used to obtain the Fourier transform, it is easy to perform many scans to improve the signal-to-noise ratio (noise adds up as the square root of the number of scans, whereas signal adds linearly). Digital subtraction (that is, point-by-point subtraction of the separate spectra by a computer) can also be used to produce good difference spectra. This method has great advantages in obtaining infrared spectra in aqueous solutions.

PHYSICAL BASIS OF INFRARED SPECTRA

Molecular Vibrations

A simple way of relating energetics and structure is provided by the example of the potential-energy curve of a diatomic molecule (with atoms A and B). The curve in Figure 3.4 shows how the energy of a molecule changes as the distance between its nuclei changes.

At the minimum of the potential-energy curve the attraction due to various dispersive forces is balanced by the forces from charge repulsion. The stretching or contracting of the bond is then similar to the behavior of a spring, since there is a restoring force. The vibrations of the bond can in the first instance be analyzed in terms of the molecule's undergoing simple harmonic oscillations about its equilibrium bond length. The vibrational frequency, ν_{vib}, is given by (see Appendix II)

Figure 3.4   A molecular potential energy curve for a diatomic molecule. R_e is the *equilibrium bond length* of the molecule. The parabola represents the potential energy of an ideal molecule that performs simple harmonic oscillation about R_e.

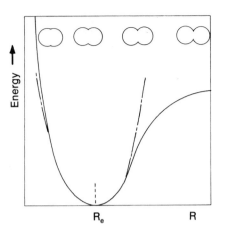

$$\nu_{vib} = \frac{1}{2\pi} \sqrt{\frac{k}{\mu}}$$

where k is the force constant of the bond and μ is the reduced mass of the molecule AB, which is given by

$$\frac{1}{\mu} = \frac{1}{M_A} + \frac{1}{M_B}$$

where M_A and M_B are the atomic masses of A and B.

Worked Example 3.1   Meaning of Reduced Mass

a. In a hypothetical diatomic molecule AB, the mass of A is infinitely large. Which of A or B will determine the frequency of the vibration?

b. How will this alter if A and B are the same mass?

Solution

a. In effect, the molecule is like a single atom B attached by a spring to a brick wall A. The frequency is determined by the smaller mass, B.

b. If A and B are the same mass, we should expect them to contribute in equal proportions.

The **reduced mass** is a mathematical way of expressing this concept. In (a) this is given by $1/\mu = \frac{1}{\infty} + 1/M_B = 1/M_B$. Therefore $\mu = M_B$, while in (b) $M_A = M_B$, so $1/\mu = 1/M_A + 1/M_A = 2/M_A$. Therefore, $\mu = M_A/2$. These values of μ determine the vibrational frequency as given by the equation

$$\nu = \frac{1}{2\pi} \sqrt{\frac{k}{\mu}}$$

Worked Example 3.2   Changing the Reduced Mass by Isotopic Substitution

Calculate the reduced mass of an $O-H$ "molecule" and an $O-D$ "molecule" What is the ratio in vibrational frequency between these two?

Solution

The reduced mass μ_{OH} of the OH "molecule" is given by

$$\frac{1}{\mu_{OH}} = \frac{1}{16} + \frac{1}{1} = \frac{17}{16}$$

$$\therefore \mu_{OH} = \frac{16}{17}$$

Similarly the reduced mass of the $O-D$ "molecule" is given by

$$\frac{1}{\mu_{OD}} = \frac{1}{16} + \frac{1}{2} = \frac{9}{16}$$

$$\therefore \mu_{OD} = \frac{16}{9}$$

Assuming that the force constant remains the same each time, the equation gives

$$\frac{(\nu_{vib})_{OH}}{(\nu_{vib})_{OD}} = \sqrt{\frac{\mu_{OD}}{\mu_{OH}}} = \sqrt{\frac{9}{17}} \approx \frac{1}{\sqrt{2}}$$

The effect of the isotopic substitution of D for H is to decrease the vibrational frequency by a factor of $\sqrt{2}$.

The Morse Curve

Real molecules do not obey the laws of simple harmonic motion exactly. The potential-energy curve shown in Figure 3.5 is clearly more complex than that for a simple harmonic oscillator (which has the parabolic shape shown in Figure 3.4). Morse derived an empirical expression to fit the experimentally observed curve for a diatomic molecule.

Vibrational Energy Levels

Quantum mechanics tells us that only certain values of the energy are allowed. These values are represented by the horizontal line on the Morse curve in Figure 3.5. Values of energy in between these energy levels are not permitted and the spacing between energy levels becomes smaller at higher energy values. As the energy increases, the atoms move farther from their equilibrium positions and hence enter a nonparabolic region of the Morse curve. Therefore, the vibrations can no longer be treated as simple harmonic but are described as **anharmonic** oscillations.

Figure 3.5   The Morse potential energy
curve of a molecule. The horizontal lines
represent the allowed vibrational energy
levels.

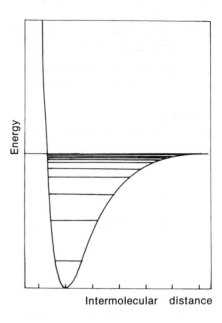

Intermolecular   distance

Zero-Point Energy

The energy of a molecule in the lowest vibrational energy level is often called the
zero-point energy (ZPE). The ZPE is given by $\frac{1}{2} \cdot h \, \nu_{vib}$. For a frequency of 1200
cm^{-1}, this would be 1.2×10^{-20} J per molecule (equivalent to 7.2 kJ $\cdot$ mol^{-1}).
Even at absolute zero temperature, the molecule possesses this amount of energy;
thus, its position can never be precisely defined (as expected from the Heisenberg
uncertainty principle).

Vibrational Energy Level Transitions

At room temperature most of the molecules are in the ground vibrational state
(see Chapter 2). The absorption of radiation of the correct frequency will cause
transitions from the ground state. There is a selection rule that allows transitions
only between adjacent levels. For a vibration of a diatomic molecule to be spec-
troscopically active, the dipole moment must change during the vibration (see
Chapter 2). This is the main selection rule that we need to consider for transitions
between vibrational levels.

Transitions from the ground state to the first excited state are called **fundamentals** or **first harmonics.** Since only fundamentals are strictly allowed, these
constitute the dominant transitions. Those transitions to the higher levels, unlike
the fundamental ones, arise only as a result of anharmonicity and are called **second harmonics** or **overtones.** Most overtones are not observed at all. However,

they can sometimes be observed when hydrogen bonding is present—since the vibration is then anharmonic—and the selection rules depend on having simple harmonic motion.

POLYATOMIC MOLECULES

Number of Vibrations

In a diatomic molecule there is only *one* mode of vibration—the stretching of the bond. In polyatomic molecules we have to allow for bonds to stretch *and bend* as the molecule distorts. For a molecule that consists of N atoms, there are $(3N - 6)$ ways in which the molecule can vibrate, or $(3N - 5)$ if the molecule is linear, for the following reasons:

In order to describe the motions of a molecule of N atoms, a total of $3N$ coordinates is required (x-, y- and z-coordinates for each atom). Three of these are required to specify the position of the center of mass of the molecule. Three more are required to specify the rotational motions (only two are required to specify rotation in a linear molecule). This means that relative atomic positions at a given time can be fully specified by $(3N - 6)$ coordinates, or $(3N - 5)$ in a linear molecule. Any overall motion of the molecule can be represented by a superposition (or combination) of these fundamental modes of vibration.

Worked Example 3.3   Number of Fundamental Vibrations

How many vibrational modes will the following molecules have: (a) H_2O, (b) CO_2, (c) the tripeptide glutathione ($COOH-CH(NH_2)-(CH_2)_2-CONH-CH(CH_2SH)CONH-CH_2-COOH$).

Solution

a. The number of vibrational modes of a nonlinear molecule like H_2O is $(3N - 6)$. Here $N = 3$, so the number is $(3 \times 3 - 6) = 3$.

b. Again $N = 3$ here but CO_2 is linear. The number of vibrational modes is then $(3 \times 3 - 5) = 4$.

c. Here $N = 37$. The number of vibrational modes is $(3 \times 37 - 6) = 105$.

Vibrational Spectra of Polyatomic Molecules

Transitions between the energy levels corresponding to fundamentals and (more weakly) second harmonics, and so forth, are expected for each vibration. A large molecule is therefore expected to have a very complex vibrational spectrum. Fortunately, many normal modes of vibration are localized because they involve a large displacement of just two bonded atoms with little interaction from any

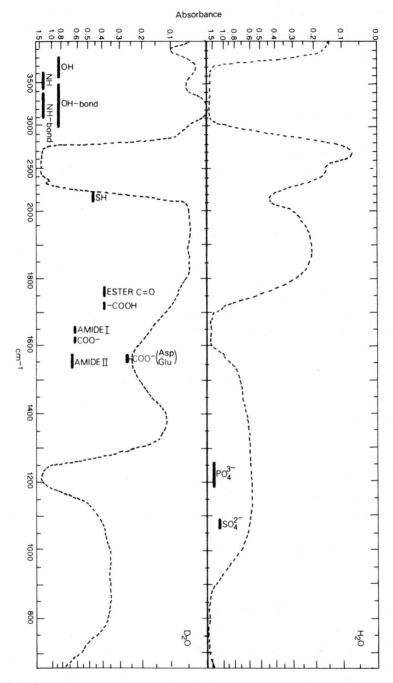

Figure 3.6   Some representative characteristic group frequencies. The infrared curves of H_2O and D_2O shown are those for the typical balance obtainable in a double beam spectrometer.

other vibrations. This forms a basis for characteristic frequencies of the compound. Some representative ones are shown in Figure 3.6, while some less common ones are given in the examples in this chapter.

Solvent Effects on Spectral Transitions

Associated with each vibrational energy level are many closely spaced rotational energy levels. However, the rotational levels are "blurred" or unresolved in solution because of their short lifetimes (see Chapter 2). Vibrational transitions in solution are therefore correspondingly broad.

The solvent may also cause shifts in the vibrational frequencies of a molecule because specific interactions with the molecule lower or raise some energy levels. For instance, we might expect polar solvents to stabilize excited vibrational states if these are more polar than the ground state. This would result in a lowering of the energy of the transition and therefore a shift in the frequency of absorption. At any instant the solvent "cage" may be arranged differently around the molecule because of molecular motion. The corresponding transitions will all have slightly different energies, resulting in a broadening of the absorption band. (This effect is very much more marked and important for transitions between electronic states; see following chapters).

Worked Example 3.4   Solvent Interactions Can Broaden Spectral Bands

SCN^- is one of several anions that stabilize the dead-end complex of creatine kinase. SCN^- in aqueous solution has a strong infrared absorption at 2066 cm^{-1}—the "triple bond" region of the spectrum, which arises from the $C-N$ stretching. When SCN^- is bound to Mn (II), the absorption occurs at 2093 cm^{-1} and has a bandwidth of 31 cm^{-1} at half the maximum height ($\Delta\nu_{1/2}$). When the anion is bound to the enzyme Mn–ADP–creatine kinase–creatine complex, the band position is unaltered, but $\Delta\nu_{1/2}$ is reduced to 19 cm^{-1}. Comment on these observations.

Solution

The unaltered position of the band in the complex suggests the SCN^- is bound to the divalent metal in the complex. The reduced bandwidth in the enzyme complex may indicate that the anion is partially screened from the solvent when bound to the enzyme (see G. H. Reid, C. H. Barlow, and R. A. J. C. Burns, *J. Biol. Chem.*, 253 (1978):4153–4158.

SOME BIOLOGICAL EXAMPLES

Analysis of the characteristic group frequencies forms the basis of most applications of infrared spectroscopy in biological systems.

Worked Example 3.5   Mode of Binding Oxygen to Iron in Oxyhemoglobin

In complexes of transition metals, either both oxygen atoms (I) or only one oxygen atom (II) may be bound to the metal.

I   II
800–900 cm^{-1} 1100–1150 cm^{-1}

The range of O—O stretching frequencies in each case is indicated. The infrared difference spectrum of hemoglobin has a unique band at 1107 cm^{-1} when the sample contains $^{16}O_2$. When the $^{16}O_2$ is exchanged with $^{18}O_2$ the band position shifts to 1065 cm^{-1}. Comment on the assignment of the 1107 cm^{-1} band.

Solution

The band position at 1107 cm^{-1} probably arises from the O—O stretching, and binding to the iron in hemoglobin is best represented as structure II. The assignment is confirmed by the sensitivity in band position to changes in oxygen isotope. The isotope shift is given by

$$\frac{(\nu_{vib})^{18}O_2}{(\nu_{vib})^{16}O_2} = \sqrt{\frac{\mu^{16}O_2}{\mu^{18}O_2}} = \sqrt{\frac{8}{9}} = 0.94$$

(For further details see C. M. Barlow et al., *Biochem. Biophys. Res. Commun.*, 55 (1973):91–95.) From consideration of the reduced mass of the ^{18}O isotope, the stretching frequency would therefore be shifted to 1043 cm^{-1} for the *isolated* molecule. That it shifts only to 1065 cm^{-1} in the protein indicates that it is not quite a simple O—O stretch, but may be modified by being attached to the iron atom.

Worked Example 3.6   Hydrogen Bonding of Peptide Group

How would you expect hydrogen bonding to affect the following vibrations in the infrared spectrum of the peptide grouping?

(a) the C=O stretching (amide I band),
(b) the N—H stretching, and (c) the N—H bending motions.

Solution

In (a) and (b), hydrogen bonding should make it easier for the carbonyl oxygen to move toward its donor and the N—H hydrogen toward its acceptor. Less energy is required for a vibration in the H-band direction and the bands should shift to *lower* wavenumber. (c) Hydrogen bonds are usually linear, so the N—H bending motion becomes more difficult with the hydrogen now "tied" down. The band associated with this vibration will move to *higher* wavenumber.

Worked Example 3.7   Hydrogen–Deuterium Exchange

Hydrogen–deuterium exchange studies on proteins give information about the probability of exposure of amide protons to the solvent. Infrared spectroscopy can be used to monitor the gross exchange process.

The protein content of retinal outer segment (ROS) membranes is 85% rhodopsin. The degree of deuterium incorporated into rhodopsin was monitored using the amide spectral region from an ROS preparation. The two spectra shown in Figure 3.7 represent (1) that obtained after exposure for 24 h to D_2O (solid line) and (2) that obtained after denaturation in D_2O (dotted line).

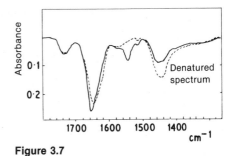

Figure 3.7

How would you interpret these spectra given the data in Table 3.1 and how would you follow the extent of deuterium incorporation at any time?

Table 3.1

Band Position	Assignment
$1736 \ cm^{-1}$	Phospholipidester carbonyl stretching
$1652 \ cm^{-1}$	Amide I (C=O stretching of peptide bond) α-helical form
$1643 \ cm^{-1}$	Amide I (random-coil form)
$1632 \ cm^{-1}$	Amide I (β-sheet form)
$1515 \ cm^{-1}$	Tyrosine band
$1546 \ cm^{-1}$	Amide II (N—'H bending and C—N stretching)
1430/55	Amide II band when N–D

Solution

From Table 3.1 we can assign the bands at $1736 \ cm^{-1}$ and $1515 \ cm^{-1}$ directly. The band at $1655 \ cm^{-1}$ can be assigned to the amide I vibration, which probably implies that rhodopsin has a high α-helical content. The contribution to the absorption at $1546 \ cm^{-1}$ from the amide II band completely disappears on deuteration, to be replaced by a new band at $1455 \ cm^{-1}$. This represents the subtraction of the N—H component from the amide II band. Denaturation leads to complete exchange and a shift in the amide I band to $1643 \ cm^{-1}$, as expected.

The extent of deuteration can be followed by measuring the decrease in the amide II absorbances (after appropriate background correction), or the increase in the "deuterated" band at $1455 \ cm^{-1}$. (To obtain absolute values of the number of unexchanged protons requires a determination of the 0% and 100% values.)
Note: This example illustrates that we can follow the gross changes. A drawback is that individual atoms cannot be followed.

Worked Example 3.8   Ionization of Carboxyl Groups

The ionization of carboxyl groups can be monitored using infrared absorption. The ionized residues have a band at about 1570 cm^{-1} (C=O stretching), which changes to about 1710 cm^{-1} on protonation.

In proteins, these bands appear as shoulders on the much more intense Amide I band. The problem of following the carboxyl bands can be overcome by the use of difference spectra. (A difference spectrum is generated by subtracting two spectra that are obtained before and after perturbation of the system).

a. Typical direct and difference (pD 4.4 as reference) spectra of lysozyme in D$_2$O are shown in Figure 3.8(a) and (b). Why is there no amide II band (1546 cm^{-1})?

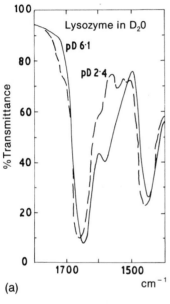

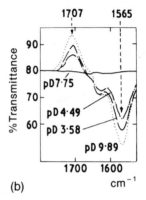

(b)

Figure 3.8

(a)

b. The difference spectra are characterized by bands at 1565 cm^{-1} and 1707 cm^{-1}, respectively. To what do these correspond?

c. How would you measure the pK$_a$ of the ionizable carboxyls and how would you interpret the results?

Solution

a. The absence of the amide II band indicates extensive deuteration. This is confirmed by the "deuterated" amide II band at 1455 cm^{-1} in Figure 3.8(a).

b. The band at 1565 cm^{-1} corresponds to COO$^-$ and that at 1707 cm^{-1} to COOD.

c. One could measure the intensity of the 1565 cm^{-1} or 1707 cm^{-1} band as a function of pH (see, e.g., Problem 6). A plot will give the pK$_a$ values. If more than one ionizable group is present, some other knowledge is then needed to assign those pK$_a$ values to individual groups (for further details see S. N. Timasheff and J. A. Rupley, *Arch. Biochim. and Biophys.*, 150(1972):318–323).

Worked Example 3.9   FTIR "Probes" the Purple Membrane During the Photocycle

The purple membrane of *Halobacterium halobium* contains a single membrane protein bacteriorhodopsin (bR) crystalized in the bilayer plane. This protein is involved in transporting protons across the membrane. One step in this process is light driven.

Figure 3.9(a) shows the FTIR difference spectra of a film of purple membrane obtained at 77 K by subtracting a reference spectrum recorded in the dark from a spectrum of the same film recorded immediately afterward under constant 500-nm illumination. Figure 3.9(b) shows similar data for a D_2O-humidified film.

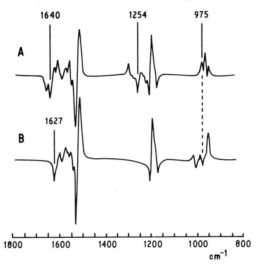

Figure 3.9

a. Suggest an explanation for the positive and negative peaks in these spectra.

b. Why are there some peaks at quite different positions in Figure 3.9(a) and (b)?

c. One of the key roles in the proton transport is thought to involve deprotonation of a Schiffs' base between a lysine residue on the protein and retinal, a light-absorbing group that is attached to bR. The peak at 1640 cm^{-1} in Fig. 3.9(a) is assigned to the C=N stretch of a Schiffs' base. Might this be protonated?

Solution

a. The negative peaks in the FTIR difference spectra reflect a loss in absorption due to the depletion of bR, and the positive peaks reflect the production of some other intermediate.

b. The main changes arise from deuterium–hydrogen exchange. (For example, a negative peak appears near 975 cm^{-1} in Figure 3.9(b) where there was a positive peak previously, and the peak at 1254 cm^{-1} is absent).

c. Yes, there is a small isotope shift for this band from 1640 cm^{-1} to 1627 cm^{-1} in Figure 3.9(b), which suggests that the base is protonated (for further details, see K. J. Rothschild and H. Marrero, *Proc. Natl. Acad. Sci. USA.*, 79(1982): 4045–4059.

INFRARED SPECTRA OF ORIENTED SAMPLES

One of the important points we stressed in Chapter 2 was the directional property associated with the transition dipole. In oriented samples this directional property can be detected in the spectra by using infrared light polarized along a particular direction with respect to the oriented samples.

Worked Example 3.10   Direction of the Transition Dipole in the Peptide Group

As we discussed previously, the following types of vibrations and their frequencies have been identified in the planar peptide group: (i) N—H stretching, (ii) C=O stretching and (iii) C—N stretching. These are shown schematically. Indicate the direction of oscillation of the transition dipole moment.

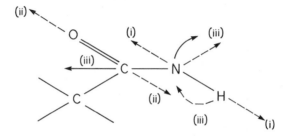

Solution

In (i) and (ii), the direction of oscillation of the transition dipole moment is parallel to the N—H and C=O bonds. We can say that the transition dipole moment is polarized parallel to the N—H and C=O bonds.

In (iii), this is more complex because we have to consider two motions. The change in dipole moment for the C—N stretching is parallel to the C—N bond. The change in dipole moment for the NH bending motion shown is at right angles to the N—H bond. The net resultant transition dipole moment is then polarized at a slight angle to the C—N bond.

CORRELATION OF THE DIRECTION OF THE TRANSITION MOMENT OF ABSORPTION BANDS WITH MOLECULAR STRUCTURE

In the α-helix of Figure 3.10(a), the N—H $\cdots$ C=O peptide hydrogen bonds are oriented parallel to the long axis. The C—N bond is almost at right angles to the long axis. Polarized infrared radiation will therefore be preferentially absorbed

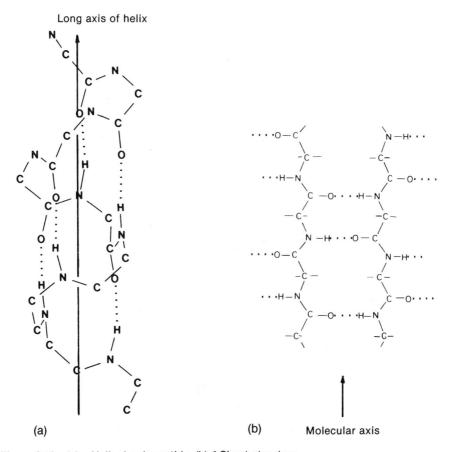

Figure 3.10 (a) α-Helical polypeptide. (b) β-Sheet structure.

by the N—H and C=O stretching motions when it is parallel to the long axis. The motions corresponding to the amide II band will show preferential absorption (dichroism) when the polarized infrared radiation is almost perpendicular to the long axis. Inspection of Figure 3.10(b) shows that opposite effects will be expected for the sheet conformation. The characteristics of the three main peptide absorptions bands are listed in Table 3.2.

Infrared dichroism is used mainly for fibrous proteins and other polymers, such as polysaccharides, that can be oriented. The result can be used to indicate whether a particular bond is oriented parallel or perpendicular to the fiber axis. In this way α- and β- (as well as collagen) configurations have been detected in oriented samples.

Table 3.2 Characteristics of some infrared absorption bands of the peptide group[a]

Vibration	Direction of Oscillation of Transition Dipole	α-Helix Frequency (cm^{-1})	Hydrogen-bonded Forms		
			Dichroism (relative to long axis)	β-Sheet Frequency (cm^{-1})	Dichroism
N—H (stretch)	↔	3290–3300	∥	3280–3300	⊥
Amide I (C=O stretch)	↔	1650–1660	∥	1630	⊥
Amide II C—N (C—N stretch) ╲ (N—H bend) H	↖	1540–1550	⊥	1520–1525	∥

[a]Adapted from J. A. Schellman and C. Shellman in *The Proteins*, 2nd ed., vol. 2, ed. H. Neurath (New York: Academic Press, 1962).

Worked Example 3.11   Dichroism of an Oriented Sample

The infrared spectrum of a thin film of poly-γ-benzyl-L-glutamic acid is shown in Figure 3.11. The sample was oriented by stretching the film. The dotted spectrum is obtained when the radiation is polarized perpendicular to the long axis of the fiber while the continuous spectrum is obtained when the radiation is polarized parallel to the long axis. Use Tables 3.1 and 3.2 to explain these observations.

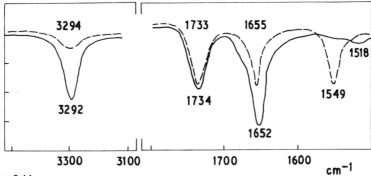

Figure 3.11

Solution

   This is an example of IR dichroism. Those vibrations with transition dipoles parallel to the electric vector of the radiation will show preferential absorption. From Table 3.2, we can suggest that the polymer is in the α-helical form. We expect that when the radiation is parallel to the long axis, we shall have preferential absorption both in the N—H stretching region at 3292 cm^{-1} and in the amide I region at ~ 1652 cm^{-1}. Similarly, when the polarization is perpendicular to the long axis, we expect preferential absorption in the amide II band (1549 cm^{-1}). The band at about 1733 cm^{-1} can be assigned to the C=O stretching of the carboxyl groups (see Table 3.1). This band does not show any dichroism. Two explanations could account for this: (1) the carboxyl groups are not ordered in the fiber, so their transition dipole moments are randomly oriented, or (2) the transition dipole moment is oriented at an angle to the long axis such that its perpendicular and parallel components are equal. (The dichroism of a band ($A_{\parallel}/A_{\perp}$) is approximately given by $2\cos^2\theta/\sin^2\theta$). When $\theta = 54°$, the ratio is approximately unity (for further details, see M. Tsuboi, *J. Polym. Sci.* 59(1962):139–153). Although it is possible, dichroic ratios have not often been used to determine θ because of experimental errors related to the spectrometer, the resolution of the spectrum, and the nature of the sample, all of which have to be allowed for. FTIR instruments should be better in this respect.

POLYMER SPECTRA CAN BE COMPLEX BECAUSE OF INTERACTIONS BETWEEN ELECTRIC TRANSITION DIPOLES

The dichroic properties referred to in Table 3.2 are useful empirical guides for identification of the conformation of polymers. However, the situation is more complex because of interactions between neighboring transition dipoles. The bands may be split or shifted as a result of the interactions between neighboring peptide groups (see Appendix IX).

The approach to calculating what will be observed when there is an interaction between transition dipoles essentially reduces to identifying the relative directions (or phases) of the dipoles and then considering the interactions between them, just as for classical dipoles (see Appendix X). The allowed transitions can be deduced from inspecting the vector sum of the dipoles along appropriate axes, which depend on the geometry of the system. It may be expected, therefore, that the calculation of a polymer spectrum will be very complex. However, in an extended polymer of many similar units, symmetry considerations simplify the problem.

Worked Example 3.12   Spectrum for an Extended Polypeptide Chain

In a single extended polypeptide chain with a β-sheet structure, two consecutive peptide units are regarded as a "unit cell" in order to consider the vibrationally active infrared transitions. These are (1) Those in which the transition dipoles have a finite (vector) sum *parallel* to the axis of the chain, and (2) Those in which the transition dipoles have a finite sum *perpendicular* to the axis of the chain. State whether the transition dipoles are in or out of phase for the motions shown in Figure 3.12, and deduce the number of bands in the spectrum by considering their vector sums.

Figure 3.12   (a)   (b)

Solution

The directions of motions giving rise to the transition dipoles are indicated by the arrow. In (a), the motions are in the same sense and therefore in phase. Note too, that the vector sum is obtained from the diagram,

(a)   (b)

from which it is clear that the vector sum is nonzero only parallel to the axis. (The perpendicular components will cancel out). For (b) the directions of motion are out of phase and, as expected, the vector sum of the dipoles now has a nonzero component only perpendicular to the axis. Each characteristic band for a monomer is split into two bands because of this interaction, one of which is polarized parallel to and the other perpendicular to the molecular axis.

PROBLEMS

1. CO_2 is a linear molecule and is expected therefore to have four fundamental vibrational modes. These are shown in Figure 3.13. The vibration at 1340 cm^{-1} cannot be detected in the infrared spectrum of CO_2. Comment.

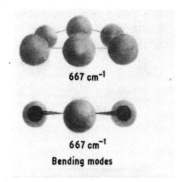

667 cm^{-1}

667 cm^{-1}

Bending modes

2. Alanine has a band at 1308 cm^{-1} assigned to the —CH deformation. In deuteroalanine this band is absent, but a new band appears as 960 cm^{-1} Why?

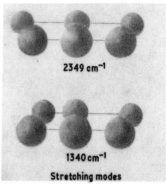

2349 cm^{-1}

3. Infrared spectra have been used to examine deuteroalanines produced during transamination by glutamate–alanine transaminase. Deuteroalanine has bands at 1451 cm^{-1}, 1356 cm^{-1}, and 1018 cm^{-1}, which are associated with the vibrations of the CH_3 group. During a transamination reaction in D_2O with alanine, these bands disappear. Suggest a reason for this.

1340 cm^{-1}

Stretching modes

Figure 3.13

4. Infrared spectroscopy has been used as a method for determining base pairing. The standard difference spectra of poly (A + U) and poly (G + C) against their hydrolysates are shown in Figure 3.14. The hydrolysis difference spectrum of *E. coli* RNA is shown in Figure 3.14(b) and can be simulated by using a mole fraction of 0.27 of the base-paired (A + U) and a mole fraction of 0.33 of the base-paired (G + C). What can you conclude from this?

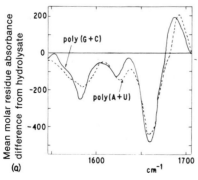

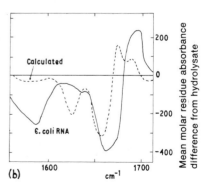

Figure 3.14

5. E. coli ribosomes consist of protein and RNA. The infrared spectra (of a 1%
   solution in a 1-cm cell) of *E. coli* 70S ribosomes and native RNA, both in D_2O,
   are shown in Figure 3.15. When the appropriate amount of the spectrum of RNA
   is subtracted from the *E. coli* spectrum, the resulting spectrum is that of a typical
   protein. What can you conclude from this?

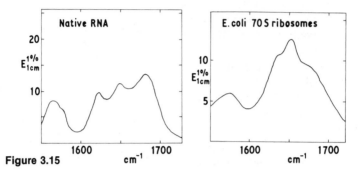

Figure 3.15

6. Heparin is predominantly an alternating copolymer of L-iduronic acid and D-
   glucosamine. The ionization state of iduronic acid was monitored by following
   the change in absorbance of the carboxylate asymmetric stretch at 1614 cm^{-1} as a
   function of pH^+ (the experiment was performed in D_2O). The protonated form
   has an absorbance of the carboxyl at 1730 cm^{-1}). Deduce the ionization states of
   heparin under physiological conditions (pH = 7).

Absorbance: 0.156 0.156 0.156 0.152 0.135 0.090 0.063 0.049 0.032 0.023 0.020
      pH: 6.5   5.75 5.25 4.85 4.30 3.75 3.20 2.70 2.15 1.55 1.00

7. Hydrogen–deuterium exchange data are generally presented as plots of the fraction of unexchanged protons versus $\log(k_0 t)$, where k_0 is the psuedo-first-order
   rate constant for exchange of peptide groups of polypeptide chains and t is the
   time. Two such plots are shown in Figure 3.16 for an intact antibody (IgG) and
   an antibody where the interheavy chain disulfide has been reduced. Suggest a
   reason for the displacement of the plots.

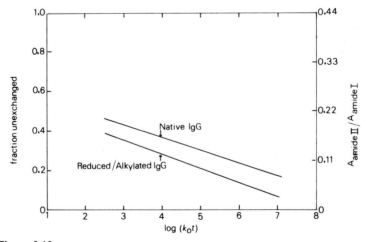

Figure 3.16

8. The protein bacteriorhodopsin (bR) is the light-driven proton pump in the purple membrane of *Halobacterium halobium*. An infrared dichroic study of oriented dried purple membranes shows that the amide I (essentially C=O stretching) is polarized perpendicular to the long axis of the protein. What does this suggest about the conformation of bR in the membrane.

9. Cellulose I possesses a unit cell containing two parallel chains. The repeating unit in the chain is cellobiose (Figure 3.17). The infrared spectrum of the crystalline form of cellulose has a number of bands in the OH region but none of them correspond to free OH stretching frequencies. Dichroism studies have shown that the polarization of one of these bands is perpendicular and two are parallel to the long axis. On this basis, can you suggest a hydrogen-bonding scheme involving the protons shown?

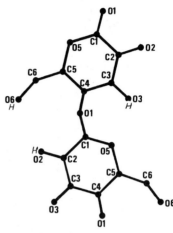

Figure 3.17

10. Studies on oriented samples have shown that the bands expected for a monomer can be further split into two or more bands in a polymer. These bands can have different polarizations. Would you expect polarization effects and the splitting to remain in solution?

11. CO binds covalently to the heme group in myoglobin (Mb). The heme is planar and its Fe atom lies in the heme plane (state a). The bond between Fe and CO can be broken by light. After photodissociation, the heme group buckles and the Fe moves out of the mean heme plane (state b). CO ultimately rebinds and the system returns to state a (Figure 3.18).

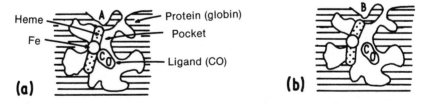

Figure 3.18

The stretching frequency of free CO is 2140 cm^{-1} for $^{12}C^{16}O$. When bound to Mb, this alters to 1945 cm^{-1}
(a) What would be the stretching frequencies for the $^{13}C^{16}O$ isotope when free and when bound?

(b) The rebinding of CO can be monitored simultaneously for two isotopes, $^{12}C^{16}O$ and $^{13}C^{16}O$, by measuring the growth of the absorption spectra at Mb-bound CO stretching frequencies with time-resolved infrared spectroscopy, as shown in Figure 3.19. By inspection, what can you deduce about the isotope effect on the rate of rebinding. Does this fit with the usual assumption from kinetics that the rate for the ^{13}C isotope should be slower by a factor $\exp(\Delta E/kT)$, where ΔE is the difference in the zero-point energies of the Fe—CO stretching frequencies of the two isotopes? There is a difference of 5 cm^{-1} in the Fe—CO stretching frequencies of the two isotopes ($k = 1.38 \times 10^{-23}$ J·K^{-1}).

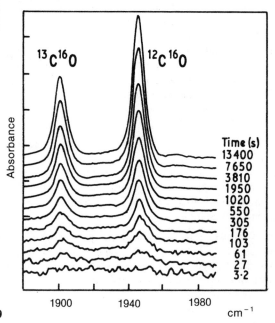

Figure 3.19

CHAPTER 4

Ultraviolet and Visible Absorption Spectroscopy

OVERVIEW

1. Electronic spectra involve transitions between different electronic energy states.

2. The accessible regions for observing these are 200–400 nm (ultraviolet) and 400–750 nm (visible).

3. The term **chromophore** describes the group giving rise to the electronic transition. In biological systems there are relatively few chromophores in the accessible range. Examples include aromatic amino acids, nucleic acid bases, NADH, hemes, polyunsaturated centers, and some transition metal ions.

4. Two parameters characterize an absorption band: the position of the maximum (λ_{max}) and the extinction coefficient (ϵ). The relationship between extinction coefficient and concentration is given by the **Beer-Lambert law.**

5. The electronic energy levels are described by molecular orbitals. For molecules with unsaturated bonds, the important molecular orbitals are π, π^*, and n. Transition metal complexes have molecular orbitals involving the d-electrons of the metal ion. Charge transfer between groups can sometimes occur and this can give intense bands.

6. The main selection rules involve spin and displacement of charge. Transitions are allowed (1) if the spin orientations of the promoted electrons remain unchanged and (2) if they involve unsymmetrical movement of charge. (The extent of charge displacement is related to the electric transition dipole moment.) Selection rules are not rigorously obeyed.

7. The different time scales for an electronic transition ($\sim 10^{-15}$ s) and molecular rearrangement ($\sim 10^{-12}$ s) form the basis of the **Franck-Condon principle.**

8. The sensitivity of electronic spectra to the solvent environment of the chromophore leads to shifts in λ_{max} and ϵ and is the basis of solvent perturbation spectra.

9. Important biological applications of ultraviolet (UV) spectroscopy are measurements of concentrations and of the interaction of ligands (including H^+) with macromolecules.

10. Transition metal ions can give spectra characteristic of the nature and stereochemistry of the coordinated ligands.

11. The electric transition dipole moment has a direction associated with it. In oriented samples this can lead to **dichroism**.

12. Neighboring electric transition dipole moments can interact. This leads to **hyperchromism** and **hypochromism**, which are particularly important in studies of nucleic acids.

INTRODUCTION

Electronic spectra involve transitions between different energy levels (states) of the electrons in molecules. Electromagnetic radiation causes the electrons to oscillate in response to the incident electric field.* If the induced frequency of oscillation coincides (i.e., is in **resonance**) with the difference in energy between two different electronic states, the probability of a transition becomes high.

Electronic transitions give rise to the familiar manifestations of the interaction of electromagnetic radiation with matter: color, vision, and the conversion of sunlight into energy by plants. In addition, electronic spectroscopy experiments are relatively simple and inexpensive to perform, yet they can be done with great precision.

The energies involved in electronic transitions correspond to the absorption of photons in the visible (400–750 nm) and ultraviolet (200–400 nm) regions of the electromagnetic spectrum. Below 200 nm is the far- or vacuum-ultraviolet region, so called because absorption by oxygen of the air is considerable in this region and spectra can be obtained only if the spectrometer is evacuated. Routine instruments therefore extend only to about 185 nm.

The most general application of electronic spectroscopy is to measure concentrations—for example, of proteins, NADH, and nucleic acids—because in most cases there is a direct and simple relationship between the number of molecules present and absorption. In some cases, however, the absorption depends on the molecular environment in a significant way. The change in spectrum with environment can then be interpreted in terms of biological events, such as the unfolding of nucleic acids or the binding of coenzymes to protein. The great utility of such "probes" of environment is the experimental simplicity and sensitivity and the fact that these experiments can be carried out in solution. It is usually not

---

*The extent of these induced oscillations depends on the *polarizability* of the molecule.

possible, however, to make measurements in intact systems, because radiation in this range is scattered and so cannot penetrate tissues (see Chapter 8).

In a large number of molecules, the absorption of a photon arises from the excitation of the electrons of a small group of atoms. The term **chromophore** (Gr. "color bringer") is used to describe the unit giving rise to the selective absorption in question.

The electronic absorption spectra of molecules enables us to account for their color. The **color** of a compound is determined by the light it does *not* absorb. Copper sulfate solution appears blue in white light because it absorbs light in the red region.

PARAMETERS OF ELECTRONIC SPECTRA

The electronic spectrum is usually presented as a plot of absorbance versus the wavelength of irradiation (usually expressed in nanometers). Two parameters are used to characterize a particular band, its position at the maximum (λ_{max}) and its intensity.

The intensity of light falls off exponentially as it passes through the absorbing sample. The experimental measure of intensity at a particular wavelength is the **extinction coefficient** (ϵ), which is given by the Beer-Lambert law

$$\log_{10}\left(\frac{I_0}{I_t}\right) = \epsilon c l$$

where I_0 is the radiation incident on the compound, I_t is the radiation transmitted by the compound. In other words, $I_0 - I_t$ is the radiation absorbed, c is the concentration of the compound, and l is the length of the cell through which radiation actually travels (that is, the path length).

The extinction coefficient is a measure of the "absorbing power" of the compound and is clearly related to the transition probability at a given wavelength. The variation of ϵ with wavelength constitutes the absorption spectrum. Spectra that contain several bands are characterized by the values of λ_{max} and the corresponding value of ϵ (often called ϵ_{max}). The Beer-Lambert law is often expressed in the form

$$A = \log_{10}\left(\frac{I_0}{I_t}\right) = \epsilon c l$$

where A is called the **absorbance**, or **optical density**, of the compound. (A is dimensionless). A plot of A versus the wavelength is an alternative way of presenting the spectrum.

Worked Example 4.1   Determination of Concentration

If 20.8% of the 340-nm radiation incident on a given solution of NADH is transmitted and if the extinction coefficient of NADH at 340 nm is $6.22 \times 10^6 \text{ cm}^2 \cdot \text{mol}^{-1}$, what is the concentration of NADH in the solution? (The path length is 1 cm).

Solution

$$I_t = \frac{20.8}{100} I_0$$

The absorbance is then

$$A = \log_{10}\left(\frac{I_0}{I_t}\right) = 0.682$$

From the Beer-Lambert law, $A = \epsilon c l$; hence,

$$c = \frac{0.682}{6.22 \times 10^6 \times 1} = 0.11 \times 10^{-6} \text{ mol} \cdot \text{cm}^{-3}$$

$$= 0.11 \times 10^{-3} \text{ mol} \cdot \text{dm}^{-3}$$

Worked Example 4.2   Practical Considerations for Measurement of Absorbance

Calculate the absorbance of (a) a $1\text{-mmol} \cdot \text{dm}^{-3}$ solution and (b) a $1\text{-}\mu\text{mol} \cdot \text{dm}^{-3}$ solution of NADH in a cell of path length 1 cm. Comment on the percentage of transmitted radiation in these two cases.

Solution

$$A = \epsilon c l$$

Since ϵ is in units of $\text{cm}^2 \cdot \text{mol}^{-1}$, c must be expressed in $\text{mol} \cdot \text{cm}^{-3}$. (To convert $\text{mol} \cdot \text{dm}^{-3}$ to $\text{mol} \cdot \text{cm}^{-3}$, we must divide by 10^3).

a. $A = (6.22 \times 10^6) (1 \times 10^{-3} \times 10^{-3}) (1) = 6.22$

$$\log_{10}\left(\frac{I_0}{I_t}\right) = 6.22$$

$$I_t = 6.01 \times 10^{-5}\% \ I_0$$

b. $A = (6.22 \times 10^6)(1 \times 10^{-6} \times 10^{-3})(1) = 0.00622$

$$I_t = 98.6\% \ I_0$$

The value of I_t in both (a) and (b) shows that it will be extremely difficult to measure the concentration of either of these solutions. In (a) I_t is negligibly small, and in (b) the amount of radiation actually absorbed is very small. In practice, I_t should be between 10% and 90% of I_0, (that is, values of A should be between 1.0 and 0.05). The most accurate measurements will be when $I_t \sim 50\% \ I_0$. These conditions could be met by altering c, the path length, or changing the wavelength of the radiation (and hence ϵ). In practice, c is most commonly varied.
    The units of ϵ depend on those chosen for c and l. If c is given in $\text{mol} \cdot \text{dm}^{-3}$ and l is given in cm, then ϵ has units of $\text{mol}^{-1} \cdot \text{dm}^3 \cdot \text{cm}^{-1}$.*

---

*Molar (M) is often used instead of $\text{mol} \cdot \text{dm}^{-3}$ in texts in which S.I. units are not insisted upon.

ELECTRONIC ENERGY LEVELS

The electronic energy levels of molecules are described by molecular orbitals (see Appendix V). When an electron undergoes a transition, it is transferred from one molecular orbital to another. However, there are cases where we can consider the excitation to be localized (approximately) to a particular bond (e.g., $C=O$) or group of atoms. To illustrate the principles involved, it is sufficient to consider the molecular orbitals associated with these chromophores.

For this chapter we shall need to know about three molecular orbitals that are associated with unsaturated centers in molecules. These are the π bonding orbital, the π^* antibonding orbital, and the n nonbonding (lone-pair) orbital. The shapes and relative energies of these are shown in Figure 4.1(a) for the $C=O$ group.

In transition metal complexes, the electronic energy levels may also be described by molecular orbitals formed between the metal and ligands. However, it is conceptually convenient to consider the energy levels as being essentially those of the d-orbitals of the central atom (see Appendix VI). In the absence of ligands, all five d-orbitals are degenerate (have the same energy). In the presence of ligands, the orbitals are split into different energies. The exact splitting depends on the stereochemical arrangement of the ligand. That for an octahedral arrangement is shown in Figure 4.1(b).

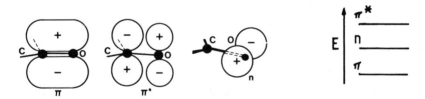

(a) π, π^*, and n molecular orbitals of the $>C=O$ group and their relative energies.

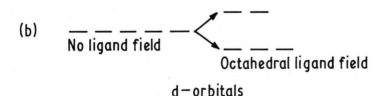

(b)   No ligand field

Octahedral ligand field

d−orbitals

Figure 4.1   (a) π, π^*, and n molecular orbitals of the $-C=O$ group and their relative energies. (b) Splitting of the five d-orbitals of a transition metal ion in an octahedral ligand field.

ELECTRONIC TRANSITIONS: SELECTION RULES

Symmetry Considerations

Transitions between electronic energy levels are allowed if they result in an unsymmetrical movement of charge. The magnitude of this movement is reflected in the electric transition dipole.

Worked Example 4.3   Allowed and Forbidden Transitions

a. Which of the following electronic transitions between the energy levels indicated in Figure 4.1 are allowed: (i) $\pi \rightarrow \pi^*$ and (ii) $n \rightarrow \pi^*$?

b. Are any $d \rightarrow d$ transitions allowed?

Solution

a. (i)   The $\pi \rightarrow \pi^*$ transition results in an unsymmetrical displacement of charge and is an allowed transition.

   (ii) The $n \rightarrow \pi^*$ transition can be thought of as occurring in two stages: a rotation and a displacement of charge. The rotation part interacts with only the magnetic component of the radiation. This interaction is very weak and is classified as forbidden (see Chapter 2).

b. Since the shapes of all d-orbitals are equivalent (see Appendix VI) all $d \rightarrow d$ transitions must be forbidden, since they will not result in an unsymmetrical displacement of charge.

Forbidden Transitions

The symmetry selection rule was derived by considering isolated molecular orbitals. In a real molecule, the exact shapes of the orbitals will be influenced by their environment. Distortions in these shapes are caused by mixing or "contamination" with other orbitals; this can make forbidden transitions allowed. Vibrational motions can also cause relaxation of the selection rules. The name **vibronic** has been given to this latter mechanism since a transition becomes allowed because it involves both a vibration and an electronic excitation simultaneously. Of the two mechanisms, contamination with other orbitals would be expected to be more efficient than the vibronic.

Examples of forbidden transitions that become allowed through these mechanisms are the $n \rightarrow \pi^*$ transitions that occur at about 290 nm in carbonyl compounds (with an extinction coefficient $\epsilon = 20$ cm$^{-1} \cdot$dm$^3 \cdot$mol^{-1}) and the $d \rightarrow d$ transitions that are responsible for the color of transition metal complexes.

Spin Considerations

In dealing with electrons, we have to remember that they also have the intrinsic property of spin. A molecular orbital may contain up to two electrons, but they must have opposed spin orientations. One selection rule states that during an electronic transition, the spin orientation of the promoted electron must remain unchanged (that is, opposed to the spin orientation of the electron left behind). Spin-orientation transitions (or flips) can occur only as a result of interaction of the spin with a magnetic field. This means that a transition in the ultraviolet or visible involving a spin reorientation will require *both* electric and magnetic radiation.* However, the magnetic transition moment is very much smaller than the electronic one (about 10^{-5} times). This is why such transitions are weak.

Worked Example 4.4   Spin-Allowed Transitions

Which of the following transitions between the energy levels in Figure 4.2 are spin allowed?

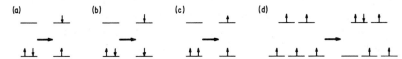

Figure 4.2

Solution

a. The orientation of the spin does not change and the transition is allowed.

b. This is a spin-forbidden transition. The orientation of the spin changes.

c. This is a trick. The two electrons in a molecular orbital must have opposed spins. (This is known as the **Pauli principle**). The initial state cannot exist.

d. This is a spin-forbidden transition. The five energy levels in the initial state could correspond to those from the five *d*-electrons of a transition metal (such as Fe(III) or Mn(II)) when it is in a ligand field. In this case the ligand field would have split the *d*-orbitals into two sets. The transition would also be symmetry-forbidden and hence very weak.

THE TIME FOR AN ELECTRONIC TRANSITION

Electronic transitions between two states occur when a transition dipole oscillates between the two states involved. The resonance condition $E = h\nu$ means that the frequency of oscillation is ν. The time taken to oscillate between the states—i.e., for the transition to occur—is thus $1/\nu$. A typical electronic transition at 420 nm (equivalent to a frequency of 7.14×10^{14} Hz) will therefore take $(1/7.14 \times 10^{14})$ s, or 1.4×10^{-15} s.

---

*The mechanism by which the spin-selection rule is relaxed is a magnetic one and arises because the electron orbiting in the electrostatic field from the nucleus produces a fluctuating magnetic field that can cause the spin reorientations (see Chapter 2).

Worked Example 4.5   Electronic Transitions are Faster than Nuclear Vibrations

Compare the time for an electronic transition at wavelength 420 nm with that for a typical nuclear vibration at 1500 cm^{-1}

Solution

From the preceding text, we know that the electronic transition takes 1.4×10^{-15} s. For the vibration, we have to convert 1500 cm^{-1} to frequency. Since $v' = v/c$, we obtain $v = 1500 \times 3 \times 10^{10}$ s$^{-1} = 4.5 \times 10^{13}$ s^{-1}. The time for a vibration is therefore $1/(4.5 \times 10^{13}) = 2.2 \times 10^{-14} = 22 \times 10^{-15}$ s. This suggests that the nuclei have no time to move appreciably during the electronic transition.

The previous worked example provides the basis of the **Franck-Condon principle,** which states that *an electronic transition occurs so rapidly that during it the nuclei are static.* Since vibrational transitions by definition require a change in internuclear distance, and since this cannot occur over the time scale of an electronic transition, it follows that electronic transitions are not accompanied by vibrational transitions.

Figure 4.3 shows the Morse curves for the ground and first excited electronic states. The horizontal position of the upper curve depends on the bond length in

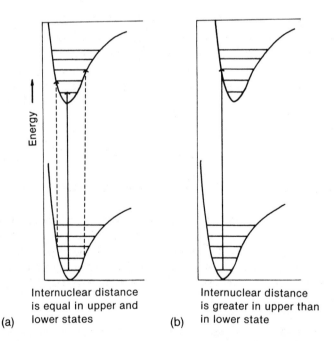

Internuclear distance is equal in upper and lower states	Internuclear distance is greater in upper than in lower state
(a)	(b)

Figure 4.3   Morse curves for the ground and first excited electronic state for a diatomic molecule. The horizontal lines represent the allowed vibrational levels. In (a) the internuclear distance is equal in the upper and lower states, while in (b) the internuclear distance is greater in the upper state than in the lower state.

the excited state compared with the band length in the ground state. The two examples in the figure have bond lengths in the excited state that are either the same or slightly longer than in the ground state. We can assume initially that all the nuclei are in the ground vibrational level of the ground electronic state. An electronic transition in which the nuclei do not move can be represented by a vertical line in the diagram. Depending on the displacement of the two curves, this vertical line will meet the upper curve at different vibrational levels. Since a *molecule* will be vibrating, there will be a whole series of such lines (usually unresolved in solution) from different relative displacements of ground and excited states.

ABSORPTION RANGE OF CHROMOPHORES

We can divide the transitions from those chromophores we expect to observe into two main categories. In the first group are those involving transitions to π^* orbitals, which are generally associated with unsaturated centers. These occur from about 190 nm through to the visible, depending on the extent of conjugation. In the second group are those transitions involving d-electrons in transition metal complexes, which usually occur in the visible region.

We shall consider absorption from three main categories of molecules: (1) the peptide bonds and amino acids in proteins; (2) purine and pyrimidine bases in nucleic acids, and (3) highly conjugated (double bonds) systems.

Peptide Bonds and Amino Acids

The $n \rightarrow \pi^*$ and $\pi \rightarrow \pi^*$ transitions of the peptide group occur in the far-ultraviolet (~ 200 nm). (The π-electrons are delocalized over the carbon, nitrogen, and oxygen atoms). The $n \rightarrow \pi^*$ forbidden transition is typically observed at 210–220 nm ($\epsilon_{max} = 100$), while the main $\pi \rightarrow \pi^*$ transition occurs at ~ 190 nm ($\epsilon_{max} = 7000$).

A number of amino acid side chains, including His, Arg, Glu, Gln, Asn, and Asp, have transitions around 210 nm. However, these cannot usually be observed in proteins because they are "swamped" by the absorptions from the more numerous and intensely absorbing amide backbone groups, which have significant absorption up to around 230 nm.

The most useful ultraviolet range for proteins is at wavelengths greater than 230 nm, where there are absorptions from the aromatic side chains of phenylalanine, tyrosine, and tryptophan (see Table 4.1). Note that the absorbance of phenylalanine in this range is very low (the absorbance around 257 nm with $\epsilon_{max} = 200$ arises from a $\pi \rightarrow \pi^*$ spin-forbidden transition). If tyrosine and tryptophan are present in a protein, then the phenylalanine side chain will con-

Table 4.1   Absorption maxima (λ_{max}) and extinction coefficients (ϵ) for selected amino acids at neutral pH

Molecule	λ_{max} (nm)	ϵ at λ_{max} ($\times 10^{-3}$) ($cm^2 \cdot mol^{-1}$)
Tryptophan	280	5.6
	219	47.0
Tyrosine	274	1.4
	222	8.0
	193	48.0
Phenylalanine	257	0.2
	206	9.3
	188	60.0
Histidine	211	5.9
Cystine	250	0.3

tribute very little to the absorbance in this region. Because Trp has the most intense absorbance, it is often used as the basis for protein concentration measurements. However, its absorption between 240 and 290 nm consists of at least three (unresolved) electronic transitions. The tyrosine spectrum is also complex. Assignment of individual transitions in proteins where there are many Trp or tyrosine residues in a variety of environments is virtually an impossible task.

Of the remaining side chains of interest, the disulfide group of cystine has a weak absorption ($\lambda_{max} = 250$ nm, $\epsilon_{max} = 300$). This chromophore is important in generating optical activity (see Chapter 10).

Purine and Pyrimidine Bases in Nucleic Acids

The absorption of nucleic acids also arises from $n \rightarrow \pi^*$ and $\pi \rightarrow \pi^*$ transitions. The spectra of the constituent purine and pyrimidine bases occur between 200 and 300 nm. Once again the deceptively simple absorption bands are composites of several transitions. Some representative spectral values of purine and pyrimidine bases and their derivatives are listed in Table 4.2. The absorption spectra of these bases when they are in polymers is very much influenced by electronic interactions between bases (see hyperchromism).

Highly Conjugated Systems

Finally, in molecules that contain many unsaturated groupings that are all conjugated, we note that the molecular orbitals containing the electrons in the systems will extend over these groupings. The resulting high degree of delocalization of the electrons means that the energy required for a transition decreases, with the result that the spectrum is often in the *visible* region.* For example, the porphyrin

---

*This idea, that energy levels are closer together as the electrons are less restricted, is nicely predicted by simple wave mechanics—the particle in a box (see page 10, "What Is Matter?").

Table 4.2 Absorption maxima (λ_{max}) and extinction coefficients (ϵ) for purine and pyrimidine bases and their derivatives

	λ_{max} (nm)	ϵ at λ_{max} ($\times 10^{-3}$) ($cm^2 \cdot mol^{-1}$)
Adenine	260.5	13.4
Adenosine	259.5	14.9
NADH	340	6.23
	259	14.4
NAD^+	260	18
Guanine	275	8.1
Guanosine	276	9.0
Cytosine	267	6.1
Cytidine	271	9.1
Uracil	259.5	8.2
Uridine	261.1	10.1
Thymine	264.5	7.9
Thymidine	267	9.7
DNA	258	6.6
RNA	258	7.4

ring system in Figure 4.4(a) is mainly responsible for the color in heme proteins. The spectrum of the zinc porphyrin molecule shown in Figure 4.4(b), which arises from $\pi \to \pi^*$ transitions, has bands in the ultraviolet and visible regions. (The most intense band is called the *Soret* band, after its discoverer.)

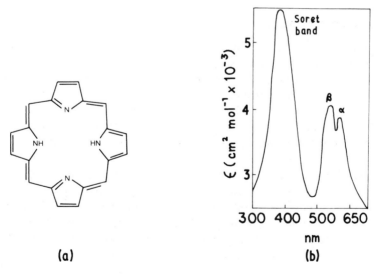

(a) (b)

Figure 4.4 (a) The skeleton of the porphyrin molecule. (b) The spectrum of a conventional metal porphyrin in the visible and ultraviolet regions.

TRANSITION METAL SPECTRA

$d \rightarrow d$ Transitions

The specra of transition metal complexes can be interpreted in terms of $d \rightarrow d$ transitions. The intensities and band positions depend on the metal, the ligand, and the arrangement (stereochemistry) of the ligands. For example, $[Co(H_2O)_6]^{2+}$, which is octahedral, has absorption bands at 513 nm ($\epsilon_{max} =$ 10 cm$^2 \cdot$mol^{-1}) and is pink, while $[Co(Cl)_4]^{2-}$, which is of lower symmetry and therefore has much more intense bands, at 667 nm ($\epsilon_{max} = 600$ cm$^2 \cdot$mol^{-1}) is blue.

The difference in energy between the ground and first excited states (and hence the position of the absorption bands) depends on the **ligand field splitting** of the d-orbitals (see Appendix VI). The ligands may be placed in a **spectrochemical series** according to their field strength. A shortened version of this series is I$^-$ < Br$^-$ < Cl$^-$ < SCN$^-$ (S-bonded) < F$^-$ < OH$^-$ < oxalate < H$_2$O < NCS$^-$ (N-bonded) < NH$_3 \simeq$ pyridine < ethylenediamine < dipyridyl < o-phenanthroline < NO$_2^-$ < CN $\simeq$ CO From left to right across the series, the effect of the ligands on the spectrum is to shift the absorption bands to shorter wavelengths (higher energy). The band positions in spectra of complexes will therefore reflect the position of the ligands in this series. In complexes with different kinds of ligands, the absorption spectrum generally appears at a position corresponding to the average ligand field. In this way a fingerprint may be established that makes it possible to infer the nature of the binding groups.

The theory of the spectra of transition metal ions has been quite well worked out and details can be found in most inorganic texts. Figure 4.5 illustrates the relationship between the colors and $d \rightarrow d$ absorption spectra of some inorganic complexes.*

Charge-Transfer Spectra

In addition to the relatively weak $d \rightarrow d$ transitions, many transition metal complexes exhibit very intense absorptions ($\epsilon_{max} = 10^3 - 10^4$ cm$^2 \cdot$mol^{-1}) that are often referred to as charge-transfer spectra. In these, the electronic transition can be thought of as one in which an electron is removed from one atom and "transferred" to another. A well-known example is the Fe(III) thiocyanate ion, Fe(SCN)$^{2+}$, which is intensely red and in which an electron from the thiocyanate is transferred to the metal. Another example is permanganate, MnO$_4^-$, which is purple.

Charge-transfer bands most often appear in the ultraviolet and tail into the visible. However, one important exception is the Fe(III) porphyrin systems, which have bands at ~1,000 nm, well into the infrared. The position of charge-transfer bands depends on both metal ion and ligand and on the relative ease of oxidation or reduction of these species. It is not surprising, therefore, that many metal-containing oxidases and electron-transfer enzymes show strong charge-transfer bands.

*For further details, see C. S. G. Phillips and R. J. P. Williams, *Inorganic Chemistry,* vol. 2 (Oxford: Clarendon Press, 1966) chap. 28.

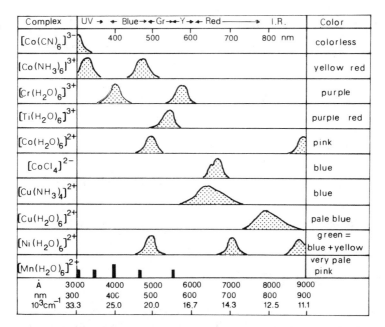

Figure 4.5 The relationship between the colors and the $d \rightarrow d$ absorption spectra of some inorganic complexes (from C. S. G. Phillips and R. J. P. Williams, *Inorganic Chemistry,* vol. 2. (Oxford: Clarendon Press, 1966), p. 377).

Worked Example 4.6   Charge Transfer Is Important in Heme Proteins

The color of most heme proteins is due to the porphyrin ring. The spectrum of an Fe(III) hemoprotein is shown in Figure 4.6.* The band positions are at 1,000; 625; 590; 530; 500; and 400 nm. Refer to Figure 4.4(b) and suggest an interpretation for these transitions.

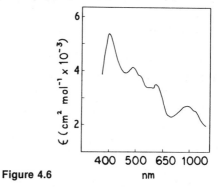

Figure 4.6

Solution

From Figure 4.4(b), the bands at 590, 530, and 400 nm arise from transitions associated with the porphyrin ring. The other bands, which are quite intense, must arise from the presence of the transition metal. The $d \rightarrow d$ transitions, which are orbitally forbidden and therefore weak, are even more unlikely for Fe(III) because here they are also spin forbidden (see "Spin Considerations"). The most likely explanation is that they are charge-transfer bands.

---

*Each of the five *d*-orbitals can contain up to two electrons. Fe(III) can therefore have five unpaired *d*-electrons or one unpaired *d*-electron. These cases are called **high spin** or **low spin**. In this example, Fe(III) is high spin (see Appendix VI).

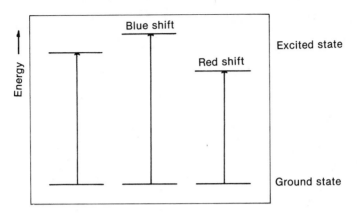

Figure 4.7   Solvent interactions with chromophores can result in red shifts or blue shifts.

SOLVENT EFFECTS ON SPECTRA

In solution, the molecules in each electronic state will be surrounded by a solvent "cage." The solvent can interact with the chromophore and alter its energy levels (that is, electronic distribution) by various mechanisms, such as electrostatic or hydrogen bonding. As a result, absorption spectra are often sensitive to the nature of the solvent. If the solvent interacts differently with the ground and first excited states, the energy gap between them will alter. An increase in the energy gap will cause a blue shift in the absorption band, while a decrease will cause a red shift (see Figure 4.7).

Because of the motions of the solvent molecules, their interactions with the chromophore will change with time. In a solution containing many molecules, a given chromophore will therefore "sample" many slightly different solvent environments (each, of course, corresponding to slightly different energy levels). The time scale for motion of the solvent molecules is around 10^{-11} s, which is very much slower than that for an electronic transition (10^{-15} s).* The energy of the electronic transition depends on the momentary positions of neighboring molecules and hence is slightly variable. This causes a broadening of the absorption band. Figure 4.8 shows the absorption spectra of tryptophan and tyrosine in 20% dimethyl sulfoxide. The shifts are very small, so accurate direct determination is difficult; measurement is easier from the **difference spectrum** between the two solvents. The experimental procedure is illustrated in Figure 4.9.

---

*We note that the average solvent cage will be different in the different electronic states. Because the rearrangement time is slower than time for the electronic transition, the excited state will instantaneously have the solvent cage of the ground state. This is discussed again in the chapter on fluorescence.

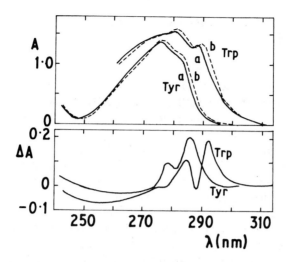

Figure 4.8   Absorption spectra of tryptophan and tyrosine in water (solid line), and 20% dimethyl sulphoxide (dotted line) and the resulting difference spectra.

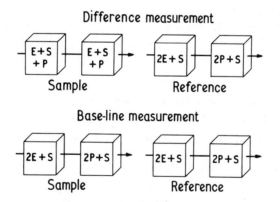

Figure 4.9   Experimental procedure for difference spectra. To obtain good difference spectra the use of pairs of tandem cells (or a single cell with two compartments) is recommended. The contents in the sample beam in the base-line measurement are mixed and the difference between this and the unmixed reference sample is recorded. For solvent perturbation measurements, the combinations of solvent and perturbants would be as shown. (E = protein concentration, P = perturbant concentration, and S = solvent.)

APPLICATIONS OF ULTRAVIOLET SPECTRA TO PROTEINS

The spectrum of a protein containing several chromophores looks like a single spectrum rather than a readily separable sum of individual spectra because the absorption bands are broad. So although the electronic spectra of proteins can be used to indicate the type of aromatic amino acids present, they do not give any information about structure.

Probably the most universal use of absorption spectra is the measurement of protein concentrations. Other applications of the ultraviolet spectra of proteins usually involve monitoring the perturbations (such as changes in pH, temperature, solvent, and denaturation) of the tyrosine or tryptophan absorption bands.

Chemical modification of tyrosine or tryptophan can alter their absorption spectra. Observation of the absorption of ligands (extrinsic chromophores) extends the range of applications considerably. (Perhaps the best known is the use of NAD^+ and NADH in coupled assays.) In addition, covalently attached probes can be used on "reporter groups."

In contrast with electronic spectra of protein chromophores, the electronic spectra of the transition metal "chromophore" can provide structural information in addition to information about the nature of the ligand. However, the use of metal ions is limited to those systems in which they occur naturally or can be used as probes. For instance, zinc can often be replaced by cobalt, and magnesium by manganese. The structural information is obtained by comparing the spectrum from the metal in the protein with spectra from model compounds of known stereochemistry. Difficulties arise in finding suitable model compounds because the metal may have unusual coordination geometries in proteins.

Worked Example 4.7   NADH Absorption Can Be Used to Monitor a Reaction

NADH absorbs at 340 nm ($\epsilon = 6.2 \times 10^6$ cm$^2 \cdot$ mol^{-1}) and at 260 nm ($\epsilon = 14 \times 10^6$ cm$^2 \cdot$ mol^{-1}), while NAD^+ absorbs only at 260 nm ($\epsilon = 18 \times 10^6$ cm$^2 \cdot$ mol^{-1}). Suggest a method for following the kinetics of the reaction

$$AH_2 + NAD^+ \underset{\text{enzyme}}{\overset{\text{Dehydrogenase}}{\rightleftharpoons}} A + H^+ + NADH$$

Solution

Since NADH has an absorption at 340 nm, while NAD^+ does not, the reaction can be followed by monitoring this wavelength.

Worked Example 4.8   Ligand Binding

The spectra of solutions ($29.7\mu M$) of NADH, both free and when bound to an excess of the tetrameric enzyme glyceraldehyde-3-phosphate dehydrogenase, are shown in Figure 4.10(a).

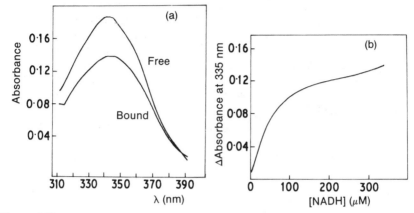

Figure 4.10

a. Calculate the extinction coefficient at 340 nm for NADH when bound to the enzyme. (Path length = 1 cm).

b. Figure 4.10(b) is a titration of the difference in absorbance at 340 nm between NADH and the same concentration of NADH in the presence of a fixed concentration of enzyme ($24.2 \, \mu M$). The curve increases linearly until a coenzyme-enzyme ratio of 2 is obtained. Interpret this finding and calculate the concentration of NADH bound to the enzyme when the absorbance is 0.04 and 0.10.

Solution

a. From Figure 4.10(a), we see that the absorbance at 340 nm for the bound form is about 0.14. Hence from $A = \epsilon cl$, we obtain

$$0.14 = \epsilon(29.7 \times 10^{-6} \times 10^{-3})(1)$$
$$\epsilon = 4.7 \times 10^{6} \, cm^{2} \cdot mol^{-1}$$

This refers to the bound extinction coefficient, which we shall call ϵ_b.

b. In the presence of the enzyme, the absorbance comes from NADH in two possible environments. If we use subscripts to indicate these bound and free states, we have, in the cuvette with enzyme,

$$A = \epsilon_b c_b l + \epsilon_f c_f l \tag{4.1}$$

Since c_T, the total concentration, is the sum of c_b and c_f, we can write

$$A = (\epsilon_b - \epsilon_f)c_b l + \epsilon_f c_T \tag{4.2}$$

(continued)

Worked Example 4.8   (continued)

In the cuvette without enzyme, the absorbance is $\epsilon_f c_T$. Thus the difference spectrum is

$$\Delta A = (\epsilon_b - \epsilon_f)c_b l \tag{4.3}$$

The titration curve in Figure 4.10(b) is thus a measure of binding of NADH to the enzyme (c_b). The curve suggests that two molecules of NADH bind very tightly to the enzyme while the third and fourth bind less tightly, presumably indicating "negative" cooperativity. From the above formula, we can calculate the amount of bound NADH at any point on the titration curve.

Since $\epsilon_b - \epsilon_f = 1.5 \times 10^6 \text{cm}^2 \cdot \text{mol}^{-1}$, then when $\Delta A = 0.04$, $c_b = 27\ M$ and when $\Delta A = 0.1$, $c_b = 67\ M$ (see J. D. Bell and K. Dalziel, *Biochim. Biophys. Acta,* 391 (1975): 249–358).

Worked Example 4.9   Probing Catalytic Subunits of Aspartate Transcarbamylase

Ultraviolet difference spectroscopy has been used to study the catalytic subunits of aspartate transcarbamylase. The binding of succinate and carbamyl phosphate to the enzyme, both individually and together, are shown in Figure 4.11. Comment on the magnitude of the spectrum of the ternary complex.

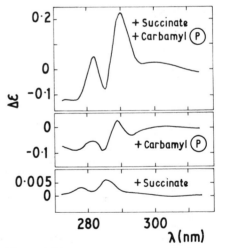

Solution

The intensity in the difference spectrum for the ternary complex is substantially greater than the sum of the intensities in the spectra of the binary complexes formed by each substrate. This suggests that a structural change in the enzyme, involving some aromatic amino acid side chains, has occurred in the ternary complex (for further details, see K. D. Collins & G. R. Stark, *J. Biol. Chem.,* 244 (1969): 1869–1877).

Figure 4.11

Worked Example 4.10   Solvent Perturbation to Determine the Accessibility of Tryptophan Residues in Lysozyme

Solvent perturbation is often applied to proteins to determine whether tyrosine or tryptophan amino acids are located on the surface or buried within a protein. In general, the spectra from those residues on the surface will be perturbed by changes in the physical properties of the solvent (e.g., hydrogen bonding or electrostatic) in their immediate environment. Those residues deeply buried within the protein will be shielded from the effects of the solvent. The experimental method is to use a solvent mixture of 80% water and 20% of a less polar solvent, such as ethylene glycol, glycerol, or dioxane. The number of exposed versus buried chromophores can be calculated from the ratio of the intensity in the difference spectra between the native protein and a standard (which is assumed to be 100% exposed to solvent).

a. What assumptions are implicit in this method?

b. Lysozyme has six Trp and three Tyr residues. A "model" mixture of the esters of these two amino acids in water had a maximum absorbance at 281 nm. The difference spectrum between water and a 20% ethylene glycol solution of this mixture had a maximum absorbance at 292 nm (see Figure 4.12). The ratio of the absorbance in the difference spectra ($\Delta\epsilon_{292}$) to that in water (ϵ_{281}) was 0.042. When lysozyme was used, this ratio was 0.034. What can you deduce from this?

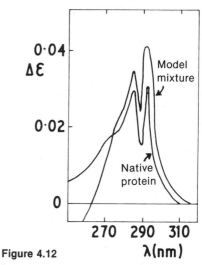

Solution

a. There are three assumptions implicit in the solvent perturbation method: (1) Only exposed chromophores are affected; (2) No preferential solvation of the protein occurs—that is, the solvent cage surrounding the protein has essentially the same composition as the bulk solvent; and (3) The perturbant does not alter the native conformation of the protein.

Figure 4.12

b. The peak at 292 nm in the solvent difference spectrum for lysozyme arises predominantly from the Trp chromophores (see Figure 4.8). The ratio of the absorbances (0.034/0.042 suggests that the relative exposure of Tryptophan residues in lysozyme is 0.8 (that is, about five tryptophan residues).

Note: An important point is that this method measures only the *average exposure.* All six tryptophans may in fact be exposed some of the time. Note also that one could choose other standards, such as the denatured protein, for the 100% exposed standard. The solvent perturbation method should always be interpreted with caution (for further details, see, for example, E. J. Williams and M. Laskowski, Jr., *J. Biol. Chem., 240* (1965): 3580–3584).

Worked Example 4.11   Ionization of Tyrosine

The spectrum of tyrosine (λ_{max} = 274 nm, ϵ_{max} = 1400 cm$^2 \cdot$ mol^{-1}) changes on addition of base (λ_{max} = 295 nm, ϵ_{max} = 2600 cm$^2 \cdot$ mol^{-1}) (see Figure 4.13.)

The ionizations of four tyrosine residues in a 20-μM solution of a protein were followed by monitoring the absorbance at 295 nm as a function of pH. The following results were obtained in a cell of 1-cm path length:

pH:	8	8.5	9.0	9.5	10.0	10.5	11.0	11.5	12.0	12.5
A_{295}: ($\times 10^3$)	0.32	0.94	2.5	5.2	8.0	9.6	10.4	11.2	16.0	20.0

Interpret these data. (Assume that ϵ_{295} = 2600 cm$^2 \cdot$ mol^{-1}.)

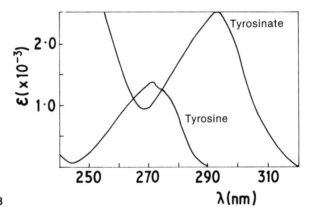

Figure 4.13

Solution

The plot of absorbance versus pH is clearly biphasic (see Figure 4.14). The abrupt increase above pH 11.5 suggests that the protein is becoming denatured. The changes in absorbance up to this pH indicate tyrosines with a pK_a of about 9.5. The total change in absorbance between pH 7 and pH 10.5 is (0.0112 − 0.0003) = 0.0109. From $A = \epsilon cl$, we obtain

$$0.0109 = 2600cl$$
$$c = 41.9 \times 10^{-6}M$$

Since the protein concentration is $20 \times 10^{-6}M$, this means that we are observing the titration of two tyrosines with the same pK_a value. This emphasizes one of the limitations of this technique: It can be used to follow changes, but it gives little structural information in helping to resolve and assign which tyrosines have ionized.

(continued)

Worked Example 4.11   (continued)

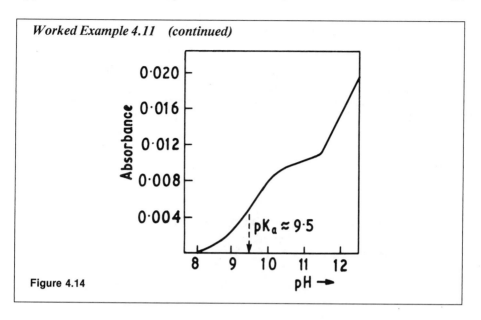

Figure 4.14

Worked Example 4.12   Deducing the Environment of a Transition Metal Ion

The visible absorption spectra of Co(II) in an octahedral $[Co(H_2O)_6]^{2+}$ and tetrahedral environment $[Co(Cl)_4]^{2-}$ are shown in Figure 4.15 (a). Compare these with the spectrum of the cyanide inhibitor complex of carbonic anhydrase and deduce the environment of cobalt in the enzyme. Why does the spectrum change markedly in the absence of inhibitor (see Figure 4.15(b))?

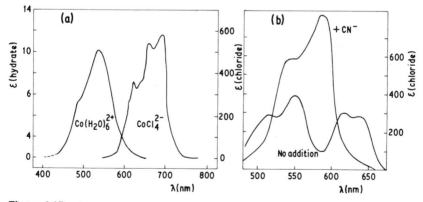

Figure 4.15

(continued)

Worked Example 4.12   (continued)

Solution

   The intensity of the Cyanide–Co(II) carbonic anhydrase spectrum suggests that it is in a tetrahedral environment. (The exact position of the bands will depend on the ligands attached to the metal.) In the absence of inhibitor, the spectrum indicates that a change in the Co(II) environment has occurred. (In fact, this sort of spectrum has been interpreted by some to indicate the presence of a pentacoordinate Co(II) environment on the basis of studies with model compounds.)

PROPERTIES ASSOCIATED WITH THE DIRECTION OF THE TRANSITION DIPOLE MOMENT AND INTERACTIONS BETWEEN THEM

There are three main properties associated with the direction of the transition dipole moment: (1) linear dichroism, (2) exciton splitting, and (3) hypochromism (and hyperchromism).

Linear Dichroism of Oriented Samples

We recall that the comparison of the absorption of parallel- and perpendicular-polarized light by an oriented sample is known as **linear dichroism.** The transition dipole associated with a given transition (for example, $\pi \to \pi^*$) has a definite orientation, which causes selective absorption of light polarized along this orientation. This property was the basis of infrared dichroism, which we discussed in the previous chapter and showed its use in determining the orientation of particular bonds with respect to molecular axes.

Determination of the Orientation of the Transition Dipole.   Just as we assigned bands in infrared to various vibrations and then calculated their expected polarization, we can in principle do the same with electronic transitions in the ultraviolet region. For instance, the $\pi \to \pi^*$ transition (see Figure 4.16) of an isolated C=O group will be polarized along the C=O axis, while the forbidden $n \to \pi^*$ transition is polarized perpendicular to it.

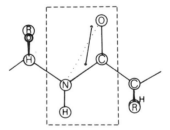

Figure 4.16

Worked Example 4.13   Direction of the $\pi \to \pi^*$ Transition Dipole in a Peptide

From measurements on myristamide, for which the structure is known, the orientation of the $\pi \to \pi^*$ transition dipole was found to lie at an angle of $9.1°$ to the N—O axis toward the C—O axis rather than along the —C—O axis (see Figure 4.16). Suggest a reason for this.

Solution

In a peptide group, the π-electrons are delocalized to some extent over the nitrogen atom as well as over the carbon and oxygen atoms, thus altering the orientation of the transition dipole.

Determining the Orientation of a Group.   In small molecules, linear dichroism measurements have been used quite successfully to determine the orientations of groups for which the direction of the transition dipole is known.* However, in polymer structures, far less has been done with electronic spectra than with infrared, for two main reasons: (1) There are experimental difficulties in making measurements in the region in which the peptide chromophore absorbs; and (2) The differences in the absorption of the peptide chromophore in various conformations are not so well resolved as in the infrared. Other factors, like hyperchromism and hypochromism, further complicate the ultraviolet spectra; nevertheless, some qualitative information can be obtained.

Worked Example 4.14   Orientation of Base Pairs in DNA

The main 260-nm $\pi \to \pi^*$ absorption of nucleic acid bases has a transition dipole in the plane of the bases. The absorption of polarized light in a sample of oriented DNA is polarized mainly perpendicular to the long axis. What can you deduce about the orientations of the base pairs?

Solution

The plane of the bases must be perpendicular to the long axis.

Resolving Ultraviolet Bands.   In general, different transitions will have different dichroisms. This can help to resolve ultraviolet spectra, which often consist of overlapping bands. However, the effect of polarized light on the absorption spectrum of an oriented sample may be difficult to observe directly if there are overlapping bands. A more sensitive method is to measure the dichroic ratio (d) of the spectrum, which is defined as

$$d = \frac{A_\| - A_\perp}{A_\| + A_\perp}$$

---

*Once again, though, (see Worked Example 3.12) there are corrections that involve both the extent of the orientation and the instrumentation that can make these measurements quantitatively unreliable.

where $A_{\parallel}$ and $A_{\perp}$ are the absorbance of light polarized parallel and perpendicular, respectively, to the long axis of the oriented molecule. The dichroic ratio of bands with transition dipoles parallel to the long axis will be positive while that for bands with transition dipoles perpendicular to the long axis will be negative.

Worked Example 4.15   Plots of Dichroic Ratio Versus Wavelength Resolve Ultraviolet Spectra

Typical polarized ultraviolet peptide absorption spectra of the peptide chromophore of an oriented sample (thin film) of the α-helical form of poly-L-glutamic acid are shown in Figure 4.17. These clearly resolve two bands that have been assigned to $\pi \rightarrow \pi^*$ transitions. From the data given for the parallel and the perpendicular absorbances, plot the dichroic ratio at each wavelength to see if this resolves any other transitions. Can you suggest why there are two $\pi \rightarrow \pi^*$ transitions?

λ (nm):	180	190	204	210	222	230	200
$A_{\parallel}$:	0.53	0.63	0.53	0.47	0.38	0.27	0.18
$A_{\perp}$:	0.45	0.46	0.53	0.56	0.38	0.29	0.18

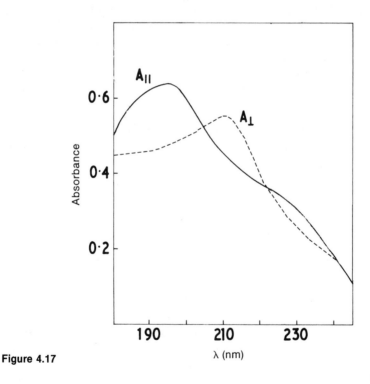

Figure 4.17

(continued)

Worked Example 4.15   (continued)

Solutions

The dichroic ratio d at each wavelength can be calculated from the equation $d = (A_\parallel - A_\perp)/(A_\parallel + A_\perp)$.

λ (nm):	180	190	204	210	227	230	240
d:	0.08	0.15	0	0.09	0	0.04	0

Plotting the dichroic ratio clearly resolves the spectrum into three bands (see Figure 4.18). The third band (near 230 nm) is perpendicularly polarized and is not so obvious in the spectrum in Figure 4.17. This band, which is weak, arises from the forbidden $n \rightarrow \pi$ transition of the C=O bond. In an α-helix, the C=O bonds are oriented parallel to the long axis of the molecule, so the n $\rightarrow \pi^*$ transition is expected to be perpendicularly polarized. As for the $\pi \rightarrow \pi^*$ transitions, we note that in a monomer we should expect to observe one $\pi \rightarrow \pi^*$ transition polarized essentially parallel to the axis. However, the perpendicular-polarized $\pi \rightarrow \pi^*$ transition, which is clearly resolved, arises because in the spectra of polymers we have to consider the interaction between neighboring chromophores (as we had to previously in dealing with the infrared spectra of polymers).

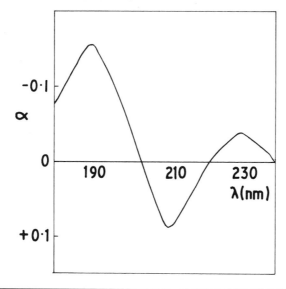

Figure 4.18

Exciton Splitting

In the chapter on infrared, we saw that in a helix the transitions expected in a monomer will be split into two bands, one of which is polarized parallel (and red shifted with respect to the monomer) and the other, polarized perpendicular with respect to the helix axis. Historically this splitting was observed in ultraviolet absorption spectra before infrared spectra. This phenomenon was called **exciton splitting** because it is equivalent to similar optical properties first noted in crystals where a collective excitation takes place—the excitation "hops" (migrates) from molecule to molecule through the crystal.

The concepts that were used to discuss interactions between transition dipoles in dealing with infrared vibrational spectra apply equally here. The only difference is that the transition dipoles here arise from the electronic transitions. As before, once the relative directions (phases) of the dipoles have been worked out, the dipole-dipole interaction can be calculated (see Appendix X).

Again, we recall that in a polymer with n residues, exciton effects split each monomer excited level into n levels, but only a few of these transitions are allowed.

Worked Example 4.16   Exciton Splitting

Can you see any evidence of exciton splitting in the spectra in Figure 4.19 of the α-helix, β-sheet, and random conformations of poly-L-lysine?

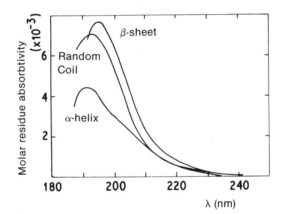

Figure 4.19

Solution

The only spectrum that shows evidence of exciton splitting is that for the α-helix which has two bands—those at 192 nm and 205 nm—assigned to $\pi \rightarrow \pi^*$ exciton splittings. The reduction in intensity of the α-helix spectrum could arise from very strong exciton interactions that shift other bands into other regions of the spectrum. However, calculations show that, in general, exciton effects are quite small. In fact, the reduction in the intensity of the α-helix form and also the slight increase for the β-sheet form compared with the random form arise from other effects, namely, **hypochromism** (less color) and **hyperchromism** (more color).

Hypochromism and Hyperchromism

The term hypochromism signifies that the absorption intensity of a sample is less than the sum of its constituent parts. The phenomenon is most familiar in the spectra of nucleic acids. For instance, the absorbance of ordered double-helical DNA is less than that of the free bases or that of denatured DNA, which implies that hypochromism reflects a structural interaction.

The explanation of the effect is complex, but again arises from interactions between transition dipoles. In exciton splitting, we considered only interactions between two identical transition dipole moments. With hyperchromism and hypochromism, we have to consider additional mechanisms, including the interaction between nonidentical neighboring transition dipoles.

One very qualitative way of describing this interaction is to assume that the transition dipole μ_1 polarizes the electrons in neighboring chromophores and thereby induces instantaneous dipoles μ_2. Hypochromism arises when there is a parallel arrangement of dipoles (see Figure 4.20(a)). Note that the direction of the induced dipoles μ_2 will be opposite to that of the inducing dipole μ_1. The net result will appear as a reduction in the magnitude of μ_1.

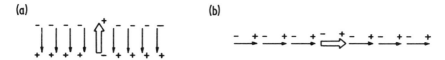

(a)   **(b)**

Figure 4.20 Arrangement of transition dipoles to account for (a) hypochromism and (b) hyperchromism.

If the dipoles form a head-and-tail arrangement, we have the arrangement shown in Figure 4.20(b). In this situation, the net effect will appear as a lengthening of μ_1 and an increase in intensity (hyperchromism).

The μ_2 dipoles are equivalent to transition dipoles on neighboring chromophores that correspond to different transitions from those characterized by μ_1. The effects of μ_1 and μ_2 are complementary—the overall spectral intensity remains constant; therefore hypochromism in the transition characterized by μ_1 will, in principle, be matched by hyperchromism in the transition characterized by μ_2 (which may not always be observed because it may not be in an accessible region). Hypochromic effects are seldom used to analyze structural details. Rather, they are used as a quantitative method of monitoring ordered structure and changes therein.

Worked Example 4.17   Nucleic Acids Show Hypochromism

The melting of r-RNA can be followed by monitoring the optical density at 260 nm. Figure 4.21 shows the temperature variation of this absorbance (relative to that at 20°C) for a solution of 2-mM r-RNA and 1-mM Mn(II) ions. Comment on the shape of the curve and the use of this method.

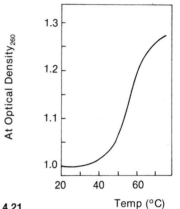

Solution

The increase in absorbance results from the unstacking of the bases as the r-RNA melts. This leads to the separation of chromophores, with a consequent decrease in hypochromic-

Figure 4.21

ity (as the interactions between the chromophores decrease). The smooth shape of the curve suggests a continuous disruption of base pairs over the temperature range. This optical assay allows the determination of the stability of r-RNA (and also DNAs) under a large variety of conditions to be monitored.

PROBLEMS

1. Calculate the value of the absorbance to be used to measure most accurately the ultraviolet properties of a solution of a protein.

2. The absorption spectra of dinitrophenyl (Dnp) ligands undergo a blue shift when the ligands are transferred from water to a less polar solvent, but when they are bound to a hydrophobic anti-Dnp-antibody binding site, a red shift is observed in the spectrum. Can you suggest a reason for this?

3. To a good approximation, N-bromosuccinimide-oxidized tryptophan has, unlike tryptophan, no solvent perturbation spectrum at 292 nm. Lysozyme has six tryptophan residues. Assume that solvent perturbation studies indicate that three are exposed. (a) If the oxidation of three tryptophan residues completely elimi-nates the solvent perturbation, what can you conclude? (b) If the solvent per-turbation disappears only after all six residues are oxidized, how is your conclu-sion altered?

4. A method for obtaining the number of base pairs in RNA in based on obtaining the fully thermally denatured ultraviolet spectra (between 230 and 290nm) of A–U and G–C base pairs. A denaturation difference spectrum can be obtained by subtracting the native spectrum from the denatured spectrum (see Figure 4.22). Why are these difference spectra positive? How would you use these spectra to obtain the number of base pairs in a given RNA?

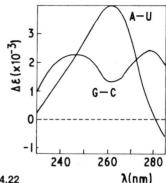

Figure 4.22

5. The absorption spectrum of Co(II)-substituted β-lactamase (a natural Zn(II) enzyme) is shown in Figure 4.23. Comment on the intensities of the bands and the possible stereochemistry of the metal site (see Figure 4.15).

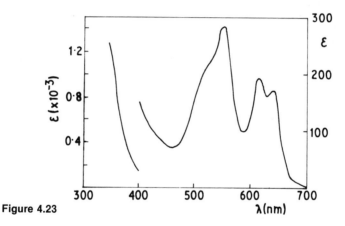

Figure 4.23

6. The tyrosine residues of an antibody fragment (molecular weight 25,000 daltons) were reacted with tetranitromethane. Nitrotyrosine has an absorption band at 428 nm, which, in this protein, has an extinction coefficient of 4100 cm² · mol⁻¹ at pH 10. At this pH, the absorbance is 0.154 for a 4×10^{-5}-mol·dm⁻³ solution of protein in a cell with a path length of 1 cm. Determine the number of tyrosines modified and the pK_a values.

A_{428}:	0.065	0.067	0.069	0.079	0.084	0.117	0.126	0.134	0.142	0.156
pH:	5.47	5.86	6.09	6.41	6.54	7.26	7.42	7.69	7.92	8.5

7. Many biological reactions occur rapidly. The analyses of these rapid reactions can yield valuable information about biological processes. A common experimental technique is to observe the reaction with a spectrophotometer immediately after rapid mixing of the reactants (for example, enzyme and substrate). Figure 4.24 shows the change in OD, observed at 340 nm, after 1-mmol·dm⁻³ pyruvate was added to a sample containing 2.5-μmol·dm⁻³ lactate dehydrogenase to

which stoichiometric amounts of NADH were bound. The enzyme, in this case, consisted of four chemically identical subunits, so the bound-NADH concentration was 10 μmol·dm^{-3}. Comment on the results.

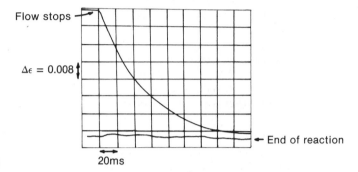

Flow stops

$\Delta\epsilon = 0.008$

← End of reaction

Figure 4.24   20ms

8. The interaction of chlorophyll molecules with one another is important for an understanding of photosynthesis. Studies of electronic spectra have been used to provide information relevant to the structure of the dimers. For example, bacterio-chlorophyll can form dimers in carbon tetrachloride, and a simple monomer-dimer equilibrium exists. The absorption spectra of monomers and dimers are shown in Figure 4.25. Comment on the shape of the band at the longest wavelengths.

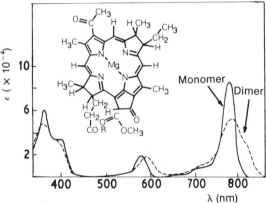

Figure 4.25

9. In principle, what information can be obtained from linear dichroism measurements on oriented samples?

10. The dichroic ratio for an oriented sample is independent of wavelength. What can you deduce about the number of overlapping bands making up the absorption?

11. The increase in intensity in the absorption spectrum of viral DNA on denaturation is much smaller than the increase on denaturation of the replicative form of this DNA (found in host bacteria). Explain this.

12. Suggest a simple reason for the intensity of charge-transfer bands in terms of the strength of the electric transition dipole.

CHAPTER **5**

Fluorescence

OVERVIEW

1. Fluorescence is the emission of radiation that occurs when a molecule in an excited electronic state returns to the ground state. It involves excitation (10^{-15} s) followed by emission. There is a finite lifetime (10^{-6}–10^{-9} s) in the excited state.

2. A physical picture of fluorescence comes from considering the energy levels involved in electronic transitions using Morse curves and the Franck-Condon principle. There are various processes by which the excited state can lose energy—only one of which is fluorescence. The emission spectrum is red shifted with respect to the excitation spectrum.

3. The measurable parameters are the excitation and emission spectra and the intensity, lifetime, and polarization of the fluorescence.

4. The term **fluorophore** describes the molecular group giving rise to fluorescence. The relatively few naturally occurring fluorophores in biology include tryptophan, the Y-base of t-RNA, NADH, and chlorophyll. Synthetic fluorescent probes are widely used.

5. The fluorescence intensity and λ_{max} (the position of the maximum of the emission) are very sensitive to environment.

6. The quantum yield, Φ_F, is the fraction of molecules that becomes deexcited by fluorescence. It is defined as $\Phi_F = \tau/\tau_F$, where τ is the observed lifetime of the excited state and τ_F the lifetime that it would have if fluorescence was the only deexcitation process.

7. There are several important deexcitation processes that can be characterized by first-order rate constants

$$(1/\tau) = k = k_F + k_E + K_q[Q] + \Sigma_i ki$$

where $k_F = 1/\tau_F$, k_E represents energy transfer to a neighboring chromophore on the same molecular complex, $k_q[Q]$ represents deexcitation by an added solute $[Q]$, and $\Sigma_i k_i$ represents other processes.

8. Molecular events on the time scale of the fluorescence lifetime can be investigated by analyzing quenching processes and depolarization.

9. Distances between fluorophores can be obtained from energy-transfer processes.

10. Applications include ligand binding (including H^+) to macromolecules, environmental probes (including polarity and conformational changes), molecular dynamics (including molecular tumbling and oxygen quenching), distance r between fluorophores (energy transfer $\sim 1/r^6$), and various assays (including those using fluorescent antibodies).

11. Phosphorescence is the emission of radiation observed when a forbidden transition occurs from an excited state (usually a triplet state) to the ground state. The lifetime in these excited states is 10^{-2} to 10^2 s. Phosphorescence is rarely observed in solution.

INTRODUCTION

In discussing absorption processes between electronic energy levels, we have mainly been concerned with the excitation of a molecule from a ground state to a higher energy level. In many instances, the excitation energy is lost as heat to the surroundings as the molecules relax (return) to the ground state. However, in some cases, reradiation (emission) occurs. This process is called **fluorescence.**

Fluorescence, therefore, involves two processes: absorption and subsequent emission. Each process occurs in the time scale given by the inverse of the transition frequency (about 10^{-15} s), but there is a time lag of about 10^{-9} s when the molecule exists in the excited state. The lifetime of the molecule in the excited state depends on competition between the radiative emission and any radiationless process, such as the transfer of the excitation energy to the surrounding medium. These nonradiative processes provide an alternative mechanism for the excited molecules to relax back to the ground state, and their presence will result in a diminution or a **quenching** of the fluorescence intensity.

As we shall see, fluorescence occurs at a lower frequency than that of the incident light. Since the detection frequency is different from the incident frequency, sensitivity is high because there is no background signal from the excitation source. It is often possible to measure fluorescence at concentrations in the $10^{-8}M$ range, which is about two orders of magnitude below those generally used for absorption spectra.

Many of the uses of fluorescence are very similar to those described in the section on ultraviolet—for example, measuring binding, monitoring conformational changes, or following a reaction. In addition, fluorescence is a particularly powerful technique because there are many reactions, solvent rearrangements, and molecular-motion processes that take place on the same time scale as the lifetime of the excited state. The resulting sensitivity of fluorescence to this time scale and to environment is the basis of many of its applications to biochemistry. By contrast, we note that absorption of light in the ultraviolet occurs in about 10^{-15} s—a time scale during which the chromophore and its environment are essentially static.

Worked Example 5.1   Fluorescence in Green Plants

In photosynthesis, green plants harvest light by trapping energy from electromagnetic radiation. This energy is then utilized in a series of reactions to produce ATP. These reactions can be blocked, e.g., by a weedkiller (see Figure 5.1(a)). Explain why the fluorescence of the receptor molecule is then much increased (Figure 5.1(b)) in the presence of weedkiller.

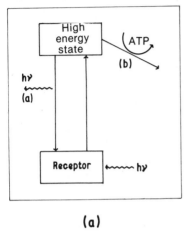

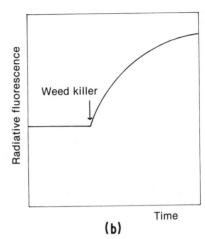

(a)   (b)

Figure 5.1

Solution

   There are at least two routes by which energy can be removed from the initial reaction in photosynthesis—energy transfer and fluorescence. In the absence of weedkiller, the fluorescence is quenched as the excitation energy is used to make ATP. In the presence of weedkiller, the energy-transfer reactions leading to the formation of ATP are shut off. The molecules can then relax back to their initial state by giving up their excitation energy, using the only "open" route—fluorescence.

PHYSICAL PICTURE

To obtain a simple physical picture of fluorescence, we return once again to a discussion of the energy levels involved in electronic transitions. In general, at room temperature a molecule will be in its ground electronic state and its lowest vibrational level. Absorption of the appropriate energy results in excitation into the upper vibrational levels of the singlet excited state.

   The spectral transitions between the vibrational levels of different electronic states are governed by the Franck-Condon principle. On the energy diagram (see Figure 4.3), we illustrate this by a vertical transition in which the internuclear separation is the same in both electronic states. Different vibrational levels in the excited state will receive the transition depending on the relative positions of the ground- and excited-state energy curves (see Figure 4.3). In solutions, when molecules have reached the Franck-Condon excited state, several events can occur as the molecules reestablish (or relax back to) their equilibrium populations (mainly in the lowest vibrational energy level in the ground electronic state). The first event is that the molecules in vibrational levels above the lowest one in the excited state lose their excess energy (usually as heat) and return to the lowest level (thus establishing a Boltzmann population distribution in the excited state). The process takes about 10^{-12} to occur. From this excited state, **spontaneous emission** can occur.* The emitted light is fluorescence.

EXCITATION AND EMISSION SPECTRA

The dependence of the fluorescence intensity on the wavelength of the *exciting* light is referred to as the **excitation spectrum**. Conversely, the **emission spectrum** describes the variation of the fluorescence intensity with the wavelength of the *emitted* light. The position of the maximum in the emission spectrum (λ_{max}) is sensitive to the polarity of the environment and the mobility of the fluorophore. There is often a mirror-image relationship between the excitation and emission spectra.

Worked Example 5.2   Fluorescence Occurs at Longer Wavelengths than Absorption

Figure 5.2 illustrate the processes leading to fluorescence. Explain why the fluorescence spectrum is shifted to longer wavelengths than those of the absorption spectra and why there is a mirror-image relationship between them. Does the shape of the fluorescence spectrum depend on the wavelength of the exciting light?

Solution

   The fluorescence usually appears at longer wavelength (lower frequency) than the incident light because the energy of the emitted electromagnetic radiation differs from that absorbed by an amount equivalent to the vibrational energy lost to the surroundings. Only one transi-

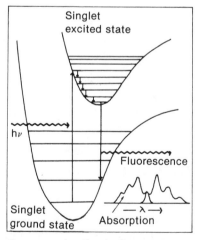

Processes leading to fluorescence
Figure 5.2

(continued)

---

*Stimulated emission does not occur, since there is usually no applied electromagnetic radiation at the correct frequency.

> ### Worked Example 5.2   (continued)
>
> tion between the lowest vibrational levels (termed 0–0) coincides in the two processes of the absorption and emission, namely the one involving the lowest vibrational states of the ground *and* excited states. Transitions probable in absorption are also probable in emission because the transition dipole connecting the two states is the same each time, hence the mirror-image relationship. Since the fluorescence originates from molecules in the lowest vibrational level of the excited state, the shape of the spectrum is independent of the wavelength of the exciting light. (However, as we shall see later, the intensity of the spectrum does depend on the wavelength of the exciting light.)

TRANSITION PROBABILITY AND LIFETIME

In absorption, the probabilities of electronic transitions, together with the populations of the energy levels, determine the intensities of the corresponding absorption bands. The same factors apply in fluorescence. However, other radiationless processes can deplete the excited state, thus reducing its population and decreasing the observed fluorescence intensity. The intensity will therefore depend on the relative rates of the competing processes. If fluorescence is the only means of depopulating the excited state, then this process is kinetically a first-order one, in which the rate constant k_f is that for spontaneous emission.* The inverse of the rate constant in a first-order process is defined as a **relaxation time.**† In this instance, it is referred to as the **radiative lifetime** τ_F, such that $\tau_F = 1/k_F$. Since the absorption probability is related to the emission probability, τ_F can often be related to the molar extinction coefficient (ϵ_{max} in units of cm^2 $mol^{-1} \equiv (mol \cdot dm^{-3})^{-1} cm^{-1}$). For example, the relationship

$$\frac{1}{\tau_F} \simeq 10^4 \epsilon_{max}$$

can be used to obtain a very approximate value of τ_F in seconds.

   In discussing the various time scales mentioned so far, we must be careful not to confuse lifetime (or relaxation time) with the time for a transition. The latter is given by the reciprocal of the frequency of the transition (see Worked Example 4.5). The lifetime refers to a bulk property that is a measure of how long the molecules exist in a particular state. Figure 5.3 summarizes some of the pathways that are discussed in this chapter with typical values of their rate constants. The wavy lines indicate nonradiative pathways, which are often termed **internal conversion.** As we shall see, measurement of lifetimes can be used in a number of ways to obtain dynamic and structural information about molecules.

---

*Spontaneous emission, like that from radioactive decay, is a first-order process, since it does not involve collisions between molecules.

†Strictly speaking from elementary kinetics, relaxation time is the correct definition of the inverse rate constants, rather than the more commonly used *lifetime*. The relaxation time is the time for the process to fall to $1/e$ of its initial value.

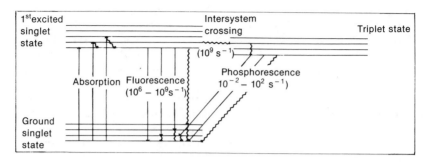

Figure 5.3   Some pathways of relaxation from the excited state.

Worked Example 5.3   The Radiative Lifetime Provides a Time Scale for Events in the Excited State

In the photosynthetic unit, hundreds of chlorophyll molecules transfer their energy to a site at the reaction center. The chlorophyll molecule at this center has unique properties because of its environment, which lowers the energy of its excited state. This makes it an energy trap (see Figure 5.4). The reaction centers mediate the change of light into chemical energy. Can you give a lower limit for the rate of energy transfer? Chlorophyll has an extinction coefficient of about 10^5 $(mol \cdot dm^{-3})^{-1} \cdot cm^{-1}$.

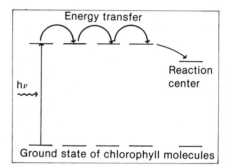

Figure 5.4

Solution

   We can estimate the lifetime of the excited state from $1/\tau_F \sim 10^4 \epsilon_{max}$, which gives $\tau_F \sim 10^{-9}$ s. To be effective, energy transfer must depopulate the excited state faster than does the fluorescence process. The time of energy transfer must therefore be faster than 10^{-9} s. (Values of about 10^{-10} s or less would ensure that most of the energy found its way to the reaction center.)

QUANTUM YIELD

The last worked example illustrated that other processes can compete with fluorescence to depopulate the excited state. These competing processes will cause a more rapid decay of the excited molecules than expected if fluorescence is considered to be the only process by which the molecules can return to the ground state. The overall rate constant (k) for the depopulation of the excited state is obtained by summing the individual rate constants for all the competing processes (which are also assumed to be first-order processes). That is,

$$k = k_F + \Sigma \, k_i$$

where k_i represents the various competing radiationless processes. The lifetime (τ) or relaxation time, for the overall process is given by

$$\tau = \frac{1}{k} = \frac{1}{k_F + \Sigma \, k_i}$$

The fraction of excited molecules that becomes deexcited by fluorescence is called the **quantum yield** (Φ_F) (or the **fluorescence efficiency**) and is given by the equation

$$\Phi_F = \frac{k_F}{k_F + \Sigma_i \, k_i}$$

Substituting for τ and τ_F, we also obtain that

$$\Phi_F = \frac{\tau}{\tau_F}$$

Absolute values of Φ_F are difficult to measure experimentally because instrumental correction factors have to be known. In practice, they are obtained by comparison with a standard sample for which the quantum yield is known. For instance, quinine sulfate in $0.5\text{-mol} \cdot \text{dm}^{-3}$ H_2SO_4 has $\Phi_F = 0.7$.

The quantum yield is sensitive to the immediate surroundings of the fluorescent chromophore (**fluorophore**) and also to specific quenching processes. Changes in quantum yield are accompanied by changes in fluorescence intensity. Measurements of intensity are usually sufficient if only relative values of quantum yield are to be studied.

FLUORESCENCE INTENSITY

The quantum yield (fluorescence efficiency) Φ_F will reflect all the processes that compete with fluorescence for the depopulation of the excited state. The fluorescence intensity (F_λ) will therefore depend on the initial population of the excited state (I_A) multiplied by the quantum yield Φ_F.

$$F_\lambda = I_A \Phi_F$$

This expression refers to the total fluorescence intensity emitted in all directions. In practice, only a small amount is collected by the spectrometer. The equation has to be multiplied by an instrumental factor (Z) that allows for this.

Worked Example 5.4   *Fluorescence Intensity Depends on the Excitation Wavelength*

Using the Beer-Lambert law, deduce how the intensity of the fluorescence spectrum will depend on the excitation wavelength.

Solution

The intensity of the fluorescence is given by $F_\lambda = I_A \Phi_F Z$. The population in the excited state depends on the amount of light absorbed, I_A, which can be calculated from the Beer-Lambert law as

$$I_A = I_t - I_0 = I_0\{1 - \exp[-(2.3\epsilon(\lambda_A)cl]\}$$

where $\epsilon(\lambda_A)$ is the extinction coefficient at the exciting (absorbing) wavelength, c is the concentration of the absorbing molecules, and l is the path length. For small absorbances, $\exp[-(2.3\epsilon(\lambda_A)cl)] << 1$, so we can replace the exponential by $(1 - 2.3\epsilon(\lambda_A)cl)$. Thus

$$I_A = I_0\{1 - [1 - 2.3\epsilon(\lambda_A)cl]\}$$
$$= I_0(2.3\epsilon(\lambda_A)cl)$$

The intensity of the fluorescence F(λ) is thus given by

$$F_\lambda = I_0(2.3\epsilon(\lambda_A)cl\Phi_F Z)$$

From this equation, if the intensity I_0 of the exciting light is kept constant, and if we assume that Φ_F and Z are independent of the wavelength of the exciting light, then for a given solution

$$F_\lambda \propto \epsilon(\lambda_A)$$

That is, the intensity in the fluorescence spectrum will depend on the extinction coefficient at the wavelength of the absorbing species. Since the extinction coefficient also reflects the probability of a transition, this result is expected because transitions probable in absorption are also probable in emission. Thus, while the *shape* of the fluorescence spectrum is independent of the exciting wavelength, the *intensity* is not.

FLUORESCENCE POLARIZATION

Measurements of fluorescence polarization can give information on rotational motions in proteins.

If the chromophore is excited with plane polarized light and the fluorescence is observed through analyzing polarizers, then it is found that the degree of polarization of the fluorescence usually decreases. This phenomenon is called **fluorescence depolarization**, and the reasons for its occurrence are discussed later in the chapter. The polarization is conventionally defined as

$$P = \frac{I_\parallel - I_\perp}{I_\parallel + I_\perp}$$

where $I_\parallel$ and $I_\perp$ are, respectively, the fluorescence intensities resolved in directions parallel and perpendicular to the direction of the exciting beam. In solutions in which the molecules are randomly oriented but immobilized, the value of P is called the **intrinsic polarization** (P_0). If the molecules in that system undergo motion during the lifetime of the excited state, then P may differ from the P_0 value.

Another relationship, similar to P, is sometimes used—the **anisotropy factor***

$$A = \frac{I_\parallel - I_\perp}{I_\parallel + 2I_\perp}$$

Measurements of P and A are sometimes carried out under **steady-state** conditions, that is, using constant illumination. In such cases only the average motion of the system is observed. However, nanosecond pulses of polarized light can be used to measure $I_\parallel$ and $I_\perp$ separately and as a function of time. This technique often enables different motions to be probed or **time resolved**.

NATURAL FLUOROPHORES AND FLUORESCENT PROBES

The main fluorophores used in biochemistry can be classified into natural fluorophores and fluorescent indicators, or **probes**.

Natural fluorophores include the aromatic amino acids, flavins, vitamin A, chlorophyll, and NADH. Nucleic acids do not have appreciable fluorescence, with the exception of the Y-base in t-RNA. Typical fluorescence characteristics of some of these are listed in Table 5.1. The relative experimental sensitivity will be governed by the product of ϵ_{max} and Φ_F.

Worked Example 5.5   Protein Fluorescence Arises from Tryptophan Residues

Use Table 5.1 to explain the statement in the title of the problem.

Solution

From Table 5.1, the relative sensitivity of Trp suggests that the fluorescence from this residue will dominate the fluorescence spectrum of a protein. (Additionally, in proteins the fluorescence of Tyr residues is usually quenched in the presence of Trp residues because of the transfer of excitation energy from Tyr to Trp residues by the Förster resonance-transfer mechanism; see later).

Because there are so few natural fluorophores, many of the applications of the fluorescence technique in biochemistry involve adding a fluorescent probe or labeling reagent to the system. These probes may be bound either covalently or

---

*The value of $I_\parallel + 2I_\perp$ expresses the total incident light—parallel to the incident axis and the *two* directions at right angles to it.

Table 5.1

Fluorophore	Conditions	Absorption λ_{max} (nm)	ϵ_{max} ($\times 10^{-3}$)	Fluorescence λ_{max} (nm)	ϕ_F	τ_F (ns)	Sensitivity $\epsilon_{max}\phi_F$ ($\times 10^{-2}$)
Trp	H_2O pH 7	280	5.6	348	0.20	2.6	11.
Tyr	H_2O pH 7	274	1.4	303	0.1	3.6	1.4
Phe	H_2O pH 7	257	0.2	282	0.04	6.4	0.08
Y-base	Yeast t-RNA^{Phe}	320	1.3	460	0.07	6.3	0.91

Table 5.2 Typical fluorescent probes[a]

Probe	Uses	Absorption λ_{max} (nm)	Absorption ϵ_{max} ($\times 10^{-3}$)	Emission[b] λ_{max} (nm)	Emission[b] ϕ_F	τ_F (ns)	Sensitivity $\epsilon_{max}\phi_F$ ($\times 10^{-2}$)
Dansyl chloride	Covalent attachment to protein: Lys, Cys	330	3.4	510	0.1	13	3.4
1,5-I-AEDANS	"	360	6.8	480	0.5	15	34
7-Chloro-4-nitrobenzo-2-oxa-1,3-diazole (NBD)	Lys, Tyr	345	9.5	—	~1		
Fluorescein isothiocyanate (FITC)	Covalent attachment to protein: Lys	495	42	516	0.3	4	116
8-Anilino-1-naphthalene sulfonate (ANS)	Noncovalent binding to proteins	374	6.8	454	0.98	16	67
Pyrene and various derivatives	Polarization studies in membranes	342	40	383	0.25	100	100
Ethenoadenosine and various derivatives	Analogues of nucleotides bind to proteins. Incorporate into nucleic acids	300	2.6	410	0.40	26	10
Ethidium bromide	Noncovalent binding to nucleic acids	515	3.8	600	~1	26.5	38
Proflavine monosemicarbazide	Covalent attachment to RNA 3'-ends	445	15	516	0.02	—	30

[a]After C. R. Cantor and P. R. Schimmel, *Biophysical Chemistry*, (New York: W. H. Freeman and Company, Publishers, 1980).

[b]Values shown for ϕ_F and τ_F are near maximum typically observed in biological samples at ambient temperature. Other (considerably smaller) values are often found.

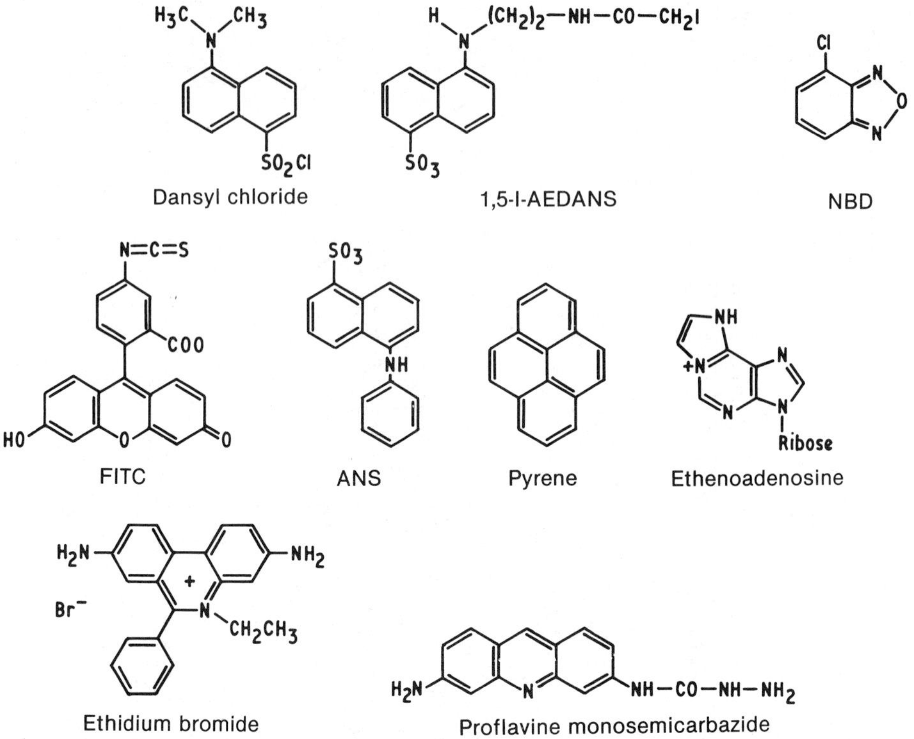

Figure 5.5   Structures of some fluorescent probes.

noncovalently, and much ingenuity has gone into designing site-specific probes. Some of the more common probes are listed in Table 5.2 and their structures are shown in Figure 5.5.

Transition metal ions are not usually fluorescent, however, a number of the lanthanide ions do have observable fluorescence. These can often be used as probes for nonfluorescent ions, notably calcium and, sometimes, magnesium.

EFFECT OF ENVIRONMENT ON FLUORESCENCE PARAMETERS

Fluorescence is very sensitive to environment and the various parameters (e.g., λ_{max}, ϕ_F, τ_F) can be affected in a variety of ways. Some examples of these are given next.

Environmental Effects on λ_{max}

In general, a molecule in the first excited electronic state will have a charge distribution different from that in the ground state. Interactions of the chromophore with the surrounding solvent molecules may occur prior to emission. These interactions will alter the energy of the excited state and also the frequency of the fluorescence emission. This can result in the nonequivalence of the absorption and emission transitions between the lowest vibrational levels in each state (the 0–0) and the breakdown of the mirror-image relationship.

This sensitivity of the fluorescence to environment can be used in a number of ways, of which one of the best known is to estimate polarity. In general, the excited state will be more polar than the ground state; therefore, the excited molecules will tend to interact with a polar solvent (or environment) so as to align the solvent dipoles. This alignment decreases the energy of the excited state and causes the emission spectrum to shift toward the red.

The position of λ_{max} as a measure of polarity must, however, be used with caution. A blue shift (with respect to the probe's fluorescence) can arise if the molecules do not have time to undergo rearrangement (and hence lower the energy of the excited state) during the lifetime of the excited state. This phenomenon is known as **orientation constraint** and emphasizes the ambiguities that may arise in using λ_{max} as a measure of the polarity of an environment.

Worked Example 5.6   Fluorescent Probes of Environmental Polarity

The fluorescent spectra of 1-anilino-8-naphthalene sulfonate (ANS) shifts toward the blue, and the quantum yield increases as the solvent polarity decreases in the order: ethylene glycol, methanol, *n*-propanol, and *n*-octanol (see Figure 5.6). ANS binds to apomyoglobin with a 1:1 stoichiometry and is displaced by the heme. When ANS binds, λ_{max} shifts from 515 nm (in water) to 454 nm and is accompanied by an increase in quantum yield from 0.004 to 0.98. What can you deduce about the ANS binding site?

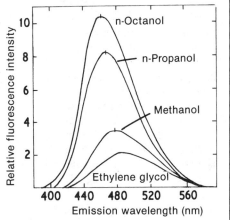

Figure 5.6

Solution

   The shift toward the blue suggests that ANS binds in a (nonpolar) hydrophobic environment. The increase in quantum yield also supports this conclusion. The displacement of ANS by the heme suggests that ANS binds in or near the heme pocket (which one might expect to be hydrophobic).

Environmental Effects on Quantum Yield

The immediate surroundings can modify the fluorescence. A large number of applications of fluorescence in biochemistry simply involve monitoring the intensity of the fluorescence of an intrinsic or extrinsic probe.

In general, the quantum yield (fluorescence intensity) increases as the polarity of the solvent (or environment) decreases. One postulated mechanism for this is that the rate of intersystem crossing (transferring to different excited states) is reduced in nonpolar solvents.

Any constraint in the excited state increases the probability of fluorescence and hence the quantum yield. Measurement of the lifetime can be used to help to distinguish between a polarity effect and an environmental constraint. The consequence of constraint will make the observed value of τ_F closer to the expected value.

Worked Example 5.7   Changes in the Fluorescence Intensity of a Probe Can Be Used to Monitor Binding

Glutamate dehydrogenase consists of six identical subunits, and its activity is allosterically regulated. The enzyme forms ternary complexes with the substrate analogue glutarate and NAD^+ in which the oxidized coenzymes are more firmly bound than in their binary complexes with the enzyme alone.

The fluorescence quantum yield of NADH bound to the enzyme is markedly *enhanced* compared with that of the free coenzyme and λ_{max} (460 nm) is blue shifted by about 10 nm compared with free NADH. With a trace amount of NADH present as a probe, the binding of NAD^+ in the presence of a saturating concentration of glutarate was monitored by measuring the fluorescence enhancement. The results are shown in Figure 5.7, where the percent saturation of NAD^+ is also shown (obtained

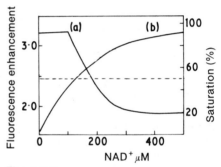

Figure 5.7

from binding studies under the same conditions). At half-saturation of the enzyme sites, the NADH probe signals a sudden change. Suggest a biochemical explanation for this.

(*Clue:* NAD^+ and NADH compete for the same site.)

Solution

At half-saturation, three of the six subunits will have NAD^+ tightly bound; the NADH probe cannot be on these subunits. The most likely explanation is that the binding of NAD^+ as a ternary complex with glutarate to three of the six subunits causes a conformational change in the other three subunits. The reduction in fluorescence quantum yield of NADH must also result from the change in environment caused by the conformational change.

> ### *Worked Example 5.8   Probing Rigidity of Environment*
>
> The radiative lifetime of Vitamin A_1 calculated from the absorption and emission spectra is about 2.4 ns. A solution of retinol has a quantum yield of 0.004 and a measured lifetime of 2.4 ns in water, but when bound to BSA has a quantum yield of 0.04 and a measured lifetime of 8.0 ns. What can you conclude about the rigidity of the retinol environment in BSA?
>
> Vitamin A_1 (all *trans*-retinol)
>
>
> ### *Solution*
>
> From $\Phi_F = \tau/\tau_F$, we obtain that in water, $\tau_F = 2.4 \times 10^{-9}/0.004 = 600$ ns, while when bound to BSA, $\tau_F = 8 \times 10^{-9}/0.04 = 200$ ns. These values are much longer than the calculated value of τ_F of 2.4 ns because Vitamin A_1 is undergoing rearrangement in the excited state. The decrease in τ_F when retinol is bound to BSA suggests that the probe is so constrained that it does not have sufficient freedom to undergo rearrangement on excitation.

Specific Quenching Processes. In favorable cases, an analysis of specific **quenching processes** can lead to useful biochemical information. A low quantum yield may also reflect the ground-state environment of the fluorophore. For instance, the low quantum yield found for some tyrosines in proteins has been attributed to hydrogen bonding with peptide carbonyl groups.

There are two types of quenching frequently studied in biological systems. The first is a result of a long-range (up to a distance of about ~5 nm) nonradiative process called **resonance energy transfer** from one chromophore to another. Analysis of this quenching allows the distance between the chromophores (which need not be the same molecule) to be measured. The second type of quenching arises from collisional processes. Here the addition of oxygen or other paramagnetic species or of heavy ions, such as I^- or Cs^+ all lead to an enhanced rate of intersystem crossing. The fluorescence quenching requires a bimolecular collision between quencher and fluorophore. Analysis of this quenching allows information on the dynamics of the collision process to be obtained. These two quenching processes are dealt with separately later. Note that collisional processes have also been used qualitatively to test the "availability" of a fluorophore in a macromolecule to the quencher.

Worked Example 5.9   Fluorescence Quenching by I⁻ Ions Gives
Information on Tryptophan Accessibility in Lysozyme

Iodide acts as a collisional quencher of tryptophan fluorescence. 80% of the fluorescence from the six tryptophans in lysozyme arises from Trp-62 and Trp-108. The tryptophan fluorescence is significantly quenched (about 60%) in the presence of I⁻ ions (about $0.2M$). When Trp-62 is oxidized with N-bromosuccinimide, which makes it nonfluorescent, the residual fluorescence is hardly quenched at all by the same concentration of I⁻ ions. Explain what this means in terms of accessibility to solvent of the two tryptophan residues.

Solution

In the modified enzyme, the fluorescence of Trp-108 is not quenched by I⁻, which indicates that this residue may not be accessible to solvent. The quenching of the fluorescence of the native enzyme suggests that Trp-62 is the more exposed to the solvent. Note, however, that the accessibility of I⁻ to the tryptophans may be influenced by interaction of its charge with any charges on the protein, thus creating a possible ambiguity in the results.

Environmental Effects on Lifetimes

If a molecule contains two or more fluorophores, their environments may be different, but their individual emissions will probably not be resolved in the fluorescence spectrum in solution. If they have different fluorescence lifetimes, however, then following the decay of the fluorescence gives another chance to resolve them. The simplest way of doing this is to excite the sample with a short pulse of light (~1 ns) and monitor the emission $S(t)$ as a function of time. $S(t)$ is related to the initial emission intensity $S(0)$ following the pulse and the lifetime by

$$S(t) = S(0)e^{-t/\tau}$$

where $1/\tau$ is the sum of the rates of the transitions from the excited state. When there are two or more fluorophores, $S(t)$ will be expressed as a sum of exponential terms. Analysis of these may resolve the individual fluorophore lifetimes, which can then be used as probes of the different environments.

Worked Example 5.10   Resolving Lifetimes of Tryptophan Residues
in Lysozyme

Lysozyme has six tryptophan residues located at positions 28, 62, 63, 108, 111, and 123. Trp-62, -63, and -108 are located in the active site. As mentioned before, however, most of the fluorescent emission arises from Trp-62 and Trp-108. The excitation pulse and fluorescence decay are shown in Figure 5.8. To analyze the fluorescence decay, a correction had to be made for the shape of the excitation pulse. The fluorescence decay was then analyzed in terms of two exponentials. Measurements were also made in samples where Trp-62 and -108 were oxidized either individually or together (to make them nonfluorescent).

(continued)

Worked Example 5.10   (continued)

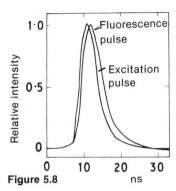

Figure 5.8

The following results were obtained:

Sample	Lifetimes (ns)	
Lysozyme	2.7	0.6
Oxidized Trp-62 lysozyme	1.5	0.3
Oxidized Trp-108 lysozyme	2.7	0.3
Oxidized Trp-108 and -62 lysozyme	—	0.3

What can you deduce about the lifetimes for the individual tryptophans?

Solution

When Trp-108 and -62 are oxidized, the single lifetime of 0.3 ns must represent the weighted mean of the lifetimes of all the other tryptophans. From the preceding results, Trp-62 must have a lifetime of 2.7 ns and Trp-108 of 1.5 ns. This latter value is contributing to the value of 0.6 ns in native lysozyme (for more details, see C. Formoso and L. S. Förster, *J. Biol. Chem.,* 250 (1975):5738–5745). (One of the limitations is the difficulty of quantitatively analyzing contributions from two or three similar exponentials. This is also a problem in the analysis of several similar rotational motions. Note also that quite a large "correction" for the pulse may have to be made, which presents a further difficulty in analysis of lifetimes of the same order as the pulse response.)

MEASUREMENTS OF MOLECULAR DYNAMICS

The lifetime of the molecule in the excited state before fluorescence occurs is finite—about 10^{-9} s. Any molecular events, such as tumbling or collisions, on this sort of time scale can lead to changes in the observed fluorescence. These changes can, in turn, be analyzed to yield information on molecular dynamics.

Dynamic Quenching: The Stern-Volmer Relationship

The effect of collisions with other molecules can shorten the fluorescence lifetime by providing additional processes that depopulate the excited state. There can also be a dynamic quenching of the fluorescence as a result of the encounters between an added quencher (such as I^- or O_2) and the fluorophore. The rate of quenching encounters is given by the product of the bimolecular quenching rate constant, k, and the quencher concentrations $[Q]$. Since the concentration of Q is often much greater than that of the excited molecule, the rate process may be regarded as essentially pseudo first order with a rate constant $k[Q]$. The quantum yield in the absence of quencher is (see the discussion of quantum yield earlier in the chapter)

$$\Phi_F = \frac{k_F}{k_F + \Sigma_i k_i}$$

In the presence of the quencher, the quantum yield will be

$$(\Phi_F)_Q = \frac{k_F}{k_F + \Sigma_i k_i + k[Q]}$$

Thus

$$\frac{\Phi_F}{(\Phi_F)_Q} = \frac{k_F + \Sigma_i k_i + k[Q]}{k_F + \Sigma_i k_i}$$

$$= 1 + \frac{k[Q]}{\Sigma_i k_i + k_F}$$

$$= 1 + k[Q]\tau$$

where τ is the lifetime in absence of quencher. This equation is often written in terms of the fluorescence intensities in the presence (I) and absence (I_0) of quencher, that is,

$$\frac{I_0}{I} = 1 + K[Q]$$

This expression is called the **Stern-Volmer** equation, which is characterized by a quenching constant K. Thus a plot of $[(I_0/I) - 1]$ versus $[Q]$ will give a straight line of slope K. Note that $K = k\tau$, where τ is the lifetime in the absence of the quencher. This emphasizes that the larger the lifetime of a fluorophore, the higher the probability of its undergoing a collision with the quencher.

The rate constant for quenching (k) depends on the probability of a collision between fluorophore and quencher. This probability depends on their rate of diffusion (D), their size, and concentration. It can be shown that

$$k = 4\pi a D \mathfrak{N}/10^3$$

where D is the sum of the diffusion coefficients of quencher and fluorophore, a is the sum of the molecular radii, and $\mathfrak{N}$ is Avogadro's number. (For example, for oxygen quenching of tryptophan fluorescence in solution, $a = 0.4$ nm, $D_{O_2} = 2.6 \times 10^{-9} \text{cm}^2 \cdot \text{s}^{-1}$, and $D_{Trp} = 0.66 \times 10^{-9} \text{cm}^2 \cdot \text{s}^{-1}$, which gives $k \approx 1 \times 10^{10}$ $(\text{mol} \cdot \text{dm}^{-3})^{-1} \cdot \text{s}^{-1}$.

The determination of the dynamic quenching constant K can thus lead to information on diffusion coefficients. Measurements of K are better obtained from lifetimes rather than intensities because there may be other processes that affect the intensity. One example is **static quenching,** in which formation of a complex between quencher and fluorophore (characterized by an equilibrium constant K_{eq}) predates the excitation. Excitation of the fluorophores will then result in instantaneous quenching of those that are complexed. Static quenching thus reduces the concentration of the excited molecules, which alters the intensity of the fluorescence but not the lifetime.

A modified form of the Stern-Volmer equation that describes quenching data when both dynamic and static quenching are operative is

$$\frac{I_0}{I} = \frac{(1 + K_{eq}[Q])(1 + K[Q])}{K[Q]}$$

The first term describes the static quenching, which predates the excitation, and the second the dynamic quenching resulting from encounters between quencher and fluorophore during the fluorescent lifetime. A static component in the quenching results in an upward curvature in the Stern-Volmer plots. (Note that in this case, $I_0/I > \tau/\tau_Q$).

Worked Example 5.11   Quenching of Protein Fluorescence by Oxygen

The quenching of tryptophan fluorescence of lysozyme by oxygen was studied by using oxygen concentration up to 0.13M (corresponding to equilibration with oxygen at a pressure of 100 atm). This pressure has no effect on the activity of the protein. A plot of the ratio of the fluorescence intensity in the absence (I_0) and presence (I) of oxygen versus [O_2] is shown in Figure 5.9. What can you deduce from this plot?

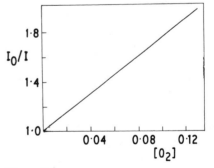

Figure 5.9

Solution

The figure shows a Stern-Volmer plot of the quenching of fluorescence by O_2. It is linear, showing no upward curvature, so the quenching is dynamic. This can be checked from lifetime measurements. A plot of τ/τ_Q versus [O_2] should have the same shape. From the slop of the curve we can calculate that K is about 7 $(\text{mol} \cdot \text{dm}^{-3})^{-1}$. The linearity of the plot also suggests that the tryptophans contributing to the fluorescence do not have significantly different Stern-Volmer quenching constants and therefore are accessible to oxygen.

Worked Example 5.12   Nanosecond Fluctuations in Proteins Detected by Oxygen Quenching of Fluorescence

In the previous example, the fluorescence lifetime was found to be 1.5 ns. The rate constant for quenching of tryptophan fluorescence by oxygen in water can be calculated to be 10^{10} $(mol \cdot dm^{-3})^{-1}s^{-1}$. What is the value of the rate constant in lysozyme? What are the implications of this result?

Solution

From the Stern-Volmer equation for dynamic quenching, $k = K/\tau = (7/1.5 \times 10^{-9}) \simeq 0.5 \times 10^{-10}$ $(mol \cdot dm^{-3})^{-1}s^{-1}$. From our previous calculations, this would imply that oxygen is diffusing through the protein at almost the same rate as in free solution. The protein must be undergoing rapid structural fluctuations on the nanosecond time scale, which permits diffusion of oxygen.

Fluorescence Depolarization

Steady-State Fluorescence Depolarization: Photoselection.   The excitation of molecules to their excited state depends on the angle θ between the plane of polarization of the incident light and the transition dipole moment of the transition. The probability of absorption is proportional to $\cos^2\theta$, and the preferential absorption that takes place with polarized excitation is called **photoselection.** The measured value of the polarization P will depend not only on (1) the orientation of the absorption transition dipole moment but also on (2) the orientation of the emission transition dipole with respect to that of the absorption. A third factor, obviously, is the ordering in the sample. In a randomly oriented sample, the $\cos^2\theta$ dependence results in the excitation of a significant number of molecules that do not have transition dipoles aligned exactly along the plane of polarization. (This number is actually proportional to $\sin\theta$.) When these nonaligned molecules emit, they will have components of both parallel and perpendicular polarized light. Therefore, P will always have a value of less than one. In fact, if the averaging over all θ is considered, it turns out that P has the characteristic value $P_0 = 0.5$ if the transition dipole moments of excitation and emission are parallel. In general, however, the absorption and emission transition dipoles are not parallel but at some fixed angle θ to each other, and this reduces P_0 further.

The term P_0 is called the **intrinsic polarization** and is given by

$$P_0 = \frac{3\cos^2\theta - 1}{\cos^2\theta + 3}$$

Thus if $\theta = 90°$, $P_0 = -\frac{1}{3}$. P_0 can take values between $\frac{1}{2}$ and $-\frac{1}{3}$.

Motional Depolarization.   Photoselection "selects," at an instant, those molecules whose chromophores have components of their transition dipole moments oriented along the direction of polarization. At that *instant,* we can regard those selected chromophores as rigid, and the system will have a characteristic value of

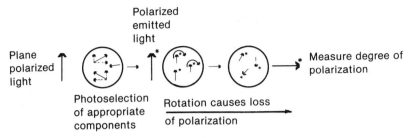

Figure 5.10 Depolarization resulting from rotation.

$P = P_0$. If in the time between absorption and emission the molecules rotate, the effects of photoselection will be lost and P will be less than P_0. Rotation thus causes depolarization (see Figure 5.10 for a simple physical picture). The relationship between P and P_0 is

$$\frac{1}{P} - \frac{1}{3} = \left(\frac{1}{P_0} - \frac{1}{3}\right)\left(1 + \frac{\tau_F}{\tau_R}\right)$$

where τ_F is the radiative lifetime and τ_R is the rotational correlation time of the molecule (see Chapter 6 for a definition of correlation time). It is important to stress that this equation, known as the **Perrin equation**, assumes that the rotational motion is isotropic.

The anisotropy of polarization (A) is also affected by motion, as shown by the equation

$$\frac{1}{A} = \frac{1}{A_0}\left(1 + \frac{\tau_F}{\tau_R}\right)$$

where A_0 is the intrinsic anisotropy (in the absence of all motion).

Measures of A and P are carried out under steady state conditions, that is, using constant illumination. The values of P_0 and A_0 can be extracted from the foregoing equations by varying τ_R or by altering either the viscosity η (by addition of sucrose and glycerol) or the temperature. Since $\tau_R \propto \eta/T$, extrapolation of plots of $1/A$ or $(1/P_0 - \frac{1}{3})$ versus T/η to infinite viscosity ($T/\eta \rightarrow 0$) will give the values of $1/A_0$ and $(1/P_0 - \frac{1}{3})$.

Disadvantages of this technique are that the addition of sucrose and glycerol may alter the conformation of the molecule and that P_0 may depend on the solvent. In addition, τ_F has to be known to calculate τ_R. Because steady-state conditions are used, the values of τ_R obtained will be average ones. If the molecule has several types of rotation (e.g., a rodlike molecule can rotate about its long axis and rotate end over end), then it is not possible to separate the different motions. This method of obtaining τ_R is now rarely used, because it is easier to obtain molecular dynamic information *directly* from pulsed fluorescence techniques. Nevertheless, measurements of P are fairly straightforward and can still give useful empirical information.

Worked Example 5.13   Fluorescence Polarization as a Binding Assay

When an antigen labeled with a fluorescent group binds to its antibody, the polarization of the fluorescence increases from 0 to 0.5. Why?

Solution

When the fluorophore is free in solution, its rotational motions must result in small values of the polarization. On binding to the antibody, the fluorophore is immobilized and the polarization increases.

Worked Example 5.14   Motion of Retinol in a Membrane

The observed fluorescence lifetime of retinol in an erythrocyte membrane is 10 ns. When polarized light was used for excitation, the polarization of fluorescence was 0.35. What can you conclude about the motion of retinol in the membrane? If you assume $P_0 = 0.5$, what other assumptions do you have to make?

Solution

From the Perrin equation, we obtain that $(1/0.35 - \frac{1}{3}) = (1/0.5 - \frac{1}{3})$ $(\times 1 + 10/\tau_R)$, from which $\tau_R \simeq 20$ ns. Tumbling of the membrane as a whole is expected to be much slower than this, which suggests that retinol has some local mobility, resulting from the "fluidity" of the membrane.

It is necessary to assume that the Perrin equation is valid here. In fact it really only applies for isotropic motion—which is unlikely in a membrane.

Time-Resolved Depolarization of Fluorescence Using Nanosecond Pulses. In time-resolved depolarization, the sample is excited with nanosecond pulses of polarized light. (The intensity of the fluorescence emitted in directions parallel $(I_\parallel)$ and perpendicular $(I_\perp)$ to the direction of excitation are measured as a function of time. There is an *initial* difference in these intensities (because of photoselection, and hence in the anisotropy A_0, which usually decays in a few nanoseconds if the molecule rotates in solution. The results are expressed as plots of the anisotropy (A) as a function of time because $A(t)$ can be *directly* related to the correlation time.* In the simplest case, for isotropic motion in which $A(t)$ decays exponentially,

$$A(t) = A_0\, e^{-t/\tau_R}$$

A plot of log A versus time is linear, with slope of $-1/\tau_R$. Note the simplicity of this method compared with the steady-state method and the fact that τ_F does not need to be known to calculate τ_R.

If there is *more* than one motion that can contribute to the decay curve, the plots of log $A(t)$ versus t will be nonlinear. The curves are then analyzed in terms

---

*The equation in terms of $P(t)$ is of the form $P(t) = C/(\exp(t/\tau_R) + D)$, where C and D are constants, so it is not so easy to use experimentally.

of two or more exponentials to give the rate constants and amplitudes of each motion. (The relative amplitudes give the weighting factors for the different types of motion to the decay). Actually, the analysis is not as straightforward as indicated because the various motions will occur on similar time scales and it is not always easy to separate two exponentials that have similar rates. Further, the response of the illumination (switching-on and -off time) is frequently of the order of 10^{-9} s, so the values of $I_{\parallel}$ and $I_{\perp}$ have to be corrected for this process before any further analysis.

These difficulties have limited the use of the technique somewhat; many applications involve using it empirically to detect changes rather than to obtain structural details.

Worked Example 5.15   Flexibility in Antibody Molecules

A schematic picture of an antibody molecule is shown in Figure 5.11. The fluorescent probe dimethylnaphthalenesulfonyl–lysine (DNS–lysine) binds to anti-DNA antibodies and is rigidly held in the combining site. The correlation time obtained from nanosecond-pulse fluorescence measurements of DNS–lysine was 100 ns. When a protein of molecular weight 50,000 is attached to the Fc part of the molecule, the value of the correlation time is unchanged. Can you suggest a reason for this?

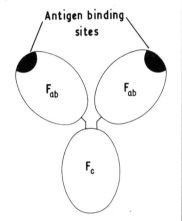

Figure 5.11

Solution

The constant value of the correlation time suggests that it is not associated with the motion of the whole molecule but rather with the Fab arms of the molecule. This indicates flexibility between the Fab and Fc parts of the molecule.

DETERMINATION OF DISTANCES BETWEEN CHROMOPHORES BY RESONANCE-ENERGY TRANSFER

The quenching of fluorescence by collisions is a very short range effect. It depends on the rate of diffusion of the molecule, which is governed by factors such as molecular size and the viscosity of the solution.

Fluorescence quenching can also occur by the transfer of excitation energy between chromophores, over much greater distances, without emission and reabsorption of radiation. This process is independent of the solution viscosity. The most common type of energy transfer is from the excited singlet state of a donor to the excited singlet state of an acceptor. In this transfer of energy, the energy donor returns from the excited state to the ground state and the energy

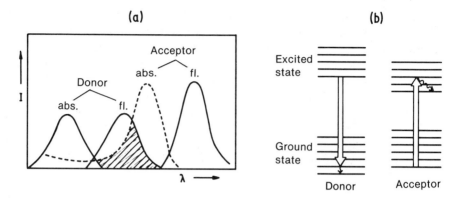

Figure 5.12   Resonance-energy transfer between donor and acceptor.

(a) The shaded area represents the region of overlap between donor fluorescence and acceptor absorption.

(b) Nonradiative relaxation to the lowest vibrational state of the ground state (by the donor) and of the first excited state (by the acceptor) leads to a loss of energy matching. Hence, continuous energy transfer back and forth between donor and acceptor cannot occur (from Cantor and Schimmel, *Biophysical Chemistry*, vol. 2 (San Francisco: W. H. Freeman and Company, Publishers, 1980)).

acceptor is simultaneously excited from its ground to its excited state. The energy separations in each case must match—that is, be in resonance.

A measure of the amount of resonance that will occur is given by the overlap of the fluorescence spectrum of the donor and the absorption spectrum of the acceptor. Figure 5.12 illustrates this resonance-energy transfer. Note that the donor and acceptor end up in excited vibrational states. Vibrational relaxation rapidly converts these to the lowest vibrational energy level in each state. The energies no longer match, so the transfer of energy back again to the original donor is unlikely. If the acceptor fluoresces, the original excitation energy reappears as acceptor emission. This is termed **sensitized fluorescence.** When the acceptor molecule is nonfluorescent, the transferred energy is dissipated by nonradiative processes.

If k_T is the rate constant for depopulation of the excited state by resonance-energy transfer, then the measured lifetime τ_T is given by

$$\frac{1}{\tau_T} = \frac{1}{\tau} + k_T$$

where $1/\tau$ is the rate of depopulation, in the absence of resonance-energy transfer.

The efficiency (E_T) of depopulation by resonance-energy transfer is then

$$E_T = \frac{k_T}{1/\tau + k_T} = \frac{1/\tau_T - 1/\tau}{1/\tau_T} \tag{5.1}$$

$$\therefore \frac{\tau_T}{\tau} = 1 - E_T$$

If Φ_T and Φ_D represent the quantum yields of the donor in the presence and absence of resonance-energy transfer, then

$$\frac{\Phi_T}{\Phi_D} = 1 - E_T$$

Förster has explained this resonance-energy transfer in terms of a dipole-dipole interaction between the donor and acceptor pair. The energy of a dipole-dipole interaction depends on $1/R^3$, where R is the intermolecular distance. The rate of energy transfer (k_T) is proportional to the square of this interaction and hence to $1/R^6$. If we define R_0 as the distance at which the energy transfer is 50% efficient, i.e., when $1/\tau = k_T$, then equation 5.1 can be rewritten as

$$E_T = \frac{R^{-6}}{R_0^{-6} + R^{-6}} = \frac{R_0^6}{R^6 + R_0^6} \tag{5.2}$$

or

$$R = R_0\left(\frac{1 - E_T}{E_T}\right)^{1/6}$$

where R_0 is a constant for each donor-acceptor pair. From a measurement of the quenching of the fluorescence (ϕ_T/ϕ_D) (or better, the lifetime τ_T/T) in the presence and absence of the acceptor, R can be calculated if R_0 is known.

Worked Example 5.16   Energy Transfer Depends on the Sixth Power of the Distance Between the Donor and Acceptor Pair

A model system for the study of the dependence of energy transfer on the separation of donor (napthyl) is dansyl-(L-propyl)n-α-naphthyl semicarbazide (see L. Stryer and R. P. Haughland, *Proc. Nat. Acad. Sci. U.S.A.,* 58 (1967): 719–726). The proline residues form a helix, the dimensions of which are known, so the distances between the napthyl and dansyl groups can be worked out. The measured efficiencies of transfer for these compounds fit well with those from (5.2) (confirming the R^6 dependence). (a) From the data in Figure 5.13, estimate R_0. (b) Very often the efficiency of transfer is calculated by measuring the quenching of the fluorescence intensity. Calculate what value of this you would expect at a distance of 6.0 nm between donor and acceptor in the system here. Comment on the use of the method in light of this.

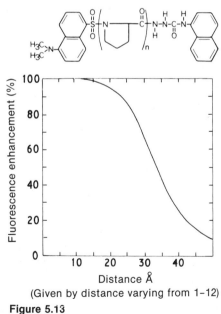

Figure 5.13
(Given by distance varying from 1–12)

(continued)

Worked Example 5.16   (continued)

Solution

a. From the graph, R_0 (the distance at which the energy transfer is 50% efficient) is about 3.4 nm.

b. When $R = 6.0$ nm, then from equation 5.2,

$$E = \frac{6^6}{6^6 + 3.4^6} = 0.97$$

Now

$$\frac{I_T}{I_D} = \frac{\phi_T}{\phi_D} = 1 - E$$

when I_T and I_D are the fluorescence intensities in the presence and absence of energy transfer, respectively.

$$\therefore \frac{I_T}{I_D} = 1 - 0.97 = 0.03$$

The fluorescence signal is quenched 3%.

This quenching is very small and suggests that distances as large as this cannot be reliably measured in this system by energy transfer. The range of distances that can be measured will depend on the value of R_0 for the system. The quantity R_0 can be evaluated from the parameters of the absorption and fluorescence spectra as shown in the next section.

Calculation of R_0.   The value of R_0 (in nanometers) can be obtained from the expression

$$R_0 = 9.79 \times 10^2 (J\bar{\nu}n^{-4}k^2\Phi_D)^{1/6}$$

where n is the refractive index of the medium, Φ_D is the quantum yield of the energy donor (in the absence of the acceptor), k^2 is the orientation factor between the donor and acceptor electric transition dipole moments (see Appendix IX), and J is the integral of the spectral overlap of the absorption spectrum of the acceptor and the emission spectrum of the donor. In Figure 5.14 J is given by

$$J = \int F_D(\lambda)\epsilon_A(\lambda)\lambda^4 d\lambda$$

The term $\epsilon_A(\lambda)$ represents the extinction coefficient of the acceptor at each value λ and therefore the absorption spectrum of the acceptor. The term $F_D(\lambda)$ is the fraction of total donor fluorescence occurring at λ.* The value J is usually calculated graphically.

---

*The emission spectrum cannot be quantified in the same way as the absorption spectrum, since the observed fluorescence intensity depends on the intensity of the light absorbed multiplied by a factor that comprises both the quantum yield and the response of the instrument to fluorescence. This means that the observed fluorescence intensity is in arbitrary units.

From the expression for R_0, two conditions favor singlet-singlet resonance-energy transfer: (1) a large spectral overlap, J, between the absorption bands of the donor and acceptor molecules and (2) a high quantum yield ϕ of the donor molecule in the absence of the acceptor. However, the major uncertainty in calculating R_0 arises from the lack of knowledge of the orientation factor (k^2). This term is equal to $(3 \cos^2\theta - 1)^2$, where θ is the angle between the electric transition dipole moments of the donor-acceptor pair. In principle, k^2 can vary between 0 (the two transition dipoles perpendicular) and 4 (parallel) (see Appendix IX). The theory will apply in macromolecules, provided the transition dipoles are rigidly fixed, but the orientation factor will not generally be known. Most authors assume a value of k^2 of $\frac{2}{3}$, which is the factor for an isotropic rapidly rotating system. If this is not the case, then some uncertainty may be introduced into the calculated distance. (For a good discussion on this problem see the article "Fluorescence Energy Transfer as a Spectroscopic Rule" by L. Stryer *Ann. Rev. Biochem.*, 47(1978):819–46.) Stryer concludes that this uncertainty is likely to be small—usually less than 20%.

***Worked Example 5.17   Graphical Evaluation of* J**

Figure 5.14 shows hypothetical fluorescence and absorption spectra for a hypothetical donor-acceptor pair. Evaluate J.

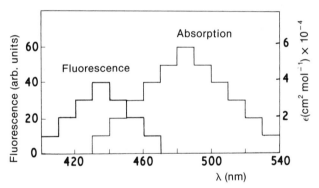

Figure 5.14

Solution

Since it is not possible to write exact expressions for $F_D(\lambda)$ or $\epsilon_A(\lambda)$ as functions of λ, we must replace the integral with a sum

$$J \simeq \sum_{i=1}^{i=4} J_i = \sum_{i=1}^{i=4} F_D(\bar{\lambda}_i)\epsilon_A(\bar{\lambda}_i)\bar{\lambda}^4 \, \Delta\lambda$$

Here we have divided the shaded overlap area into four numbered rectangles, each of width $\Delta\lambda = 10$ nm. We can then determine the values of F_D and ϵ_A at the midpoint

(continued)

Worked Example 5.17   (continued)

wavelength ($\bar{\lambda}_i$) of each rectangle. Similarly the fluorescence spectrum can be divided into seven rectangles.

First, we evaluate $F_D(\bar{\lambda}_i)$. This is given by

$$F_D(\bar{\lambda}_i) = F_i / \sum_{i=1}^{i=7} F_j \Delta\lambda$$

where F_i is the fluorescence intensity at $\bar{\lambda}_i$. Note that the sum in the denominator is taken over the entire fluorescence spectrum.

Hence

$$\Sigma F_j \Delta\lambda = (10 + 20 + 30 + 40 + 30 + 20 + 10) \cdot 10$$
$$= 1600 \text{ nm} = 1.6 \times 10^{-4} \text{ cm}$$

This value can be used in conjunction with the fluorescence intensities and extinction coefficients at the four midpoint wavelengths to construct the following table:

i	$\bar{\lambda}_i$ (cm)	$\bar{\lambda}_i^4$ (cm^4)	$F_D(\bar{\lambda}_i)$ (cm^{-1})	$\epsilon_A(\bar{\lambda}_i)$ (mol·dm^{-3})$^{-1}$·cm^{-1}	J_i cm^3·M^{-1}
1	4.35×10^{-5}	3.58×10^{-18}	25.0×10^4	1×10^4	8.95×10^{-15}
2	4.45×10^{-5}	3.92×10^{-18}	18.8×10^4	2×10^4	1.47×10^{-14}
3	4.55×10^{-5}	4.29×10^{-18}	12.5×10^4	3×10^4	1.61×10^{-14}
4	4.65×10^{-5}	4.68×10^{-18}	6.25×10^4	4×10^4	1.17×10^{-14}

Hence $J = 5.14 \times 10^{-14}$ cm^3·(mol·dm^{-3})$^{-1}$

Note: Sometimes the unnormalized form of $F_D(\lambda)$ is used directly in J, in which case J becomes

$$J = \frac{F_D(\lambda)\epsilon_A(\lambda)\lambda^4 \, d\lambda}{F_D(\lambda) \, d\lambda}$$

Worked Example 5.18   Energy Transfer as a Spectroscopic Ruler: Can Rhodopsin Traverse the Disc Membrane?

Rhodopsin is a photoreceptor protein that is in integral part of the disc membranes of vertebrate retinal rod cells. Rhodopsin consists of opsin, a protein, and 11-*cis* retinal (see Figure 5.15(a)), a prosthetic group. The

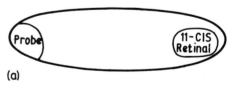

(a)

(continued)

Worked Example 5.18   (continued)

color of rhodopsin and its respon-
siveness to light depend on the presence of the 11-*cis* retinal, which is a
very effective chromophore giving a
broad absorption band at 500 nm.

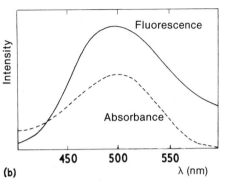

(b)

Figure 5.15

Rhodopsin was covalently la-
belled with 1.5-I AEDANS, (formula
is shown in Figure 5.5). The labeled
rhodopsin retained the 500-nm
absorption band and was regenerable
after bleaching (removal of 11-*cis*
retinal), which suggests that the
introduction of the probe had not altered the overall conformation significantly. The overlap between the
fluorescence emission spectrum of the 500-nm absorption band of 11-*cis* retinal is
shown in Figure 5.15(b). Use the data to show whether rhodopsin is sufficiently long
to traverse the membrane, which is estimated from electron microscopy to be 7.5 nm
wide.

The quantum yields in the presence and absence of 11-*cis* retinal are 0.68 and
0.75, respectively. The refractive index n of the medium is 1.33; the orientation factor $k^2 = \frac{2}{3}$. The overlap integral $J = 1.84 \times 10^{13}$ cm$^3 \cdot$(mol$\cdot$dm$^{-3})^{-1}$.

Solution

We first calculate R_0 from

$$R_0 = 9.79 \times 10^2 (Jn^{-4}k^2\phi_D)^{1/6}$$
$$= 9.79 \times 10^2 (1.84 \times 10^{13} \times 1.33)^{-4} \times 0.67 \times 0.75)^{1/6}$$
$$= 5.4 \text{ nm}$$

The ratio of the quantum yields in the presence and absence of 11-*cis* retinal is
0.68/0.75 = 0.91. The transfer efficiency is therefore 1 − 0.91 = 0.09. We use this
value to obtain

$$R = 5.4 \left(\frac{0.91}{0.98} \right)^{1/6}$$

$$= 7.9 \text{ nm}$$

Rhodopsin is thus long enough to traverse the membrane.

FLUORESCENT ANTIBODIES

The specific binding of antibodies, coupled with the sensitivity of fluorescence,
forms the basis of the use of fluorescent antibodies in various processes such as
cell sorting. After a cell population is stained with fluoresceinated antibody, the
mixed cell suspension can be examined under a fluorescence microscope (see
Chapter 11) and the number of stained cells counted.

PHOSPHORESCENCE

The presence of a second exicted state is responsible for phosphorescence. In
most biological chromophores, this other excited state is a triplet state. We recall
that the triplet state has the spins of two of its electrons arranged in a parallel,
rather than in an opposed, manner. Transitions between states of different multiplicities are theoretically forbidden, so the population and depopulation of the
triplet state proceeds as follows (see Figure 5.16):

Excitation of electrons from the ground singlet state results in transitions to
the upper singlet state. Here any excess vibrational energy in the excited state is
rapidly transferred to the surroundings as heat. As the molecule "steps down"
the vibrational energy levels, there is a finite possibility that if a vibrational
energy level of the triplet state coincides with one from the excited singlet state,

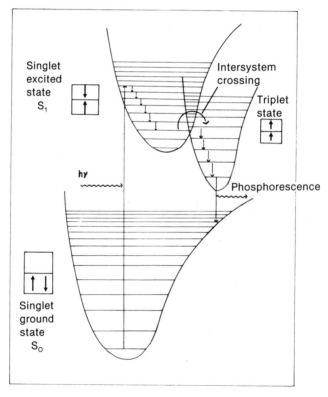

Figure 5.16   Processes leading to phosphorescence.

then the molecule will transfer to the triplet state, reversing the orientation of the spin of one of its electrons.* This switch is known as **intersystem crossing.** The reversal of the spin is characterized by a first-order rate constant, which is about 10^9 s^{-1}. Once in the triplet state, the molecule continues losing energy until it reaches the lowest triplet vibrational level. From here it can return (relax) to the ground singlet state by the spontaneous emission of radiation known as **phosphorescence.** Note that the emission wavelength is red shifted with respect to fluorescence.

The rate of depopulation of the triplet level by spontaneous emission will be very slow because the transition dipole connecting the two states is very small (forbidden transition). The lifetime of this state is therefore very long (typically 10^{-2} s to 10^2 s). The long lifetime of the triplet state makes it highly vulnerable to any quenching process that will remove energy—such as collisions with the solvent. For this reason phosphorescence is rarely observed in solution.

Worked Example 5.19   Triplet Probes of Molecular Motion

Eosin derivatives (Figure 5.17) are useful as triplet probes. In these, the nature of the group X is such that it can be covalently attached to proteins. When the triplet state is populated with molecules that have been photoselected following flash excitations of the S_0-S_1 transition with plane-polarized light, the resulting polarization of the phosphorescence is anisotropic. The time dependence of this anisotropy can be measured.

Figure 5.17

What is the main difference in the information on molecular motion available from such measurements in phosphorescence compared with fluorescence?

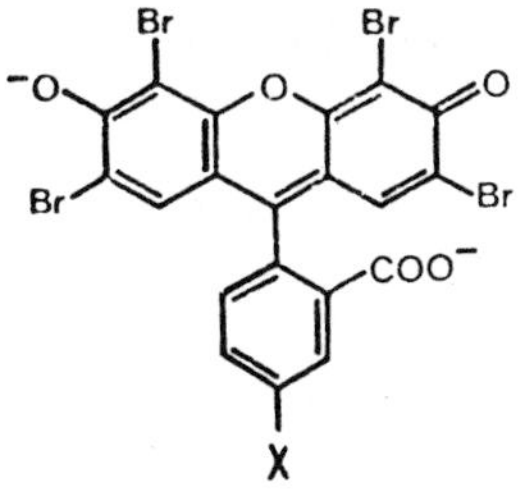

Solution

The main difference results from the lifetimes of the triplet and S_1 states. The lifetimes give the time scale over which the emission anisotropy can detect molecular motion. The lifetime of the triplet state (milliseconds) means that relatively slow molecular motions can be detected. An example would be the rotation of a protein molecule in a membrane. The fluorescence method detects motions in the nanosecond range (for an introduction to triplet probes, see R. J. Cherry, *Biochim. Biophys. Acta,* 559 (1979):289–327.)

---

*The triplet state is lower in energy than the first excited singlet state.

PROBLEMS

1. What is the difference between the time for a transition and the relaxation time (or lifetime) in fluorescence?

2. Several competing events depopulate the excited state. Write down an expression for the lifetime τ in terms of the rate constants of the following processes: (a) Intersystem crossing k_{1C}, (b) energy transfer k_T, (c) internal conversion k_C, (d) collisional quenching k_Q in the presence of a large amount of quencher Q, (e) fluorescence k_F.

3. In solution of chlorophyll, the measured lifetime is about 7 ns. However, in a photosynthetic unit, the lifetime is estimated to be about 0.1 ns. Can you suggest a reason for this difference? What are the relative fluorescence yields of chlorophyll in solution and in the photosynthetic unit? The radiative lifetime of chlorophyll is 25 ns.

4. *N*-bromosuccinimide (NBr) oxidizes tryptophan, and the product is non-fluorescent. Modification of papain by NBr showed that 2 molar equivalents modified Trp-69 and 4 molar equivalents modified Trp-69 and -177. The variation of the fluorescence spectra (excitation at 288 nm) with pH of the native, Trp-69 modified, and Trp-69 and -177 modified enzymes are shown in Figure 5.18. What conclusions can you draw from the data?

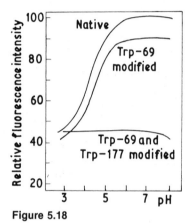

Figure 5.18

5. 1-Anilinonaphthalene-8-sulfonate (ANS) can be used as a probe for ligand binding to glutamate dehydrogenase (GDH). The ANS fluorescence alters in the presence of GDH (see Figure 5.19(a)). What conclusions might you draw from this? The addition of NADH or an inhibitor, guanosine-5'-triphosphate (GTP), alone does not alter the fluorescence of the ANS. However, in the presence of GTP, titration of the solution with NADH alters the fluorescence in a sigmoidal manner (see Figure 5.19(b)). Is ANS a useful probe for this system?

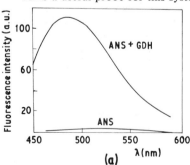
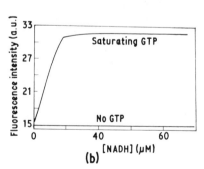

Figure 5.19

6. The intensity of ANS fluorescence increases with concentration until it reaches a maximum value, when it starts to decrease. Why?

7. Tryptophan fluorescence quenching is widely used to monitor the binding between 2,4-dinitrophenyl (Dnp) haptens and specific antibodies. Figure 5.20(b) shows the absorption and emission spectra of a purified antibody specific for Dnp ligands and the absorption spectra of ϵ-Dnp lysine. The large overlap of the absorption spectrum of ϵ-Dnp lysine with the emission spectra of the antibody results in highly efficient energy transfer. Figure 5.20(a) shows two typical titration curves for a strongly binding and a weakly binding hapten with antibody. The fluorescence quenching of a blank solution containing tryptophan (at the same concentration as the protein) and hapten is also shown. Why is it necessary to correct for the data of the blank quenching in determining the binding constant? Does this limit the utility of the method?

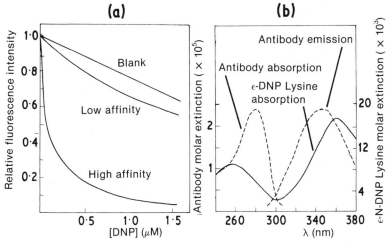

Figure 5.20

8. Thermolysin is a proteolytic enzyme (MW 37,500) that binds four calcium ions, which stabilize it against denaturation and autolysis. Two of these calcium ions are close together and can be substituted by a single terbium ion (Tb^{3+}). The active site of the enzyme also contains a zinc atom, which is essential for activity and which can be replaced by Co^{2+}. The fluorescence of terbium is partially quenched when Co^{2+} replaces zinc at the active site because of energy transfer (see Figure 5.21). Given that $R_0 = 1.63$ nm for the Tb^{3+} donor and Co^{2+} acceptor, calculate the distance between the two metal sites.

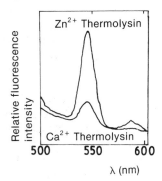

Figure 5.21

9. The kinetics of carboxypeptidase cleavage of the peptide dansyl-Gly-L-Phe were
   followed by monitoring the tryptophan and dansyl (DNS) fluorescence as a function of time after rapid mixing using stopped-flow methods.
   (a) Explain why the fluorescence of tryptophan increases while that of dansyl
       decreases.
   (b) When the zinc ion is replaced by Co^{2+}, the enzyme is active, but no dansyl
       fluorescence is observed. Suggest a reason for this and what information
       might be obtained from this observation. (The fluorescence spectra of
       tryptophan and DNS, together with the absorption spectra of DNS and Co^{2+}
       in the enzyme are shown in Figure 5.22.)

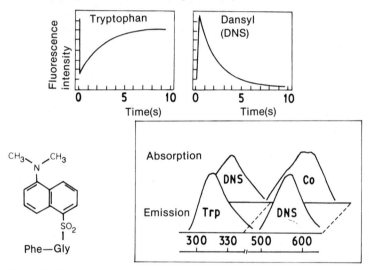

Figure 5.22

10. Figure 5.23 shows the time
    dependence of the anisotropy of the
    polarization of the fluorescence emis-
    sion of dansyl–lysine in the combin-
    ing sites of the Fab fragment and the
    $(Fab)_2$ fragment (see Figure 5.11).
    For Fab alone, a value of the rota-
    tional relaxation time of 26 ns was
    obtained from the plot. This value is
    about that expected if the Fab frag-
    ment behaved as a rigid unit. For the
    $(Fab)_2$ fragment, the curved plot of
    the results was analyzed in terms of
    two rotational times of 26 ns and 100
    ns. What does this suggest about the
    site of flexibility?

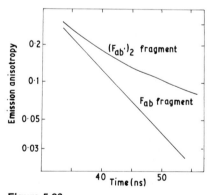

Figure 5.23

11. A protein undergoes a reversible transition between a rigid native and flexible unfolded form. A schematic plot of the predicted dependence of log A (anisotropy of the fluorescence polarization) versus time is shown in Figure 5.24(b) and (c) for two different models for the protein transition. What can you deduce about the relative rates of interconversion in each model? (Plots of log A versus time for the fully native and unfolded forms are given in Figure 5.24(a).

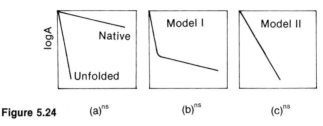

Figure 5.24   (a)^{ns} (b)^{ns} (c)^{ns}

12. Energy-transfer experiments between a covalent probe and 11-*cis* retinal have suggested that rhodopsin is at least 7.5 nm long. The calculation requires the value of the overlap integral between the emission spectrum of the energy donor and the absorption spectrum of 11-*cis* retinal. These are shown in Figure 5.25. The shaded area can be divided into approximately eight rectangular strips each 25 nm wide (2.5×10^{-6} cm). From the following data, which gives the values of the absorbance and fluorescence intensity at the midpoints of these strips, evaluate the overlap integral J.

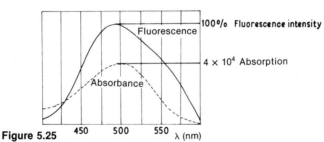

Figure 5.25

λ_i $(\times 10^5)$ (cm)	$\epsilon_A(\bar{\lambda}_i)$ $(\times 10^{-4})$ $(mol \cdot dm^{-3})^{-1} \cdot cm^{-1}$	$F_D(\lambda_i)$ $(\%)$
4.125	0.76	8
4.375	1.9	36
4.625	2.9	74
4.875	3.7	97
5.125	3.7	91
5.375	2.3	77
5.625	0.95	51
5.875	0.38	37

CHAPTER **6**

Nuclear Magnetic Resonance

OVERVIEW

1. Nuclear magnetic resonance (NMR) is a technique that detects nuclear-spin re-orientation in an applied magnetic field.

2. The parameters of NMR are chemical shift (δ), spin-spin coupling (J), area or intensity of the signal, and two relaxation times (T_1 and T_2). For molecules freely tumbling in solution, T_2 is related to the linewidth.

3. The most common method of measurement is application of a pulse of radiofrequency to the sample and detection of a transient signal from the nuclear spins. This transient is Fourier transformed to give the NMR spectrum.

4. Dipolar interactions with neighboring spins are very important in NMR. These can give rise to induced shifts and to relaxation. The dipolar interaction depends on distance ($1/r^3$), thus structural information can be deduced. In addition, Brownian motion of the molecules gives rise to a fluctuating dipolar field. This causes transitions between energy levels and relaxation. Relaxation times can be interpreted in terms of molecular motion. Very strong dipolar interactions can be induced by paramagnetic ions.

5. If a chemical group exchanges between different environments, the resulting spectra may be broadened, or the observed parameters may be an average of all the different environments.

6. The applications of NMR in biology are various and widespread. They include the measurement of the concentration and isotopic composition of molecules in intact tissue, intracellular pH, pK_a values of individual groups on macromolecules, kinetics, and dissociation constants of ligands binding to macromolecules; and structural information about molecules and molecular dynamics in a variety of systems, including membranes. The method is sensitive to environmental changes, but relatively large, concentrated samples are required because the signals produced are very weak.

INTRODUCTION

Nuclear magnetic resonance (NMR) is the spectroscopic method that is used to observe nuclear-spin reorientation in an applied magnetic field. Of all the spectroscopic methods, NMR is the most likely to give detailed information about the properties of molecules, whether they are in free solution or in a living animal. The reasons for this are that (1) individual chemical groups in macromolecules often give signals that can be resolved; (2) the NMR signals are sensitive to environment; (3) the theory is well understood and the relationship between spectral parameters and the information of interest (such as, concentration, dynamics, or structure) is relatively straightforward; (4) magnetic waves penetrate membranes and tissues with relatively little interference to the waves or the biological system. The main disadvantage of NMR is its lack of sensitivity, which arises from the small energy separation between the energy levels (see Chapter 2).

THE PHENOMENON

Certain nuclei possess the property of spin. Some of these are listed in Table 6.1. If a nucleus has a finite spin, then it has a magnetic moment μ_N that can interact with applied magnetic fields.

For many important nuclei—for example, ^1H, ^{31}P, ^{13}C—the spin I equals $\frac{1}{2}$. In such cases, only two energy levels result from an interaction with the applied magnetic field (B_0). These energy levels correspond to the spins aligned along and

Table 6.1   Properties of some magnetic nuclei

Nucleus	Spin (in units of $\hbar$)	Frequency at 10 T (MHz)[a]	Natural Abundance (%)
^1H	$\frac{1}{2}$	425.8	100
^2H	1	65.4	1.5×10^{-2}
^{13}C	$\frac{1}{2}$	107.1	1.1
^{14}N	1	30.8	99.6
^{15}N	$-\frac{1}{2}$	43.2	0.4
^{17}O	$-\frac{5}{2}$	57.6	3.7×10^{-2}
^{19}F	$\frac{1}{2}$	400.1	100
^{23}Na	$\frac{3}{2}$	112.6	100
^{31}P	$\frac{1}{2}$	172.4	100
^{35}Cl	$\frac{3}{2}$	41.7	75.5

[a]The value of γ can be obtained from the formula $2\pi \times \text{frequency} = \gamma B_0$.

Figure 6.1 Spin $\frac{1}{2}$ nuclei, interacting with
an applied field, B_0, have two possible
energy states. The populations of the two
states are N_1 and N_2.

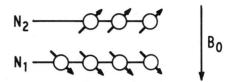

against the applied field B_0 (see Figure 6.1). The spins oriented to oppose B_0 have
the higher energy. As mentioned in Chapter 2, the spins do not align perfectly
along B_0; this gives rise to a permanent torque on μ_N. The nucleus also has the
property of angular momentum because of its spin. The net result of angular
momentum and torque is **precession**.

The frequency of precession is

$$\omega_0 = \gamma B_0 \tag{6.1}$$

where γ is a proportionality constant that is different for each nucleus (see Table
6.1) and ω_0 is called the resonant or the Larmor frequency.

Nuclei with $I > \frac{1}{2}$ also interact with magnetic fields and distribute themselves
among $2I + 1$ energy levels. These nuclei have an additional property, a **quadrupole moment** (Q) that allows the nuclei to interact with the electric fields produced by neighboring electrons and nuclei. This interaction can lead to important
modifications of the spectra (see sections on ligand binding to macromolecules
and molecular motion).

MAGNETIZATION

In NMR, we are always dealing with large numbers of spins (a typical sample
would be 1 ml of 1-mM solution, which contains nearly 10^{18} molecules). As a
result, simple classical physics can often be used to describe the properties of the
nuclear magnetization (the magnetic moment per unit volume of the sample). The
magnetization has direction (i.e., is a vector) and the components of interest are
M_z, which is defined to be along the B_0 direction, and M_x and M_y the components
at right angles to B_0. The component M_z depends on the population difference
between the spins with components parallel (N_1) and antiparallel (N_2) to B_0.

$$M_z = \gamma \hbar (N_1 - N_2) \tag{6.2}$$

The M_{xy} components arise because the spins do not align perfectly along B_0.
In an NMR sample there are many molecules, each with its spin precessing about
B_0. It is convenient to bring all these precessing spins to a common origin, as illustrated in Figure 6.2. At equilibrium, the spins are randomly distributed and the
net $M_{xy} = 0$. On application of a rotating radiofrequency field with frequency at
or near $\omega_0 = \gamma B_0$, the spins resonate and the random distribution changes into a
coherent one with net M_{xy} components. Spin systems giving rise to net M_{xy}
components are said to be **phase coherent**.

Figure 6.2   The many precessing spins in a sample can be considered to have a common origin. In (a) the equilibrium position is shown with an excess of spins oriented along the B_0 direction ($+z$), thus producing a net M_z. There is no phase coherence and $M_{xy} = 0$.

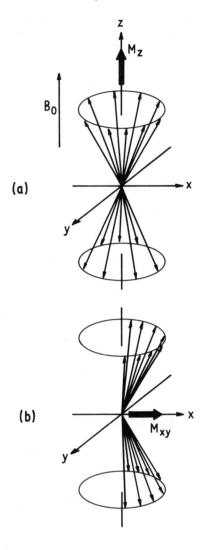

MEASUREMENT

In most forms of spectroscopy, the spectrum is measured by sweeping the applied radiation frequency. An equivalent method that can be used is induction of a transient response in the system and analysis of this transient for its component frequencies. The conventional frequency-sweep method is analogous to a piano tuner's striking and listening to each note in turn. The transient-response method is equivalent to all the notes' being struck at once. Obviously, some way of sorting the transient signal into frequencies is required. It is clear, however, that

Figure 6.3 A schematic representation of the NMR experiment. The sample is placed in a coil, which is tuned to resonate at the frequency of the applied radio-frequency field B_1.

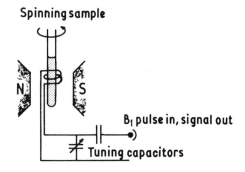

if this sorting could be done, all the frequencies could be measured very much more quickly by the transient method than by the frequency-sweep method. The former technique is also used to collect infrared spectra (see Chapter 3).

Both kinds of measurement, frequency sweep and transient response, can be used in NMR, but in most modern instruments only the transient method is used. This is the one we shall describe here, since in many ways it is easier to understand as well as being a more versatile and powerful way of collecting an NMR spectrum.

A typical experimental setup is shown in Figure 6.3. A large B_0 field (2–14 T) is applied to the sample. This field must be very homogeneous, so the sample is often spun about its vertical axis to improve the effective B_0 field homogeneity. To generate phase coherence among the spins, a rotating magnetic field is applied at right angles to B_0. This field, called B_1, is applied via a tuned coil. This generates M_{xy} components, which precess at the frequency ω_0. In addition, the phase coherence is lost by a process known as relaxation (discussed later), and the equilibrium position, where $M_{xy} = 0$, is achieved after some time. In fact, this relaxation process usually leads to an exponential decay of M_{xy} with time constant T_2. The net result is a transient signal with frequency ω_0 and decay rate $1/T_2$.

If there are many signals with different ω_0 and $1/T_2$ values, the transient, which is a superposition of all the signals, will be complicated. The transient can be resolved, however, using the mathematical method known as **Fourier transformation,** which changes the transient signal into a normal spectrum (see Appendix VII). Note that a spectrum is a plot of intensity against frequency, while the transient is a plot of intensity against time. These time and frequency "domains" are linked by the Fourier transformation process. Figure 6.4 illustrates this principle for two signals with unequal frequencies.

An important concept in using the transient method is to define the pulse of applied field B_1 in terms of angles. We discussed precession at frequency ω_0 in a B_0 field. This idea of precession can be extended to the rotating B_1 field by the expedient of using a rotating frame of reference, in which we imagine ourselves rotating at the same frequency as B_1, so B_1 effectively becomes static. At the

Figure 6.4 (a) The transient response of
two NMR signals, one with a low frequency,
the other somewhat higher. (b) The spectrum that arises from the transient response after Fourier transformation.

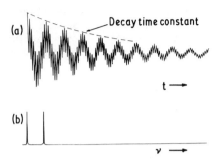

instant B_1 is applied, we have the situation shown in Figure 6.5(a), with B_1 along
the rotating axis x' and the equilibrium magnetization (M_0) along z. The component M_0 precesses around B_1 just as the magnetic moments μ_N precess around B_0.
The precession frequency of M_0 is $\omega_1 = \gamma B_1$. Since ω is an angular frequency, the
angle through which M_0 rotates is $\theta = \gamma B_1 t$, where t is the time for which B_1 is
applied. After a 90° pulse, $M_{xy} = M_0$, while after a 180° pulse, $M_z = -M_0$ and
$M_{xy} = 0$.

Figure 6.5 The effect of the application of
a rotating B_1 field on the magnetization M_0.
$x'y'$ is a reference frame rotating at the
same frequency as B_1. (b) and (c) correspond to 90° and 180° pulses, respectively.

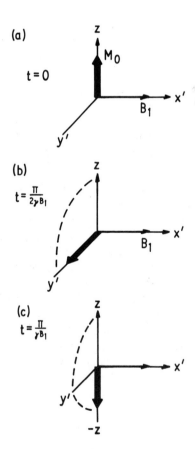

Worked Example 6.1 **B_0 and B_1 Fields**

In ^{1}H NMR experiments at 400 MHz, the 90° pulse was measured to be 10 μs. What are the magnitudes of the B_0 and B_1 fields?

Solution

From Table 6.1, $\gamma_H = (2\pi \times 425.8 \times 10^6)/10 = 2.6754$ rad$\cdot$T^{-1}.

Thus, $B_0 = 9.394$ *T.*

For a 90° pulse, $t = 10$ μs $= \pi/2\,\gamma_H B_1$; therefore, $B_1 = 5.87 \times 10^{-4}$ T.

THE SPECTRAL PARAMETERS IN NMR

There are four parameters that define the NMR spectrum of a particular group: (1) the intensity (I) or, more rigorously, the area; (2) the chemical shift (δ), which defines the position on a frequency scale; (3) the multiplet structure, which is related to the spin-spin coupling constant (J), and (4) a relaxation time (T_2), which is related to the linewidth. A spectrum of adenosine triphosphate is shown in Figure 6.6.

There is also a fifth parameter, which is not obtained directly from a spectrum—another relaxation time (T_1), which is related to the lifetime of a spin in an energy level (discussed later).

Intensity

If a system of nuclear spins is at equilibrium, then the amplitudes of the transient signals after a pulse are directly proportional to the number (concentration) of nuclei in the sample. Instrumental variables are such that accurate estimates of absolute concentration can be obtained only if a reference compound is used in

Figure 6.6   A ^{31}P spectrum of ATP. The three resonances from the α-, β-, and γ-phosphorous atoms have different chemical shifts and different multiplicities. The resonances have different linewidths, with the α-doublet appearing broader than the γ-doublet.

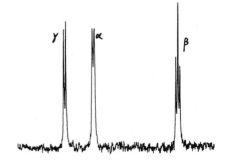

Figure 6.7 The production of fumarate, as a function of time, monitored by ^{1}H NMR in a suspension of red cells (courtesy Robert Simpson) (0.5-mL sample, concentration of fumarate ∼5 *mM*).

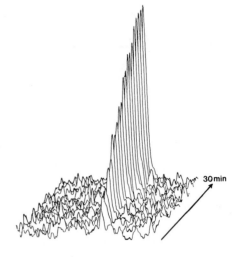

30 min

the sample. Relative concentrations can, however, often be obtained quite accurately from measurement of the resonance intensity (peak height). Figure 6.7 shows the resonance of fumarate in a ^{1}H NMR experiment carried out with a suspension of intact red cells. Malate has been added to the sample at $t = 0$, and successive spectra show the production of fumarate from the malate as a function of time.

Chemical Shift

Nuclei are surrounded by electrons, which shield them from the applied magnetic field B_0. The B_0 field induces currents in these electron clouds that reduce the effective field experienced at the nucleus because they tend to oppose B_0. The induced fields are directly proportional to B_0. Thus we can write

$$B_{\text{eff}} = B_0(1 - \sigma)$$

where σ is a **shielding constant** that depends on the nature of the electrons around the nucleus. Thus, different nuclei within a sample experience different fields, depending on their immediate chemical environment.

Rather than use a scale involving magnetic fields, it is more convenient to use a frequency scale and to normalize this scale by using a signal from a reference compound. The scale now used in NMR is called the **chemical shift** scale and is given by

$$\delta = 10^6\left(\frac{\delta_{\text{ref}} - \delta_{\text{obs}}}{\delta_{\text{ref}}}\right)$$

where δ_{ref} is the position, in hertz, observed for a reference compound and δ_{obs} is the position of the signal of interest. This scale, which is in units of parts per million (ppm), has the advantage that it is independent of B_0.

Table 6.2

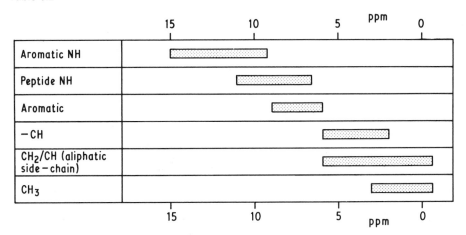

There are two distinct classes of shift that are important for our purposes: (1) an **intrinsic,** or **primary, shift,** which is characteristic of a particular chemical group and (2) an **induced,** or **secondary, shift** arising from the influence, through space, of neighboring magnetic centers.

The intrinsic shifts of different groups and of different nuclei can vary widely. Table 6.2 shows some characteristic shifts for ^1H resonances with respect to a reference compound of the type $R-Si(CH_3)_3$. Note that the shift position is often very sensitive to the *ionization state* of the molecule. The $C-(2)-^1H$ resonance of histidine, for example, shifts by about 1 ppm between the acidic and basic forms of this amino acid side chain (see the section on kinetics).

Induced (or secondary) shifts are very important in spectra of macromolecules. The spectrum of a folded globular protein is, for example, completely different from the spectrum of the unfolded protein (see Figure 6.8). The spectra are different because each chemical group in a globular protein is in a relatively well-defined environment, which causes small but significant secondary shifts of the amino acid resonances.

Some secondary shifts can be quantified, the most important of which arise from aromatic rings and paramagnetic ions. Aromatic rings give rise to **ring-current shifts** because the electrons in aromatic rings are essentially delocalized; if B_0 is applied in a direction perpendicular to the plane of the ring, a field that opposes B_0 is produced by the circulating electrons. The field produced may be considered as a dipole μ of magnitude $B_0 e^2 a^2 / 4m$, where e is the charge on the electron, a the ring radius, and m the mass of the electron. In general, the dipole μ produces a field at a point A (see Figure 6.9) proportional to $(3 \cos^2\theta' - 1)/r^3$ (see Appendix IX), where θ' is the angle between the B_0 direction and the line joining A and μ (the vector $\mathbf{r}$). In solution, the molecule tumbles and the field at A is averaged over all θ'. The induced dipole, however, will also change as the molecule

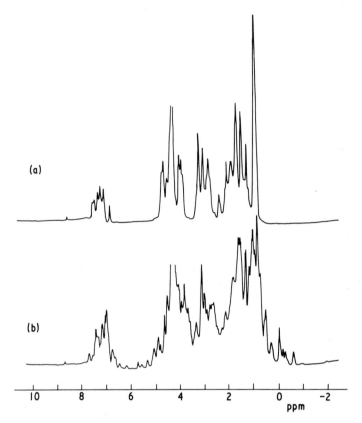

Figure 6.8   (a) The 270-MHz ^{1}H spectrum of 5-mM lysozyme at pH 4.0, 80°C (random-coil). (b) Spectrum of native protein under same conditions, but 58°C.

tumbles. The value $\mu_\perp$ when B_0 is perpendicular to the ring is greater than $\mu_\parallel$. The net field produced by these effects at the point A is given by

$$B_{\text{ring}} = \frac{(\mu_0/4\pi)(\mu_\perp - \mu_\parallel)(3\cos^2\theta - 1)}{3r^3}$$

where θ is the angle defined in Figure 6.9. The factor $\frac{1}{3}$ arises because of the averaging that results from the tumbling of the molecule.

Figure 6.9   Illustration of the geometry for discussion of ring-current shifts. The vector **r** makes an angle θ with the perpendicular to the ring and an angle θ' with the B_0 direction.

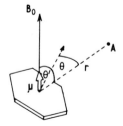

Worked Example 6.2   Ring-Current Shifts

Calculate the shifts caused by a benzene ring (a) on a group 0.3 nm above the plane of the ring and (b) on one of the hydrogen atoms attached to the ring. (The $C-C$ bond length in benzene is 0.14 nm and the $C-H$ bond length is 0.109 nm.) ($\mu_0 = 4\pi \times 10^{-7}$ H•m^{-1}, $m = 9.11 \times 10^{-31}$ kg, $e = 1.6 \times 10^{-19}$C)

Solution

There are six free electrons in the ring. If we asume there is no current induced when B_0 is parallel to the ring, $\mu_\parallel = 0$. Thus,

$$B_{\text{ring}} = \frac{\mu_0 \mu_\perp}{4\pi 3 r^3}(3 \cos^2\theta - 1) = \frac{6\mu_0 a^2 B_0 e^2 (3 \cos^2\theta - 1)}{4 \times 4\pi m \times 3r^3}$$

$$\therefore \frac{B_{\text{ring}}}{B_0} = \frac{1.4 \times 10^{-15}(3 \cos^2\theta - 1)a^2}{r^3}$$

In (a) $\theta = 0$, therefore, $3 \cos^2\theta - 1 = 2$, $a = 0.14$ nm, and $r = 0.3$ nm

$$\therefore \frac{B_{\text{ring}}}{B_0} = +2.03 \text{ ppm}$$

In (b) $\theta = 90°$, therefore, $3 \cos^2\theta - 1 = -1$, $a = 0.14$ nm, and $r = 0.249$ nm

$$\therefore \frac{B_{\text{ring}}}{B_0} = -1.78 \text{ ppm}$$

Note the change in sign and the order of magnitude of the shifts. This calculation is approximate and more sophisticated theories are possible (see, for example, S. J. Perkins, *Biol. Mag. Res.,* 4(1982):193).

Spin-Spin Coupling and Multiplet Structure

High-resolution NMR spectra exhibit multiplet structure that arises from weak interactions between magnetic nuclei (e.g., see Figure 6.6). These interactions are communicated between the nuclei by the electrons in a chemical bond. The size of the interaction is defined by the spin-spin coupling contant (J) which is expressed in hertz.

Consider, for example, a hypothetical molecule consisting of two bonded nuclei, as illustrated in Figure 6.10. Each nucleus can be oriented in two ways. These two orientations cause slightly different electron distributions, which result in small chemical shifts. These shifts cause each of the two nuclear resonances to

Figure 6.10 Two nuclei sharing a bond exhibit multiplet structure. The electron cloud communicates the direction of one nucleus to the other because it becomes slightly polarized by the nuclear spin.

Figure 6.11 Illustration of the appearance of a pair of doublets for different δ/J ratios. When $\delta/J >> 1$, the spectrum is often called an AX spectrum; when $\delta/J \simeq 1$, it is called an AB spectrum.

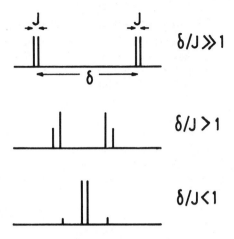

split in two, making a pair of doublets. The spectrum of a nucleus coupled to *two* equivalent nuclei is a triplet with intensities in the ratio 1:2:1 because the following orientations are possible: ↑ ↑, ↓ ↑, ↑ ↓, ↓ ↓, two of which are equivalent. Similarly, three equivalent nuclei give rise to a quartet with intensity ratios 1:2:2:1, and so on (see also Chapter 7).

The appearance of a multiplet depends also on the relative magnitudes of δ and J for the coupled nuclei. Consider a spectrum of two nuclei separated by δ and with coupling constant J. If $\delta/J >> 1$, then a simple pair of doublets is observed. However, with decreasing δ/J ratios, quantum mechanics predicts that the transition probabilities of the outer lines decrease compared with the inner ones (see Figure 6.11). When $\delta = 0$, the nuclei are said to be equivalent and no multiplet structure is observed.

Since multiplet structure arises from a through-bond interaction, the magnitude of J depends on both the nature of the bond and the number of bonds involved. There is a fairly well-defined relationship between bond angle and the observed value of J. This relationship is named after Martin Karplus, who developed the theory, and is illustrated in Figure 6.12.

Figure 6.12 The form of the Karplus equation, which relates dihedral angle ϕ to coupling constant $J_{HH'}$ (from J. Feeney, *Proc. Roy. Soc. London, Ser.*, A., 345 (1975):61).

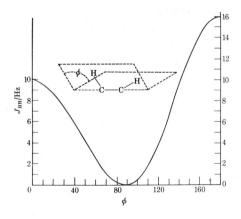

Spin-coupling multiplets can be removed if the relative orientation of coupled nuclei changes relatively rapidly compared with $1/J$ (see section on chemical exchange). Such rapid reorientation can arise if the relaxation time of one of the nuclei is very short (see section on relaxation). Another way in which coupling can be removed is by irradiation of one of the nuclei or groups of nuclei with a selective radiofrequency field. (This field is usually denoted B_2 to distinguish it from the static B_0 field and the B_1 pulse used to excite the spectrum for observation). The technique of applying another radiofrequency field is called **double resonance** and the resulting collapse of the spin multiplet is called **decoupling**. The reason that the method works is that the irradiated nuclei are stimulated to change energy levels relatively rapidly—that is, they are "stirred" by the B_2 field.

Worked Example 6.3   The Spectrum of Ethanol

Explain the appearance of the spectrum of ethanol shown in Figure 6.13.

Solution

The resonance from the CH_3 group has a relative area of three units, and because it is adjacent to two hydrogens on the CH_2 group, it is a triplet. The CH_2 resonance has a relative area of two units and is a quartet because it is next to the CH_3 group. The coupling between the OH hydrogen and the CH_2 group is removed because of rapid chemical exchange (discussed later) of this hydrogen between different ethanol molecules.

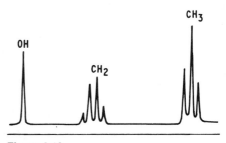

Figure 6.13

Note the asymmetry of the CH_2 quartet and CH_3 triplet because δ/J is finite.

The T_2 Relaxation Time and Linewidth

The term T_2 is defined as the time constant of the decay of the M_{xy} components (see the section on measurement). It is also related to the linewidth by the following relationship (see Worked Example 6.4 for derivation):

$$\Delta\nu_{1/2} = \frac{1}{\pi T_2}$$

Field inhomogeneity affects the observed linewidth and the transient decay because a spread of fields produces a spread of frequencies. There are, however, methods for reducing the effect of such inhomogeneities and thus obtaining T_2.

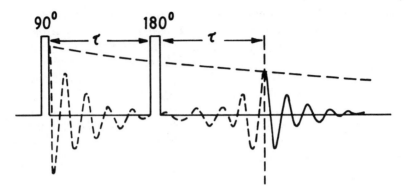

Figure 6.14   Spin echoes. Illustration of the refocusing of the magnetization in an NMR experiment by a 180° pulse after a given time. The decay of the echo as a function of τ can be used to measure T_2.

For example, a $90°-\tau-180°-\tau$ pulse sequence generates an echo at time 2τ and the decay of this echo is relatively insensitive to field inhomogeneity effects (see Figure 6.14).

The mechanisms that cause the spins to lose their phase coherence in the *xy*-plane, that is, to lead to T_2 relaxation, will be discussed in the section on relaxation.

Worked Example 6.4   Relationship Between T_2 and Line Shape

Derive the relationship $\Delta\nu_{1/2} = 1/\pi T_2$. (*Hint:*   Look at Appendix VII.)

Solution

The transient-response function for one resonance of frequency ω_0 and relaxation time T_2 is of the form (see Figure A13(b))

$$f(t) = \cos(\omega_0 t)\exp\left(\frac{-t}{T_2}\right)$$

The Fourier transform of this function is

$$F(\omega) = \frac{CT_2}{1 + (\omega - \omega_0)^2 T_2^2}$$

where C is a constant. The form of line shape given by $F(\omega)$ is known as a **Lorentzian line** and is shown in Figure A5.

The half-height of $F(\omega)$ occurs when $(\omega - \omega_0)^2 T_2^2 = 1$, i.e., when $\omega = 1/T_2$. The linewidth at half-height on the ω scale is thus $2/T_2$. Since $\omega = 2\pi\nu$, the half-height on a frequency scale $\Delta\nu_{1/2} = 1/\pi T_2$.

The T_1 Relaxation Time

The term T_1 is defined to be the time constant that describes the recovery of M_z after a perturbation. The component M_z can be perturbed in several ways. If B_1 is applied for a long time, for example, then **saturation** takes place—that is, the population difference between the levels is destroyed and $M_z = 0$ (see Chapter 2). The recovery to equilibrium can then be monitored.

In fact, there is a more practical way of carrying out the experiment that involves the use of 90° and 180° B_1 pulses. Figure 6.15 illustrates the measurement of T_1 by a 180°–τ–90° pulse sequence. The 180° pulse inverts the populations, so the magnetization component along the z-axis, M_0, becomes $-M_0$. This

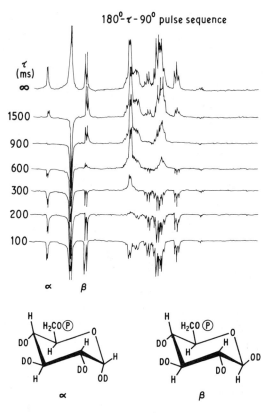

Figure 6.15   Inversion and recovery of a 270-MHz ^{1}H NMR spectrum of a mixture of anomers of glucose monitored by a 180°–τ–90° pulse sequence. The peaks labeled α and β represent the resonances of the anomeric protons. The β-resonance has a faster recovery rate (spin-lattice relaxation rate) than the α-resonance. This arises because the C-1 hydrogen of the β-anomer is close to both the C-3 and C-5 hydrogen, whereas the C-1 hydrogen of the α-anomer is close only to the C-2 hydrogen. The spin-spin coupling of the anomeric resonances arises from interactions with the C-2 hydrogen; the different values reflect the different bond angles made between the C-1 and C-2 hydrogens (see Figure 6.12) (from R. A. Dwek, *NMR in Biochemistry,* (Oxford: Clarendon Press, 1973)).

recovers along the z-axis with a first-order rate constant $1/T_1$. Thus M_z, as a function of time, is given by

$$M_z(t) = M_0(1 - 2e^{-t/T_1})$$

Since this recovery takes place along the z-axis, it cannot be observed directly because there are no M_{xy} components (no phase coherence). The recovery is therefore monitored by applying a 90° pulse at intervals, thus tipping the resultant z-magnetization into the xy-plane.

WHAT CAUSES RELAXATION?

We now consider the mechanisms for T_1 and T_2 relaxation.

T_1 Relaxation Processes

The rate at which a Boltzmann distribution of populations is set up among the energy levels is $1/T_1$. We noted in Chapter 2 that spontaneous emission is negligible in NMR. This means that all NMR transitions are caused by magnetic fields fluctuating at frequencies related to ω_0. What then causes the Boltzmann distribution to be set up? The answer is that thermal motion of molecules in the sample causes small fluctuating local fields at the nuclei of interest. Since these motions also have an energy distribution, the probabilities of inducing upward and downward NMR transitions are not equal, so a slight excess population is set up in the lower energy level.

Fluctuating magnetic fields, caused by motion, can arise from a variety of sources, but for $I = \frac{1}{2}$ nuclei, the most important one is usually the dipole-dipole interaction. The field exerted by one nuclear magnetic moment on another is given by $(3\cos^2\theta - 1)/r^3$ (see Appendix IX). If the moments μ_1 and μ_2 are on one molecule, e.g., the C-1 and C-2 hydrogens of glucose in Figure 6.15, then the separation between them, r, is fixed (see Figure 6.16). As the molecule tumbles, the spins will stay oriented in the field. Thus, θ will vary in a random way as the molecule undergoes Brownian motion, and μ_1 will produce a fluctuating field at μ_2 and vice versa.

Figure 6.16   The interaction between two dipoles in field B_0.

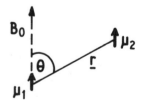

The time scale of the random motion is defined by some function $g(\tau)$. In many cases this function is exponential and can be expressed as $\exp(-t/\tau_c)$, where τ_c is a *correlation time*. This simply means that the probability of finding a correlation between the orientation of the vector **r** and some defined direction decreases exponentially with time constant τ_c.

The frequency content of a function that varies with time can be analyzed by Fourier transformation. In the case of an exponential $g(\tau)$, this transformation is given by $J(\omega) = \tau_c/(1 + \omega^2\tau_c^2)$. Note the similarity to a Lorentzian line shape (see Worked Example 6.4). The function $J(\omega)$ expresses the probability of finding a fluctuation with frequency ω for a particular correlation time τ_c. (*Note:* For a macromolecule of radius a in a solution of viscosity η, the rotational correlation time τ_c is given approximately by $4\pi\eta a^3/3kT$.)

The probability that field fluctuations will induce a transition between energy levels will depend not only on the existence of a component at the correct frequency but also on the amplitude of the fluctuation. The amplitude of the dipole-dipole interaction is $\mu_0\mu_1\mu_2/4\pi r^3$. The nuclear magnetic moments have magnitude $\gamma\hbar I$ and the transition probability depends on the amplitude squared. For two spins with the same T_1 and with $\gamma_1 = \gamma_2$, the full expression for T_1 is given by

$$\frac{1}{T_1} = \frac{3\mu_0^2\gamma_1^4\hbar^2}{160\pi^2 r^6}\left\{\frac{\tau_c}{1 + \omega^2\tau_c^2} + \frac{4\tau_c}{1 + 4\omega^2\tau_c^2}\right\} \tag{6.3}$$

Fluctuations at both ω and 2ω cause transitions; thus, there are terms for both $J(\omega)$ and $J(2\omega)$.

Worked Example 6.5   Calculations of T_1

Calculate T_1 for a pair of protons separated by 0.25 nm undergoing motion described by the correlation times (a) 10^{-8} s, (b) 10^{-9} s, and (c) 10^{-10} s. Assume the spectrometer frequency is 100 MHz.

$$(\gamma = 2.675 \times 10^8 \, \text{T}^{-1}\cdot\text{s}^{-1}, \hbar = 1.054 \times 10^{-34} \, \text{J}\cdot\text{s}, \mu_0 = 4\pi \times 10^{-7} \, \text{H}\cdot\text{m}^{-1})$$

Solution

$$\frac{1}{T_1} = \frac{3\mu_0^2}{160\pi^2}\frac{(2.675)^4 \times 10^{32} \times (1.054)^2 \times 10^{-68}}{(0.25)^6 \times (10^{-9})^6} f(\tau_c)$$
$$= 6.996 \times 10^8 \, f(\tau_c)$$

The $f(\tau_c)$ term depends on $\omega = 2\pi \times 100 \times 10^6 \, \text{rad}\cdot\text{s}^{-1}$.
Thus for $\tau_c = 10^{-8}$ s, $T_1 = 2.87$ s; for $\tau_c = 10^{-9}$ s, $T_1 = 0.63$ s; for $\tau_c = 10^{-10}$ s, $T_1 = 2.89$ s.

Note the minimum around $\tau_c = 10^{-9}$ s. This arises because the $f(\tau_c)$ term has a maximum when $\omega\tau_c \sim 1$.

Figure 6.17 Plot of the dependence of T_1 and T_2 on τ_c for two different spectrometer frequencies. Note that for long τ_c, $T_2 \ll T_1$. The relaxation is assumed to arise from a dipolar mechanism (see previous worked example).

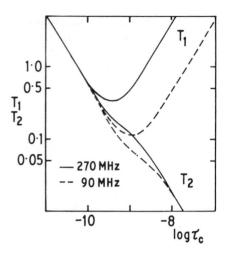

T_2 Relaxation Processes

It was shown in Figure 6.5 that a 90° pulse caused M_0 to rotate into the xy-plane, but that it also resulted in $M_z = 0$. From T_1 processes, M_z will recover to the equilibrium M_0 value, but this will also cause M_{xy} to disappear. From this simple argument, one might expect $T_1 = T_2$, and this is often observed.

There are, however, additional ways in which M_{xy} components can be destroyed. If there is a spread of ω_0 values because of local fields produced in the sample, or if there is exchange between different environments with different values of ω_0 (see section on chemical exchange), then T_2 can be decreased. Thus, in general, $T_2 < T_1$.

For dipole-dipole interactions, the T_2 equation, which corresponds to the one for T_1, is

$$\frac{1}{T_2} = \frac{3\mu_0^2\gamma^4\hbar^2}{320\pi^2 r^6}\left(3\tau_c + \frac{5\tau_c}{1 + \omega^2\tau_c^2} + \frac{2\tau_c}{1 + 4\omega^2\tau_c^2}\right) \qquad (6.4)$$

Equations 6.3 and 6.4 are plotted in Figure 6.17.

THE NUCLEAR OVERHAUSER EFFECT

In the last section we stated that the dominant relaxation mechanism for spin $\frac{1}{2}$ nuclei usually arises from dipolar interactions with other spins. In such a case, double resonance—that is, irradiation of one set of spins by a B_2 field (see section on spin-spin coupling)—can cause intensity changes in the spectra of other spins. This change in intensity results from the fact that the T_1 relaxation of one of the pair of spins is partly caused by the spin's being irradiated. For two nuclear-spin dipoles, called I and S, saturation of the S-resonance by B_2 (i.e., setting M_z of S to zero) causes a **nuclear Overhauser effect** (nOe) described by the following equation:

$$\eta = \frac{R_c}{R_I + R_x} \frac{\gamma_S}{\gamma_I} \tag{6.5}$$

where η is the fractional change in intensity of the I-resonance, γ_S and γ_I are the magnetogyric ratios of I and S, R_I is the dipolar relaxation rate of I caused by S, R_x is the relaxation rate of I from sources other than I, and R_c is a cross-relaxation rate between I and S. The term $R_c = R_I/2$ when $\omega\tau_c \ll 1$, and $R_c = -R_I$ when $\omega\tau_c \gg 1$. Note that R_I and R_c depend on τ_c and $1/r^6$ in a manner similar to equations (6.4) and (6.5).

Worked Example 6.6   Nuclear Overhauser Effects

Explain the nOe observations in the table, which were made at a field corresponding to a ^1H frequency of 100 MHz.

Molecule	S Spin (Irradiated)	I Spin (observed)	η
(a) Benzene	^1H	^{13}C	1.93
(b)	^1H (6)	^1H (5)	0.2
(c) As in (b) but bound to antibody	^1H (6)	^1H (5)	−0.8

(For (b), the molecule structure shown: a benzene ring with NO_2 at top, H (5) and H on the left/right positions, H(6) and NO_2 at lower positions, and R at the bottom.)

(Assume that for benzene $\tau_c = 5 \times 10^{-11}$s, while for the antibody $\tau_c = 10^{-8}$ s.)

Solution

(a) From Table 6.1, $\gamma_I = \gamma_S/4$. For a small molecule, $\omega\tau_c \ll 1$. Thus $R_c \approx R_I/2$ and $\eta \approx 2$ if $R_x = 0$. R_x will be small since the irradiated ^1H is on the same bond as the ^{13}C.

(b) $\gamma_S = \gamma_I$, $R_c \approx R_I/2$, and η will be 0.5 if $R_x = 0$. R_x will be significant because of other groups in the vicinity of the H (5) hydrogen.

(c) When the DNP is bound to antibody and $\omega\tau_c \gg 1$, $R_c = -R_I$, and $\eta = -1$ if $R_x = 0$.

CHEMICAL EXCHANGE

A molecule in solution may exist in different environments—for example, a ligand of a protein that exists both free in solution and bound to the protein, or an amino acid that exists in both protonated and unprotonated forms. In both of these examples, the molecule can exchange between the environments, perhaps during an NMR measurement, which usually takes about 1 s, even using the transient method. Such exchange has important consequences, both for the appearance of NMR spectra and for the information that can be extracted from the spectra.

Consider the following situation, where A and B are intercoverting forms of a molecule.

$$A \underset{k_b}{\overset{k_a}{\rightleftharpoons}} B$$

The spectral parameters of A are δ_A, J_A, $1/T_{2_A}$, and so forth, with a similar set for B. If k_a and k_b are slow, then one might expect resonances of A and B to be observed separately and independently of one another. If k_a and k_b are fast, however, then one might expect some sort of weighted mean to be observed. These two extremes are called the **slow** and **fast exchange limits**. We must now consider these limits in more detail. How fast is "fast"? How slow is "slow"? What happens between the limits?

To answer these questions, consider the situation in terms of the rotating components in the xy-plane. A 90° pulse produces two components, one for each species. These two rotate at an angular frequency $\Delta\omega = \omega_A - \omega_B$ with respect to each other. To simplify matters, let us assume $k_a = k_b = k$ and $1/T_{2_A} = 1/T_{2_B} = 1/T_2$. Without resorting to complicated mathematics, it can be seen, with the help of Figure 6.18, that if $k < \Delta\omega$, the bulk of the magnetization, either M_A or M_B, will decay as before, but there will be an additional relaxation process caused by the exchange. For example, if an A molecule changes to a B molecule, a component of M_A will rotate away from x' at a rate $\Delta\omega$. This will result in a loss of M_A at a rate of k. Thus,

$$\frac{1}{T_{2_{obs}}} = \frac{1}{T_{2_A}} + k \tag{6.6}$$

If $k > \Delta\omega$, then A and B interconvert fast enough to make the M_A and M_B components indistinguishable, and a new component, $M_B + M_B$, is observed to rotate at $\Delta\omega/2$ in the rotating frame. The observed relaxation rate in this situation is

$$\frac{1}{T_{2_{obs}}} = \frac{1}{T_2} + \frac{\Delta\omega^2}{8k} \tag{6.7}$$

Figure 6.18 Representation of M_A and M_B components in a reference frame rotating at angular frequency ω_A.

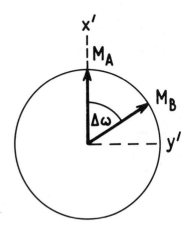

In the fast limit, the chemical shift becomes $\delta_{obs} = f_A\delta_A + f_B\delta_B$, where f_A and f_B are the fractions of the A and B species present in solution.

Thus the answer to How fast? is $k > \Delta\omega$; to How slow? is $k < \Delta\omega$; and to What happens in between? is that there is an increase in relaxation rate described by equations 6.6 and 6.7. Figure 6.19 illustrates the sort of spectra that are observed in practice if chemical exchange occurs in the right frequency range. (Note that fast and slow exchange can sometimes be relative to the linewidth; see Chapter 7.)

Figure 6.19 The effect of chemical exchange between two environments. The bottom spectrum represents two chemical environments occupied by 75% and 25% of the molecules, respectively. Their signals are 10 Hz wide and separated by 100 Hz, and their rates are shown.

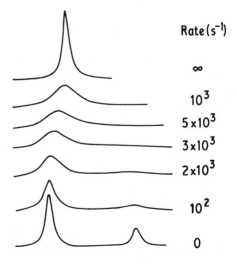

Rate (s^{-1})

∞

10^3

5×10^3

3×10^3

2×10^3

10^2

0

> ### *Worked Example 6.7   Chemical Exchange*
>
> In the exchange process
>
> $$A \underset{k_b}{\overset{k_a}{\rightleftharpoons}} B$$
>
> $\delta_A = 4.1$ ppm, $\delta_B = 3.7$ ppm, $1/T_{2A} = 3$ s^{-1}, and $1/T_{2B} = 2$ s^{-1}. With a 200-MHz spectrometer,
> (a) What would the observed shift be when $k_a = 10^4$ s^{-1} and $k_b = 5 \times 10^3$ s^{-1}?
> (b) What is the observed linewidth when $k_a = 3$ s^{-1} and $k_b = 1.5$ s^{-1}?
>
> *Solution*
> (a) The shift difference is $\delta_A - \delta_B = 0.4 \times 200 = 80$ Hz.
>
> $$\Delta\omega = 2\pi \times 80 = 502.6 \text{ rad} \cdot \text{s}^{-1}$$
>
> Since k_a, $k_b > \Delta\omega$, fast exchange is observed, and
>
> $$\delta_{obs} = f_A\delta_A + f_B\delta_B$$
>
> Now $f_A = [A]/([A] + [B])$ and $k_a[A] = k_b[B]$ at equilibrium,
>
> $$\therefore f_A = \frac{k_b}{k_a + k_b} = \frac{1}{3} \text{ and } f_B = \frac{2}{3}$$
> $$\therefore \delta_{obs} = \frac{1}{3} \times 4.1 + \frac{2}{3} \times 3.7 = 3.83 \text{ ppm}$$
>
> (b) Since k_a, $k_b \ll \Delta\omega$, slow exchange (i.e., separate lines) is observed.
> For A,
>
> $$1/T_{2obs} = 1/T_{2A} + k_a = 3 + 3 = 6 \text{ s}^{-1}$$
>
> For B,
>
> $$1/T_{2obs} = 1/T_{2B} + k_b = 2 + 1.5 = 3.5 \text{s}^{-1}$$
>
> The linewidth is given by $1/\pi T_2$,
>
> $$\therefore \Delta\nu_{1/2}(A) = 6/\pi = 1.91 \text{ Hz}$$
> $$\Delta\nu_{1/2}(B) = 3.5/\pi = 1.11 \text{ Hz}$$

PARAMAGNETIC CENTERS

The magnetic moment of an unpaired electron is much greater (658 times) than that of a nucleus; thus, the presence of a paramagnetic center in a solution can have a large effect on the NMR spectrum. These effects can be interpreted to give information about molecular motion, distance from the paramagnetic center to the observed nucleus, and binding of a paramagnetic ion to a macromolecule.

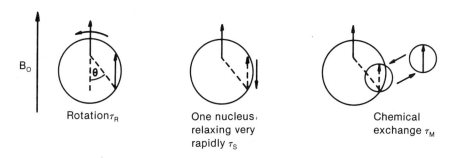

B$_0$

Rotation τ_R

One nucleus,
relaxing very
rapidly τ_S

Chemical
exchange τ_M

Figure 6.20   Illustration of some of the motions that can modulate the dipolar interaction between two dipoles. Each kind is characterized by a different correlation time (τ_R, τ_S, τ_M).

The interaction between the paramagnetic center (let us call it S) and the nucleus (I) can give rise to observed shifts or changes in the relaxation parameters (T_1 and T_2). In each case these changes arise from two kinds of mechanisms: (1) a dipolar interaction between I and S and (2) a delocalization of the electron of S, which changes the effective field experienced by I. Mechanism (2) is often called a **scalar mechanism** and is similar to the through-bond effect that causes spin-spin coupling J.

The dipolar mechanism is usually dominant and it can cause large shifts and relaxation effects. The equations for shift are essentially the same as those discussed for ring currents, since they both arise from a dipolar mechanism.

The equations for relaxation have a form similar to equations 6.3 and 6.4. In particular, by assuming that $\omega_s^2 \tau_c^2 >> 1$—a reasonable assumption in most cases—we obtain

$$\frac{1}{T_{1M}} = \frac{2}{15} \frac{\gamma_I^2 \mu_{\text{eff}}^2}{r^6} \left(\frac{3\tau_c}{1 + \omega_I^2 \tau_c^2} \right) \tag{6.8}$$

where I and S are in a macromolecular complex, r is the separation between I and S, and μ_{eff} is the effective magnetic moment of the paramagnetic center. The correlation time τ_c depends on the overall tumbling of I and S in solution, which is defined by the correlation time τ_R. The term τ_c also depends on the relaxation time of S, τ_S, which can be comparable with τ_R. There is also the possibility that S is exchanging in and out of the I–S complex with a bound lifetime τ_M (see Figure 6.20). The term τ_c is then defined by

$$\frac{1}{\tau_c} = \frac{1}{\tau_R} + \frac{1}{\tau_M} + \frac{1}{\tau_S} \tag{6.9}$$

Shift and Relaxation Probes

Different metal ions have different magnetic properties. Consider, for example, the two lanthanide ions Gd(III) and Eu(III). They are both paramagnetic, but τ_s for Gd(III) is about 10^{-9} s, while for Eu(III) it is 10^{-13} s. Consider the correlation times for these two ions when they are in a complex with a small molecule in aqueous solution with a rotational correlation time τ_R of 10^{-10} s. If $\tau_M \sim 10^{-6}$ s, for example, then equation 6.9 gives $\tau_c \simeq 10^{-10}$ for Gd(III), but $\tau_c \simeq 10^{-13}$ for Eu(III). These results mean that $1/T_{2M}$ from equation 6.8 may be 1000 times higher for Gd(III) than for Eu(III). Moreover, Gd(III) turns out to produce isotropic shifts; thus it causes no net dipolar shift (see page 136). From this argument, it is clear that Gd(III) is a *relaxation* (or *broadening*) probe, while Eu(III), which is a poor relaxation agent, is a *shift* probe.

Of the transition metal ions, Mn(II) is a relaxation probe, while Ni(II) and Co(II) are shift probes.

Worked Example 6.8   Distances from Paramagnetic Probes

The stereochemical relationship between a lanthanide shift probe and two nuclei, 1 and 2, on a molecule (see Figure 6.21) can be expressed as

$$\frac{\delta_1}{\delta_2} = \frac{r_2^3}{r_1^3} \frac{3\cos^2\theta_1 - 1}{3\cos^2\theta_2 - 1}$$

where r_1 and r_2 are the distances between the lanthanide and the nucleus, and θ_1 and θ_2 are the angles to the principal axis of the lanthanide. What shift ratio do you expect when $r_1 = r_2$, $\theta_1 = 30°$, and $\theta_2 = 60°$? What happens when $\theta = 54°44'$?

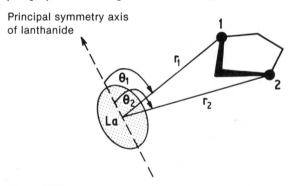

Principal symmetry axis of lanthanide

Figure 6.21

Solution

The shift ratio is $1.25/-0.25 = -5$. The negative sign means that the induced shifts on nuclei 1 and 2 are in opposite directions. The term $3\cos^2\theta - 1$ becomes zero when $\theta = 54°44'$. This is sometimes known as the *magic angle*.

APPLICATIONS OF NMR IN BIOLOGY

We have defined the NMR parameters and explained how these are affected by vmrious factors such as ring-current shifts, paramagnetic centers, double resonance, and molecular motion. How can useful information be obtained from the

NMR parameters? It is convenient to discuss the many possibilities under different headings:

1. *Analytical uses:* Compounds can be identified; concentrations can be measured, including that of the proton (pH); and biosynthetic pathways can be elucidated using the isotopes ^{13}C, ^{15}N, ^2H, or ^3H.

2. *Ligand binding and ionization states:* K_d can be measured (in the millimolar range) for a wide variety of ligands, including metal ions, anions and small organic molecules; pK_a values of ligands and individual groups on macromolecules can be measured.

3. *Kinetics:* The rates of processes can be determined over a wide range by analyzing chemical-exchange effects ($1-10^5$ s^{-1}) and by monitoring concentration changes (10^{-1} s^{-1}–10^{-3} s^{-1}).

4. *Structural information:* Structural information can be obtained from the r^6 dependence of dipolar relaxation, which can influence T_1, T_2, and the nOe. Shifts depend on $1/r^3$ and on some angle (see page 136, for example). Bond angles influence J (see section on spin-spin coupling).

5. *Molecular motion:* The terms T_1 and T_2 depend on molecular motion. In addition, information about motion can sometimes be obtained from partially averaged spectra—for example, in spectra of membranes.

6. *Spatial distribution:* Information on the spatial distribution of molecules can be obtained. In some cases, it is possible to distinguish whether a molecule is inside or outside a cell, and a method known as **NMR imaging** can be used to give good images of various objects, including the human body.

The Assignment Problem in NMR Studies of Macromolecules

Before discussing in more detail the applications of NMR in biology, we briefly discuss some of the technical problems encountered in studying macromolecules.

Even in a small protein, there are hundreds of different resonances for both ^1H and ^{13}C nuclei. It is therefore important to obtain spectral simplification whereever possible. One of the simplest ways of doing this is to use difference spectra. Such spectra, obtained by subtraction of separately collected spectra, are shown in the following examples on lysozyme and t-RNA.

There are more complicated but very powerful methods for obtaining spectral simplification. The most general of these involves presenting the spectra in two dimensions—for example, with ^{13}C shifts along one axis and ^1H shifts along the other axis. The information is obtained by collecting a series of transient signals in a time domain t_2. These transients are prepared differently by sequential changes in a time domain t_1 prior to the t_2 domain. A wide variety of experiments can be carried out by varying the method of preparing the nuclear spins in

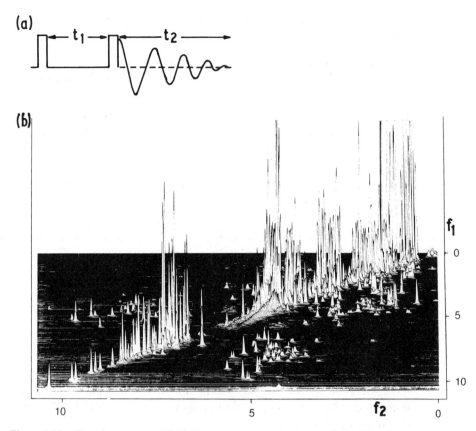

Figure 6.22   Two-dimensional NMR. If a series of transients is collected in a time domain t_2 at different values of t_1 (t_1 and t_2 are defined in (a)), then a double Fourier transformation of the data leads to a plot (b) with two frequency axes. This example shows spectra of bovine trypsin inhibitor at 500 MHz. The normal one-dimensional spectrum occurs along the diagonal, while off-diagonal peaks correspond, in this experiment, to connectivities between pairs of resonances due to spin-spin coupling (from G. Wagner and K. Wüthrich, *J. Mol. Biol.*, 155(1982):347).

the t_1 domain. After the data are collected, a double Fourier transformation is carried out so that both the t_1 and t_2 domains are presented as frequencies. An example is shown in Figure 6.22.

Even more difficult than the resolution problem is the problem of *assignment:* A resolved resonance has to be identified with a particular atom in the macromolecule. This process of identification can have several stages. In the case of proteins, the kinds of questions that arise are, Is the resonance from a $-CH_3$ or a $-CH_2$ group? Is it from a valine or an alanine? From which of several valines is it—from Val-109 or Val-92? We illustrate this general problem by discussing the assignment of the ^1H resonances of Val-109 in lysozyme and the hydrogen-bonded ring hydrogens of a G–U pair in t-RNA.

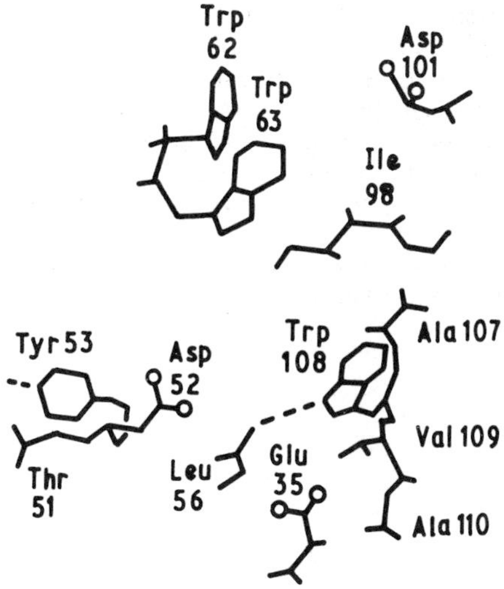

Figure 6.23 A schematic illustration of the active-site region of lysozyme determined by X-ray crystallography.

Example: **The Assignment of the Val-109 Methyl Resonances of Lysozyme.** Lysozyme is a relatively small enzyme of 14,500 daltons, but it has more than 500 resonances in its ^{1}H spectrum. The methyl resonances of valine occur at 0.96 ppm and 1.02 ppm in the isolated amino acid. In the protein, they may be shifted from these positions, but they will still have certain characteristics unique to valine—namely, that both resonances are doublets with $J \simeq 7$ Hz because of coupling to the same $-$CH group. Thus, irradiation of this one resonance will decouple both methyl groups.

The *first stage* of assignment is to identify a resonance with a particular kind of group—e.g., the valine methyl group.

The *second stage* of assignment is to distinguish between different valines—between Val-109 and Val-92, for example. At this stage, it is very useful if the protein structure is known from X-ray diffraction studies. In lysozyme, for example, it is known that metal ions bind between Asp-52 and Glu-35 at the active site (see Figure 6.23). Val-109 is the only valine in this vicinity. Therefore, one would expect a paramagnetic ion to broaden resonances of groups near its binding site. If the difference is taken between spectra collected with and without Gd(III), then the broadened resonance can be detected (see Figure 6.24). Moreover, a decoupling experiment can also be carried out to confirm the assignment of Val-109 methyls (see Figure 6.25).

Example: **The Assignment of a G–U Hydrogen-bonded Hydrogen Resonance of t-RNA.** Hydrogen atoms bonded to nitrogen and oxygen are exchangeable with solvent hydrogen atoms. If a molecule is dissolved in D_2O, these hydrogens exchange with deuterium atoms. In a macromolecule, this exchange may take milliseconds or months, depending on the position of the hydrogen. If the exchange is very fast, the resonances of the exchangeable hydrogen may be averaged with the H_2O resonance, and therefore be unobservable. If the exchange rate

Figure 6.24 The 270-MHz ^{1}H difference spectrum of lysozyme. (a) Spectrum of 5-mM lysozyme containing 50-mM diamagnetic La(III) at 55°C, pH 5.3. (b) Spectrum of the same sample after adding a 5 × 10^{-5}-M solution of the relaxation probe Gd(III). (c) Difference spectrum (a) − (b), vertical scale × 4. The peaks in the spectrum are those strongly broadened by the binding of Gd(III) to lysozyme.

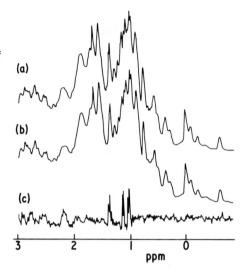

is in the millisecond-to-second range, then it is necessary to dissolve the macromolecule in H_2O rather than in the more usual solvent for ^{1}H NMR studies, D_2O. Special techniques must then be used to reduce the very large solvent resonance in the spectrum.

In t-RNA, the hydrogen-bonded NH protons of complementary base pairs can be observed in aqueous solution. In particular, the ring NH protons of G and U bases occur in a region of the spectrum (9–15 ppm) well resolved from other resonances. Figure 6.26 shows this part of the spectrum from a sample of valine t-RNA isolated from *E. Coli*. The majority of the resonances arise from the protons ringed in Figures 27(a) and (b), i.e., A–U and G–C pairs. In such cases, these protons are not close enough to give rise to an nOe effect when one member of

Figure 6.25 Lower spectrum: difference spectrum obtained as in Figure 6.24; upper spectrum: difference spectrum obtained in the same way, except that irradiation at the position indicated by the arrow was applied in both spectra. The decoupling of the two upfield doublets is clearly revealed (from I. D. Campbell, et al., *Ann. N. Y. Acad. Sci.,* 222(1973):163).

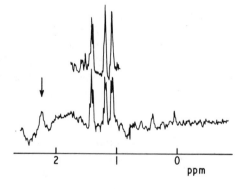

Figure 6.26 (a) A control ^{1}H spectrum of t-RNA^{val} in H$_2$O. (b) A difference spectrum obtained from the difference between the control and a spectrum collected while irradiating at 11.35 ppm (from B. R. Reid, *Ann. Rev. Biochem.*, 50(1982):969).

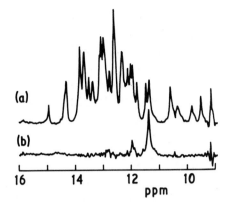

the pair is selectively irradiated. In the relatively rare "wobble" G–U base pair, however, there are two ring NH protons that do give rise to an nOe effect, as shown in Figure 6.27(c). This experiment thus allows the resonance to be assigned and the existence of wobble hydrogen-bonding in solution to be confirmed.

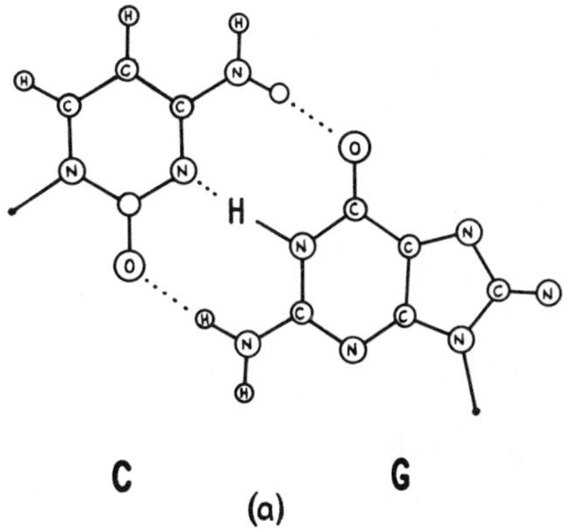

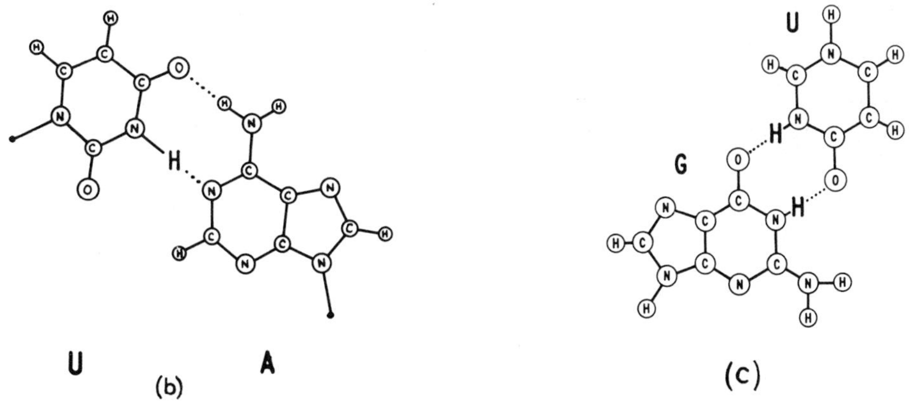

Figure 6.27 Illustration of (a) G–C (b) A–U, and (c) G–U base pairing. The observed NH protons in NMR experiments are not ringed.

Analytical Uses of NMR

NMR has become one of the most powerful methods for identifying small organic compounds and thus has become an indispensable tool for the organic chemist. This sort of analysis is no longer so important in biochemistry. However, the ability of NMR to identify molecules and measure their concentration *noninvasively* in biological tissues has important advantages. This kind of study is now being applied to a wide variety of biological systems, including human beings.

Example: Glycolysis in **E. Coli.**   It is possible to follow the fate of a particular carbon atom in glucose after its addition to a suspension of cells. Figure 6.28 shows ^{13}C spectra collected at 1-min intervals. After 6 min, the label appears in the C-3 position of lactate. At later times, the C-1 signal from fructose-1, 6-diphosphate increases and then decreases with the production of C-3–labeled alanine, C-4 valine, and C-2 ethanol and acetate. Oxygenation of the sample after these anaerobic experiments leads to the consumption of lactate and the production of glutamate labeled in the C-2, C-3, and C-4 positions. Note that label positions and metabolism can be monitored without destruction of the sample.

Ligand Binding to Macromolecules

Consider the process

$$A + B \underset{k_{-1}}{\overset{k_1}{\rightleftharpoons}} AB$$

Assume that the resonances of the A and AB species are observed with parameters δ_A, δ_{AB}, $1/T_{2A}$, $1/T_{2AB}$, and so forth. Several experimental situations are possible—for example, A and AB could be bound and free forms of ligand, bound and free forms of protein, or protonated and unprotonated forms of an amino acid side chain. Consider first the case where the exchange is fast, a situation often observed in practice. The observed parameters, call them P, are then a weighted mean of the A and AB parameters (see the section on chemical exchange). That is,

$$P_{obs} \simeq P_A f_A + P_{AB} f_{AB}$$

where $f_A = [A]/A_t$, $f_{AB} = [AB]/A_t$, and $A_t = [A] + [AB]$ and $B_t = [B] + [AB]$.
Let $\Delta_0 = P_{AB} - P_A$ and $\Delta = P_{obs} - P_A$. Since $f_A + f_{AB} = 1$, we can then write

$$\Delta = \Delta_0 f_{AB}$$

Also, at equilibrium,

$$K_d = \frac{[A][B]}{[AB]} = \frac{k_{-1}}{k_1}$$

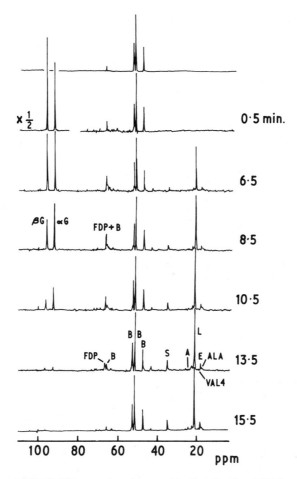

Figure 6.28   Several 90.52-MHz spectra of anaerobic *E. coli* cells at 20°C as a function of time from [^{13}C] glucose (50 *mM*) addition. Top spectrum (16,000 scans) shows the natural abundance ^{13}C peaks (assigned to the PIPES and MES buffers) detectable in the suspension prior to glucose addition. All subsequent spectra, except the last one, represent 200 transients accumulated in 1 min. The last spectrum consists of 1600 scans. The time given for each spectrum indicates the middle of the accumulation period, referred to glucose addition. Peaks labeled *B* are from buffer, *L* from lactate, *S* from succinate, *A* from acetate, and *E* from ethanol (from K. Ugurbil et al., *Proc. Nat. Acad. Sci.,* 75(1978):3742).

We can then write f_{AB} as

$$\frac{[B]}{[B] + K_d} \quad \text{or} \quad \frac{[A]B_t}{A_t \left([A] + K_d\right)}$$

If the experiment is carried out at constant A_t, and B_t is varied, then

$$\Delta = \frac{\Delta_0[B]}{[B] + K_d} \tag{6.10}$$

which is of the same form as the Michaelis-Menten equation, so K_d can be readily determined. If A_t is varied, the equations are less simple; but if $[A] >> [AB]$, then

$$\Delta = \frac{B_t \Delta_0}{A_t + K_d} \tag{6.11}$$

Let us now consider some examples of ligand-binding experiments in light of equations 6.10 and 6.11.

Worked Example 6.9   Inhibitor Binding to Triosephosphate Isomerase

At constant pH (6.0) and enzyme concentration ($1mM$), a competitive inhibitor of the enzyme triosephosphate isomerase caused the resonance of one particular histidine resonance to change as follows:

D-glycerol-3-phosphate-(mM):
0.9   1.3   3.7   3.6   4.3   5.2   6.2   13.3
Change in shift (ppm):
0.048   0.083   0.108   0.125   0.138   0.145   0.158   0.19

Use these data to calculate the dissociation constant, K_d, of the enzyme-inhibitor complex.

Solution

A plot of these data reveals a hyperbolic curve (see Figure 6.29). We can apply equation 6.10 to this situation. If Δ_0 is taken as 0.19 ppm, then at $\Delta_0/2$, $[I] = K_d$. Thus, $K_d \sim$ 2.2 mM. A more sophisticated analysis can be made using a Scatchard plot (from C. A. Browne et al., *J. Mol. Biol.*, 100 (1976):319).

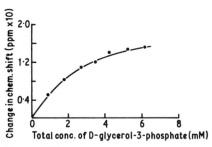

Figure 6.29

Example: Chloride Ion Binding to Cytochrome c.   The atom ^{35}Cl has $I = \frac{3}{2}$; thus, it has a quadrupole moment. Such ions often relax very strongly when they bind to proteins because of electrostatic interactions. In Figure 6.30, Δ is plotted against $[Cl^-]$; thus, equation 6.11 applies. Note that $\Delta = (1/T_{2AB} - 1/T_{2A})$, a difference between relaxation rates rather than between shifts.

The results demonstrate two classes of binding sites: a high affinity class and a low affinity class. These results, taken with other competition studies, were interpreted in terms of Cl^- binding near the heme edge. It was also found that the binding sites are different in the two oxidation states (for a review of this and other quadrupolar ion work, see S. Forsén and B. Lindman, *Meth. Biochem. Anal*, 27(1981):289).

Figure 6.30 The ^{35}Cl excess transverse relaxation rate as a function of the chloride concentration in 3-mM solutions of horse heart cytochrome c; (a) is the oxidized form and (b) the reduced form of cytochrome c.

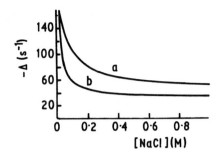

Example: Binding Constant of Paramagnetic Ions Using Water Relaxation. Some metal ions are paramagnetic; thus, they have a large magnetic moment μ_S. The relaxation rate of nuclei near these ions can be very fast because the dipole-dipole interaction becomes very large. The lanthanide ion Gd(III) and the transition metal ion Mn(II) are particularly good relaxation agents.

In Figure 6.31, at low concentrations of IgG, the Gd(III) is surrounded only by water, which has a relaxation rate of $1/T_{1A}$, in our notation. With increasing IgG concentration, more of the Gd(III) is bound to the protein. The water, which exchanges rapidly with the bound Gd(III), has a relaxation rate of $1/T_{1AB}$, and equation 6.10 applies. The reason that $1/T_{1A}$ is different from $1/T_{1AB}$ is twofold: (1) The number of water molecules around the Gd(III) changes (is reduced) as Gd(III) binds to the protein; and (2) The frequency-dependent $F(\tau_c)$ term in equation 6.3 changes because the Gd(III) tumbles more slowly when it is bound. The second effect is the more important, so the water relaxation rate is usually faster when the metal is bound because the experimental situation corresponds to the short τ_c side of the T_1 minimum shown in Figure 6.17.

Ionization States and pH

Consider the simple acid-base equilibrium $A^- + H^+ \rightleftharpoons AH$. This reaction has the same form as that analyzed in the previous section with $B = H^+$. If $K_d = K_a$ then,

$$\Delta = \frac{\Delta_0 [H^+]}{[H^+] + K_a}$$

Figure 6.31 Titration of Gd(III) with whole IgG by following the paramagnetic contribution to the water proton spin-lattice relaxation rate. The continuous curve is calculated (from K. Willan et al., *Biochem. J.*, 161 (1977):205).

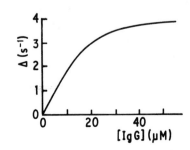

Figure 6.32   The appearance of a Δ-versus-
pH plot.

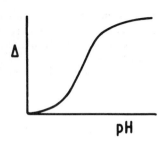

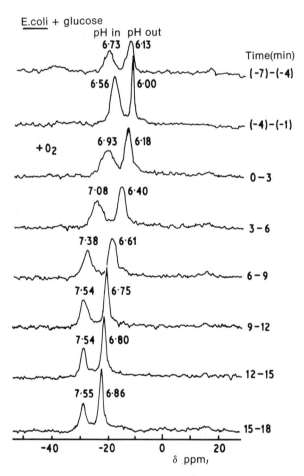

Figure 6.33 ^{31}P spectra of a suspension of *E. coli.* Intracellular pH values, obtained from the shift of the P_i resonance, are shown as a time course (from G.
Navon et al., *Proc. Nat. Acad. Sci.* U.S.A., 74(1977):888).

Since [H^+] is usually plotted in a logarithmic form as pH, the variation of Δ with pH appears as shown in Figure 6.32, rather than as a hyperbolic curve. $K_a = [H^+]$, or $pK_a = pH$ when $\Delta = \Delta_0/2$. Since large shifts are observed on ionization, pK_as can be determined by following the shift (δ) of a titrating group. Alternatively, if the titration curve is known, the pH can be determined from δ if the pK_a falls in a suitable pH range.

Example: Measurement of Intracellular pH. If a group ionizes at a pH near physiological pH, then the shift of its resonance can be a sensitive monitor of pH. Histidines and phosphates are among the relatively few groups that ionize near pH 7. In suspensions of cells, inorganic phosphate (P_i) is usually present both inside and outside the cell; thus, the shift of (P_i) can simultaneously give both intracellular and extracellular pH (Figure 6.33).

Example: The Ionization of His-β-146 in Oxyhemoglobin and Deoxyhemoglobin. Hemoglobin takes up protons on releasing O_2; this phenomenon is known as the **Bohr effect**. Perutz, in his classic X-ray studies, predicted that His-β-146 might be involved in this process due to a change in local environment around this group between the oxy and deoxy states of the protein structure (see Chapter 12). Upon ionization, the C-2 proton of histidine shifts by about 1 ppm, as shown in Figure 6.34. The pK_a of the group changes by more than 1 pH unit due to the stabilization of the charged form by Asp-94 in the deoxy structure.

Note that there are a large number of different histidines in hemoglobin, so individual resonances must be resolved and identified with a particular residue (e.g., His-146). The problems of resolution and assignment are not trivial as we have just seen. In this particular case, the C-2 hydrogen resonance of His-β-146 was identified by selectively cleaving this group, which is near the COO⁻ terminal of the β-chain, using carboxypeptidase.

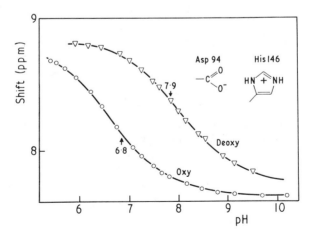

Figure 6.34   The titration of His-β-146 in hemoglobin in the oxy and deoxy form, as observed by the shift of the C-2 proton resonance (from I. D. Campbell, *NMR in Biology*, ed. R. A. Dwek et al. (London: Academic Press, 1977) p. 33).

Kinetics

NMR can be very powerful in the study of kinetic processes such as enzyme-catalyzed reactions or ligand binding. Kinetic information can be extracted in two distinct ways: (1) by analyzing the spectra from a system at equilibrium in terms of chemical-exchange theory and (2) by following changes in concentration as a function of time.

Chemical-Exchange Analysis.   We can extend the analysis given in the section on chemical exchange to the reaction

$$A + B \underset{k_{-1}}{\overset{k_1}{\rightleftharpoons}} AB$$

and apply it to kinetic events.

For a line that is an average of the A and AB resonances (for example, a ligand binding to a macromolecule), equation 6.7 for fast exchange predicts that the equation for the observed T_2 is

$$\frac{1}{T_{2obs}} = \frac{f_A}{T_{2A}} + \frac{f_{AB}}{T_{2AB}} + 4\pi^2 \frac{f_A f_{AB} \Delta_0^2}{k_1[B] + k_{-1}} \qquad (6.12)$$

Note that $k_1[B]$ and k_{-1} are the lifetimes of the A and AB species, respectively. The range of k_{-1} that can be detected depends on Δ_0, but it is usually 10^2–10^5 s^{-1}.

For the case where two lines are observed, equation 6.6 for slow exchange becomes

$$\frac{1}{T_{2obs}} = \frac{1}{T_{2A}} + 'k_1[B] \qquad \text{(for the A signal)}$$

and

$$\frac{1}{T_{2obs}} = \frac{1}{T_{2AB}} + k_{-1} \qquad \text{(for the AB signal)}$$

The range of k that can be measured with these equations is 10–100 s^{-1}, since this is related to observed linewidths ($1/\pi T_2$).

In the slow exchange case, irradiation of one resonance (e.g., A) will cause changes in intensity of the other resonances (AB) because of *transfer of saturation* from one to the other as a result of the exchange. This double-resonance method extends the range of k that can be measured to about 1 s^{-1}. The fractional change in intensity of the AB resonance is given by the equation

$$\frac{I_{AB'}}{I_{AB}} = \frac{1/T_{1AB}}{1/T_{1AB} + k_{-1}}$$

where I_{AB} and $I_{AB'}$ are the intensities of the AB resonance before and after irradiation of the A resonance, respectively.

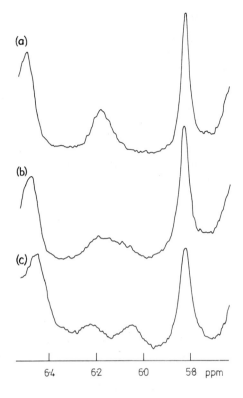

Figure 6.35 The effect of temperature on the spectrum of 8-*mM* hen lysozyme containing 4.1-*mM* (GlcNAc)$_3$ at pH 5.3. Temperatures are (a) 55°C, (b) 45°C, (c) 37°C. The passage from slow through intermediate to fast exchange conditions is observed (from C. M. Dobson, *NMR in Biology*, ed. R. A. Dwek et al. (London: Academic Press, 1977), p. 63).

Example: Inhibitor Binding to Lysozyme.   In ^{1}H spectra of lysozyme, the resonance of one of the benzenoid protons of Trp-28 occurs at about 6.25 ppm. On the addition of the inhibitor (Glc NAc)$_3$ to the enzyme, this resonance broadens and shifts. At 37°C, two resonances are observed: The 6.25-ppm resonance decreases as inhibitor concentration increases, and a new resonance appears at 6.05 ppm. When the inhibitor concentration is half that of the enzyme, two equal-intensity resonances are observed (see Figure 6.35). If this mixture is heated to 55°C, the two resonances are averaged to one because the interconversion rates between these two enzyme forms increase. Taken together with T-jump measurements, the following kinetic scheme could be derived from these experiments:

$$E + I \underset{k_{21}}{\overset{k_{12}}{\rightleftharpoons}} EI \underset{k_{32}}{\overset{k_{23}}{\rightleftharpoons}} E^*I$$

The rate observed by NMR is k_{32} ($= k_{23}$ with 50% EI) and the value obtained at 45°C was 1.3×10^2 s^{-1}. These observations were interpreted in terms of a conformational change on the enzyme.

Concentration Versus Time.   A measurement of concentration as a function of time is the method used for most kinetic studies. It usually involves a nonequilibrium reaction, such as a substrate proceeding to a product. NMR is very insensitive, and therefore slow, compared with ultraviolet spectroscopy or fluorescence in its ability to follow concentration changes. The collection of NMR data is limited to times greater than about 30 s. It can, however, follow a wide variety of nonchromophoric compounds directly without coupled assays, and it can detect these in intact tissue (see section on analytical uses).

A variation on this type of experiment is the monitoring of isotope-exchange reactions (at equilibrium), since NMR can readily distinguish between ^1H and ^2H or between ^{13}C and ^{12}C.

Example: Lactate Dehydrogenase Activity.   Lactate dehydrogenase catalyzes the conversion of pyruvate to lactate. If the pyruvate is labeled with a deuterated methyl group, then the following exchange reaction

$$
\begin{array}{ccc}
CD_3 & & CH_3 \\
| & & | \\
C{=}O & \rightleftharpoons & HCOH \\
| & & | \\
COO^- & & COO^- \\
\\
\text{Pyruvate} & & \text{Lactate}
\end{array}
$$

can be monitored by following the fractional change in labeling as a function of time, as shown in Figure 6.36. The main advantage of this type of study is that it can be carried out directly in suspensions of intact cells.

Structural Studies by NMR

Structural information can be extracted from NMR spectra in a number of ways.

1. The dependence of J on bond angle has been useful in studies of the conformation of small molecules. Conformational flexibility leads to the observation of average J values. J is difficult to measure accurately in spectra of most macromolecules.

2. T_1 and T_2 depend on r^6 for dipolar relaxation, where r is the separation between the observed nucleus (I) and the magnetic center (S) causing relaxation. In many samples, the precise nature of S is uncertain, since many nuclei, including those of the solvent, may contribute to the relaxation of I. Paramagnetic ions, however, are very useful, since their effect is very large and their contribution often dominates all others. The contribution of the paramagnetic center can be determined relatively precisely by comparing the relaxation with a diamagnetic ion (e.g., La(III)) to that with a paramagnetic ion (e.g., Gd(III)). Separations between I and S of up to about 1.2 nm can be measured with this technique.

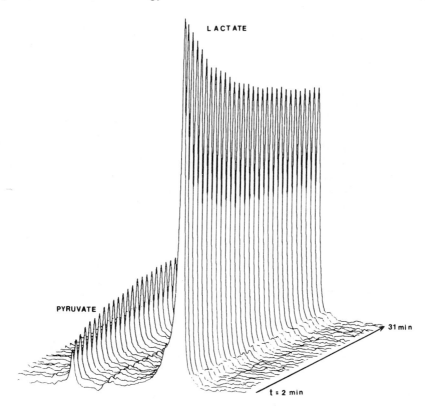

Figure 6.36 Equilibrium exchange of CD_3 pyruvate with CH_3 lactate observed by
^{1}H NMR (from R. J. Simpson et al., *Biochem. J.,* 202(1982):573).

Example: Gd(III) Binding to Lysozyme.   We have discussed the fact
that Gd(III) binds in the active site of lysozyme between the carboxyl
groups of Glu-35 and Asp-52. The observed broadening ($1/\pi T_2$) of assigned resonances can be correlated with the structure known to exist in
the crystal using X-ray diffraction. The correlation is quite good in the
0.5–1.2 nm range (see, for example, C. M. Dobson in *NMR in Biology,*
ed. R. A. Dwek et al. London: Academic Press, 1977).

3. The nuclear Overhauser enhancement (η) depends on r^6 in the same way
   as do relaxation rates. The nOe has the advantage, however, that S can
   be identified, since changes in I are monitored as various S nuclei are irradiated. This technique is very useful for detecting I–S interactions over
   distances up to about 0.4 nm.

Example: Measurement of nOe in Lysozyme.   Since many resonances
of lysozyme have been assigned, pairwise I–S interactions can be detected by observing the decrease in intensity of an I resonance when an S
is irradiated. For example, irradiation of the Trp-28 C-5 hydrogen res-

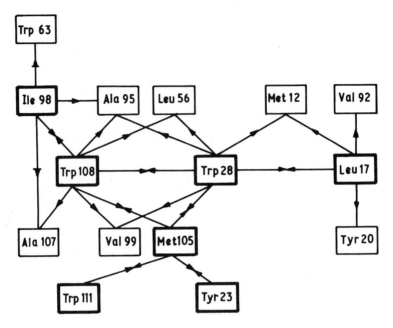

Figure 6.37 Schematic representation of the interresidue nuclear Overhauser effects observed in hen lysozyme. The residue to which an arrow points indicates that an nOe was seen on this residue on irradiation of the other residue. Residues whose resonances have been saturated are indicated by a heavy border (from F. M. Poulsen, J. C. Hoch, and C. M. Dobson, *Biochemistry,* 19(1980):2597).

onance causes the intensity of the methyl resonance of Met-105 to decrease; thus these groups are near in space. Some of the observed nOe between assigned resonances are illustrated in Figure 6.37.

4. The chemical shift of a nucleus *I* that arises from a dipole μ_s depends on $(3 \cos^2\theta - 1)/r^3$ if μ_s can be considered axially symmetrical. The shifts induced by paramagnetic ions and some aromatic rings are large enough to be quantified in some cases.

    Example: Ring-current Shifts in Lysozyme. The shifts of assigned resonances can be calculated from X-ray coordinates and compared with observed shifts, as shown in Figure 6.38.

The preceding three examples have all been studies of the enzyme lysozyme, molecular weight 14,900. The agreement between the NMR and X-ray data in each case confirms that the overall fold of the protein is the same in the crystal and in solution. Note, however, that the X-ray structural data are more precise and less ambiguous.

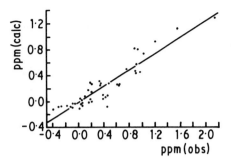

Figure 6.38 Correlation of calculated and observed ring-current shifts in lysozyme (from S. J. Perkins, and R. A. Dwek, *Biochemistry*, 19(1980):245).

At present, this sort of detailed NMR study can be done only on molecules up to about 20,000 daltons because of problems in resolving and assigning single resonances. Once assignments have been made (preferably without using X-ray data), NMR has the advantage of being able to extract structural information relatively rapidly in solutions that mimic the *in vivo* situation. NMR gives a series of *pairwise interactions;* it does not lead directly to the *shape* of the molecule in solution.

In addition to quantitative information about structure, a certain amount of qualitative information can also be obtained. For example, observed secondary shifts on an assigned resonance can be compared in related macromolecules, or the line shape of a membrane spectrum may be characteristic for a particular kind of lipid structure (see next section).

It should be remembered that NMR can also give information about molecular dynamics, ligand binding, and ionization states, which can all complement structural studies.

Molecular Motion

We have seen that T_1 depends on τ_c, the rotational correlation time for a dipole pair and that exchange between different environments can affect the observed spectra (see section on chemical exchange). Information about molecular motion is thus normally derived from analysis of T_1 or line shapes.

Example: Fluidity Gradient in Lipids. When ^{13}C T_1 measurements were made on sonicated dipalmitoyl lecithin suspensions, the results shown in Figure 6.39 were obtained. The relaxation is dominated by the dipolar interaction from the hydrogens on the same bond. Apart from the carboxyls, which have no hydrogens attached, there is an increase in T_1, both toward the CH_3 groups and toward the $N^+(CH_3)_3$ group. This means that there is increased mobility toward the ends of these chains (note that these experiments were carried out at 25 MHz, and that the motion is described by the $T_1 \simeq T_2$ region of Figure 6.17).

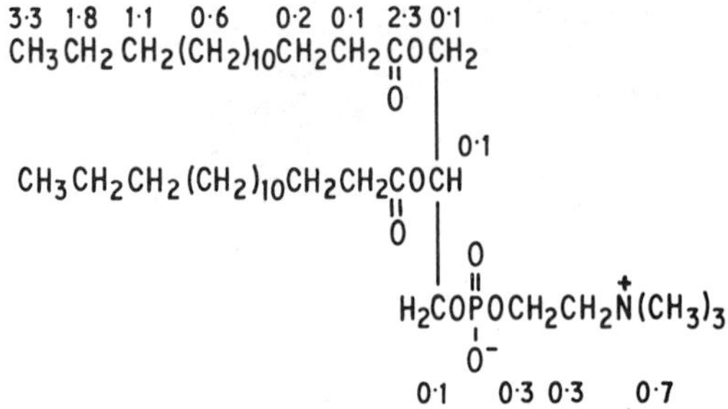

$$\underset{3\cdot3}{CH_3}\underset{1\cdot8}{CH_2}\underset{1\cdot1}{CH_2}\underset{0\cdot6}{(CH_2)_{10}}\underset{0\cdot2}{CH_2}\underset{0\cdot1}{CH_2}\underset{2\cdot3}{\underset{\|}{\underset{O}{C}}}\underset{0\cdot1}{O}CH_2$$

$$CH_3 CH_2 CH_2 (CH_2)_{10} CH_2 CH_2 \underset{\|}{\underset{O}{C}} O CH$$ 0·1

$$H_2 \underset{\|}{\underset{O}{C}} O \underset{\underset{O^-}{|}}{P} O CH_2 CH_2 \overset{+}{N}(CH_3)_3$$

0·1   0·3 0·3   0·7

Figure 6.39 The observed ^{13}C T_1 values for various parts of a dipalmitoyl lecithin chain are shown (from A. G. Lee et al., *Membr. Biol.*, 2(1976):2).

Example: "Flipping" of a Tyrosine in Cytochrome c. In general, the four ring protons of tyrosine are in different environments in a globular protein. Motion about the $C\gamma-C_\beta$ bond can cause the resonances of the two ortho and the two meta protons to be averaged—a situation that is observed in many globular proteins. This means that the motion about this bond is relatively fast (greater than about 10^2 s^{-1}). In some proteins, the motion is slower than this, so the observed line shapes are then sensitive to the rate of the motion. The constraints in the protein probably restrict this motion to 180° flips rather than permit free rotation. Figure 6.40 shows the rates of tyrosine flipping observed for one particular tyrosine in cytochrome c at different temperatures. At high temperatures, only two resonance are observed, while at low temperatures, four are observed (see Figure 6.41).

Figure 6.40 The temperature dependence of the flip rate of a tyrosine in horse ferrocytochrome as determined by line shape analysis (○) and saturation transfer experiments (■) (from I. D. Campbell, *NMR in Biology,* ed. R. A. Dwek et al. (London: Academic Press, 1977) p. 33).

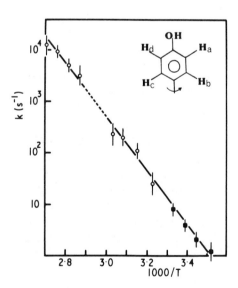

Figure 6.41 At low temperature, four resonances may be observed for a tyrosine in a protein. Rotation of the ring reduces the spectrum to two resonances.

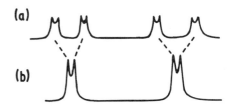

The Observation of Powder Spectra from Membranes. In several situations the spectrum that is observed is sensitive to the angle the molecule makes with the applied magnetic field B_0. This property is known as anisotropy (see also Chapter 7). This property is usually averaged out in NMR because of molecular motion in solution. When averaging occurs, the observed line shape is Lorentzian.

In some cases, where the molecular tumbling is relatively slow ($\tau_c > 10^{-6}$ s), this averaging does not occur, and a random distribution of molecules, either from a solution or a solid, gives rise to a different kind of line shape. These non-averaged or partially averaged spectra can still yield useful information. They are sometimes called **powder spectra,** since they were first observed in powdered solids. This type of spectrum is commonly observed in EPR (see Chapter 7).

Consider the example of a molecule experiencing a ring-current shift (see chemical-shift section). The shift depends on the magnitude of the dipole induced in the ring *and* on the orientation of the ring in the magnetic field. The magnitude of the dipole induced when the angle between B_0 and the ring plane (θ') is 90° ($\mu_\perp$) is greater than when $\theta' = 0$ ($\mu_\parallel$). In a crystal, where all the molecules have the same orientation, one resonance will be observed, but its position will depend on the orientation of the crystal in the applied B_0 field (Figure 6.42). In a random orientation of molecules, all values of θ' will be present; thus, an envelope of resonances will be observed. This envelope has a nonsymmetrical shape because it is more probable that the molecular axis will be more nearly at right angles to B_0 than parallel to it (see also Chapter 7).

Example: Chemical Shift Anisotropy of ^{31}P Resonances in Membranes. The shift of the ^{31}P resonances of phospholipids is sensitive to the orientation of the headgroup with respect to B_0 because of the asymmetry of the electrons in the

Figure 6.42 Representation of the spectra observed in a situation where θ' is orientation dependent: (a) the spectra from a crystal oriented such that the angle (θ') between the principal axis of the molecule and B_0 is 0° and 90°; (b) the spectrum observed from a random orientation of such molecules.

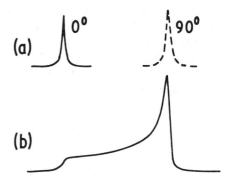

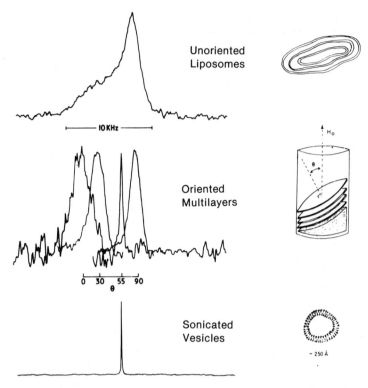

Unoriented
Liposomes

Oriented
Multilayers

Sonicated
Vesicles

Figure 6.43 The ^{31}P spectra of liposomes, oriented multilayers, and sonicated vesicles. Because the tumbling of the molecules in the liposomes is relatively slow, a powder spectrum is observed with linewidth of many kilohertz. In the vesicles, the overall tumbling is rapid enough to produce a relatively narrow Lorentzian line shape (from A. C. McLaughlin et al., in *NMR in Biology*, p. 231).

bonds. In large assemblies of lipids, this anisotropy is not averaged out and a powder spectrum is observed. The spectra observed are sensitive to both the motion of the molecules and to their orientation (Figures 6.43 and 6.44).

Example: Deuterium Spectra of Membranes. The ^{2}H (or D) nucleus has a spin I of 1; thus, three energy levels ($2I + 1$) arise on interaction with B_0. These levels are usually labeled by the quantum number m, which for $I = 1$ has the values $+1$, 0, -1. The allowed transitions (see Chapter 2) are $+1\leftrightarrow0$ and $-1\leftrightarrow0$. The energy-level diagram for such a nucleus is shown in Figure 6.45. The two transitions coincide if there is no interaction other than with B_0. An $I = 1$ nucleus also has a quadrupole moment Q, which means that the nucleus can interact with local electric fields as well as with B_0. There is, therefore, a splitting of the energy levels

Figure 6.44 Hydrated soya phosphatidyl ethanolamine adopts the hexagonal H_{II} phase at 30°C. In the presence of 50-mol % phosphatidylcholine, the bilayer phase dominates. The ^{31}P spectra of liposomes are characteristic for the phase adopted (from B. de Kruijff et al., *Trends Biochem. Sci.*, 5(1980):79–81).

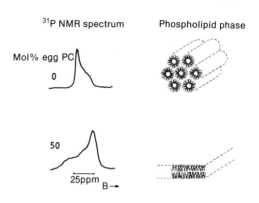

even when $B_0 = 0$. The net result is that two lines are observed in an oriented system with a splitting between the lines of

$$\Delta\nu_Q = \frac{3}{2} \frac{e^2 qQ}{h} \frac{(3 \cos^2\theta' - 1)}{2}$$

where θ' is the angle between B_0 and some axis on the molecule, often the C—D bond vector; eq is the electric field gradient to which eQ is coupled.

It is clear that $\Delta\nu_Q$ is anisotropic; thus, a random distribution of molecules will give a powder spectrum if the molecular motion is not fast enough to average the quadrupolar splitting. Since two lines occur at every value of θ', the envelope of lines is now like the sum of two envelopes of the kind shown in Figure 6.46. The splitting between the main peaks, which occurs when $\theta' = 90°$, is $3e^2qQ/4h$.

The observed powder spectra of deuterium are sensitive to the orientation of the molecules and to the extent and frequency of the molecular motion. These spectra have been used to determine the fluidity gradient in membranes by specifically incorporating ^2H at known positions along fatty acid chains (compare the

Figure 6.45 The energy levels of a system where $I = 1$. In (a) the energy levels are completely degenerate in zero applied field. In (b) the degeneracy is removed even in zero field.

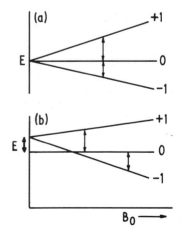

Figure 6.46   The spectra observed from a
²H nucleus (a) in an oriented sample with
$\theta' = 0°$ and 90° and (b) in a random distribution of molecules.

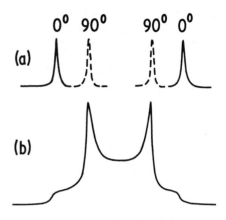

use of spin labels described in Chapter 7). Deuterium has a considerable advantage over a spin label in being a very small perturbation to the system. Figure 6.47
shows that changes in lipid order can be detected by this method when cholesterol
(33% w/w) or cytochrome oxidase (65% w/w) are incorporated into a preparation of C-6–deuterated phosphatydycholine. The cholesterol causes a large degree
of ordering, but the cytochrome oxidase has little effect.

Spatial Distribution

Biological tissues are inhomogeneous because they have compartments within
cells and many different kinds of cells. In NMR studies of intact tissues, it might
be of interest to know whether a particular resonance comes from inside or outside the cell or from an organelle, such as a mitochondrion. It would also be of
great value, especially in medicine, to obtain an image of an object, such as a
human being, using a noninvasive method like NMR. In fact, both these aims are
now possible to some extent, as is illustrated in the following two examples.

Figure 6.47   Theoretical and experimental
²H spectra of ²H-labeled lipids. The spectra
of pure phosphatidylcholine (DMPC) are in
good agreement with theory. The perturbation produced by the addition of 65% (w/w)
of cytochrome oxidase is less than that produced by 33% cholesterol (from E. Oldfield
et al., *Biochem. Soc. Symp.*, 46(1982):155).

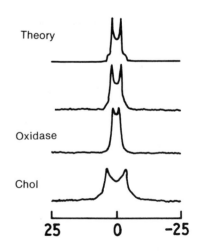

Figure 6.48   The relaxation rate of water observed in a suspension of red cells to which Mn(II) has been added. Curve *a* is from extracellular water, curve *b* is from intracellular water (from T. Conlon and R. Outhred, *Biochim. Biophys. Acta*, 288 (1972):354).

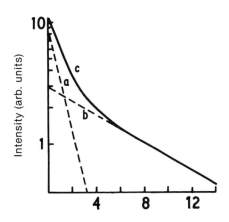

Example: Water Transport in Cells.   If a molecule could be labeled by having a different NMR parameter in different compartments of a cell suspension, then it would be possible to distinguish which compartment contained the molecule. A simple example of this technique is the use of Mn(II) to decrease the relaxation rate of extracellular water in a suspension of red cells. Since the Mn(II) does not cross the cell membrane, both the fraction of extracellular water and the rate of water transport across the cell membrane can be measured, as illustrated in Figure 6.48. The relaxation rate observed for the intracellular water is dominated by the rate of exchange across the membrane (see equation 6.6).

Example: NMR Imaging.   The usual practice in magnetic resonance is to place the sample in a *homogeneous* magnetic field in order to determine the parameters δ, J, and T_2 as accurately as possible for an assigned resonance. A rather different technique has emerged in recent years: The specimen is deliberately placed in a nonuniform field such that different parts of the structure are labeled by different NMR frequencies, thus enabling an "image" of the specimen to be determined.

Consider the example of two cylindrical tubes of water in an NMR instrument, as shown in Figure 6.49. If a large field gradient is applied in direction (1) as shown, then tube A will resonate at a lower frequency than tube B. If the gradient is applied in direction (2), then A and B resonate together. Intermediate spectra will be observed for other directions of the applied gradient. Note that the spectra from (1) and (2) are essentially the projections of the object on the direction vector of the gradient. From a series of these projections, the object can be reconstructed by a process known as **image reconstruction** (see also Chapter 11).

Several sophisticated methods have been developed to collect the NMR imaging data efficiently. An advantage of the method over X-ray tomography is that it appears to be harmless. The technique essentially measures the distribution of

Figure 6.49   Illustration of the spectra that
would be observed after the application of a
large field gradient in directions 1 and 2.

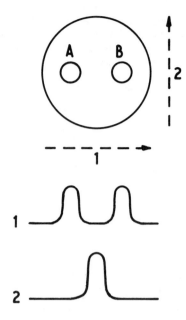

mobile water in the object, but a variety of techniques can be used to enhance the
observed contrast in the image—for example, the exploitation of variations in the
T_1 of the water in different parts of the sample. The technique has developed
rapidly in the last few years; Figure 6.50 shows the sort of picture that can now be
obtained.

Figure 6.50   An NMR image of a human
head (from F. H. Doyle et al., *The Lancet*
(July 1981):54).

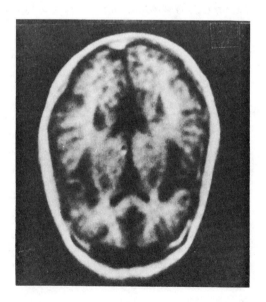

PROBLEMS

1. Describe how you might assign the aromatic ^{1}H resonances of a tyrosine in the NMR spectrum of a protein.

2. Resonances of NH groups are often observed in ^{1}H NMR spectra of macromolecules in H_2O. These hydrogens exchange rapidly with solvent water and are rarely observed for small molecules. (a) Can you say anything about the rate of exchange of the NH_3 with solvent in cases when they are observed? (b) Why is the coupling to the spin 1 of the ^{14}N nucleus not usually observed although J is greater than the linewidth (compare with the observed coupling in spin-label ESR spectra).

3. Using the data in part (a) of Worked Example 6.7, calculate the observed linewidth.

4. In a high-resolution ^{1}H NMR experiment on the enzyme triose phosphate isomerase, the chemical shift of a resonance labeled A varied with pH as follows:

pH:												
5.0	5.35	5.58	5.70	5.80	6.04	6.20	6.40	6.70	7.02	7.5	8.0	8.4
Shift (ppm):												
8.57	8.5	8.4	8.33	8.27	8.13	8.06	7.99	7.92	7.88	7.85	7.84	7.84

What can you conclude from these data given the additional information that the area of the peak corresponds to one proton? In the presence of the inhibitor D-glycerol-3-phosphate, the observed pH dependence of A was as follows:

pH:											
5.0	5.35	5.65	5.9	6.02	6.21	6.46	6.60	6.98	7.46	7.98	8.4
Shift (ppm):											
8.59	8.55	8.48	8.37	8.31	8.20	8.10	8.03	7.95	7.88	7.86	7.84

In similar titrations, it was found that the substrate dihydroxyacetone phosphate and the inhibitor 2-phosphoglycolate also perturbed resonance A in a similar way to D-glycerol-3-phosphate. In addition, three of these molecules affected another titrating resonance labeled B. Resonances A and B were assigned to amino acids separated by more than 2 nm, using the protein structure derived from X-ray crystallography. Suggest explanations for these results.

5. The smallest fragment (Fv) of dinitrophenyl-binding (Dnp-binding) antibody that retains full antigen binding activity has a molecular weight of 25,000. The ^{1}H NMR spectrum of this fragment is sufficiently complex to present problems of assignment and resolution of the resonances. When a spin-labeled Dnp ligand is added to the Fv and the resulting NMR spectrum subtracted from the original spectrum, a spin-label difference spectrum is obtained (see Figure 6.51) that contains only a few resonances. What does this subspectrum represent, and what is the physical basis for your conclusions?

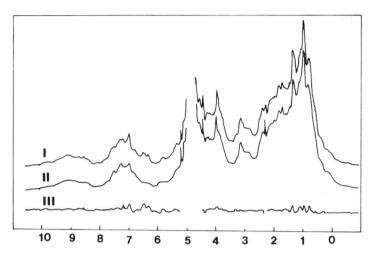

Figure 6.51   270-MHz ^{1}H spectra of the Fv fragment of an antibody: (I) without spin label; (II) with spin label; (III) a difference spectrum (I − II).

6. For the sugar β-methyl-D-glucosamine (see Figure 6.52), the distance between the H_1 and the H_4 protons is about 0.4 nm. Suppose that both H_1 and H_4 are on the principal symmetry axis of a bound paramagnetic shift probe. Calculate how far the sugar ring would have to be from the metal for the shift ratios for H_1 and H_4 to be in the ratio of (a) 8:1 and (b) 1.728:1. (c) and (d) Repeat these calculations for the case when H_1 and H_4 are on an axis perpendicular to the symmetry axis of the shift probe. What can you say about the range of distances between substrates and metal for the shift probes to be effective?

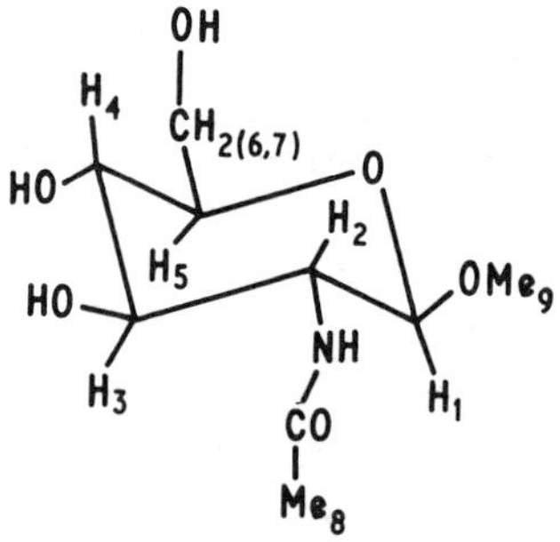

Figure 6.52 β-Methyl-D-glucosamine.

7. The following results were obtained for the spin-lattice relaxation times of the C-1 nucleus in the ^{13}C-enriched sugar β-methyl-D-glucopyranoside under various conditions—alone or in the presence of either Zn(II) or Mn (II) derivatives of the sugar-binding protein concanvalin A

Sample	T_1(s)	
Free sugar (no protein)	1.08	
+ Zn Con A	0.88	($f = 0.049$)
+ Mn Con A	0.43	($f = 0.049$)

From these data, calculate the distance of Mn (II) from the C-1 of the sugar. (f is the fraction bound and the observing frequency was 24.9 MHz.)

8. The relative sensitivity of spin $\frac{1}{2}$ nuclei is proportional to γ^3. In data accumulation, the signal-to-noise ratio improves as the square root of the time taken. Calculate the relative times required to acquire ^1H, ^{13}C, and ^{31}P spectra of a solution of ATP with the same signal-to-noise ratio.

9. In ^{31}P spectra of intact muscle, ATP and phosphocreatine are readily observed. The role of P-creatine is to regenerate ATP broken down during muscle contraction. This is carried out in the following reaction catalyzed by creatine kinase.

$$MgADP^- + \text{P-creatine}^{2-} + H^+ \rightleftharpoons MgATP^- + \text{creatine}$$

In an experiment on intact muscle, selective irradiation of the ATP resonance produced an 18% decrease in the P-creatine resonance, while selective irradiation of the P-creatine resonance produced a 30% decrease in the ATP resonance. The T_1 values for P-creatine and ATP were 3.2 s and 1.5 s, respectively, and the ratio of P-creatine concentration to ATP concentration was 6. From these data, calculate the relative flux of ATP $\rightarrow$ P-creatine and P-creatine $\rightarrow$ ATP in the intact muscle.

CHAPTER 7

Electron Paramagnetic Resonance Spectroscopy

OVERVIEW

1. Electron paramagnetic resonance (EPR) is a technique that detects unpaired electrons. Electron-spin reorientation causes absorption of incident microwave radiation when a sample is placed in a strong magnetic field.

2. In biological systems, unpaired electrons occur in free radicals and some transition metal ions. The use of synthetic, stable free radicals (spin labels) as probes extends the range of biological applications.

3. The physical basis of the technique is the interaction between the magnetic moment of the electron and the applied magnetic field.

4. EPR spectra are usually displayed as the first derivative of the absorption spectrum. A spectrum is characterized by four main parameters: intensity, linewidth, g-value (which defines position), and multiplet structure.

5. The intensity of an EPR spectrum in solution can give information on concentration, the linewidth on any dynamic process, the g-value on the immediate environment of the unpaired electron, and the multiplet structure (characterized by a hyperfine splitting constant A) on the interaction of the unpaired electron spins with nuclear spins.

6. The g- or A-values are usually antisotropic, which means that the spectra depend on the angle between the molecular axes and the applied magnetic field. In a single crystal, the spectra are sharp, but in a powder (random) sample, they are spread out because all possible angles exist. Information must then be obtained from interpretation of the shape of the spectral envelope.

7. EPR spectra of transition metal ions are characteristic of the number of unpaired electrons of the metal and the arrangement and nature of the coordinated ligands. The g- values range from about 1 to 10. Very low temperatures are often required for observation of the spectra because relaxation times are very short.

8. In molecules with more than one unpaired electron, the interaction between the spins leads to a splitting of the energy levels even in zero field (zero-field splitting, ZFS). ZFS can give rise to multiplet structure and is particularly important in understanding transition metal EPR spectra.

9. Biological applications of EPR include exploration of the ligand environment around a metal site in a metalloprotein, the detection of free-radical intermediates (spin trapping), and exploitation of the spectral anisotropy of spin labels to measure both the rate and amplitude of molecular motion in a variety of systems.

INTRODUCTION

Electron paramagnetic resonance (EPR) spectroscopy, sometimes known as electron spin resonance (ESR), is a technique that detects unpaired electrons in a sample by their absorption of energy from microwave irradiation ($\sim 10^{10}$ Hz) when the sample is placed in a strong magnetic field. (~ 0.3 T).

The vast majority of molecules contain paired electrons, but there are two important classes that contain unpaired electrons that are relevant in biological systems, namely, free radicals and transition metal ions. The widespread use of nitroxide free radicals (spin labels) as probes has enabled EPR to be applied to a wide range of biological systems that do not have intrinsic EPR signals.

THE RESONANCE CONDITION

The electron possesses a magnetic moment by virtue of its spin ($S = \frac{1}{2}$). In the presence of an applied field, the magnetic moment has two allowed orientations, which have different energies (see Figure 7.1). These orientations correspond to the two spin states of the electron ($S = \pm \frac{1}{2}$). Transitions between the two spin states can be induced if oscillating electromagnetic radiation of the appropriate

Figure 7.1   Energy-level splitting and electron resonance of an electron in an applied magnetic field.

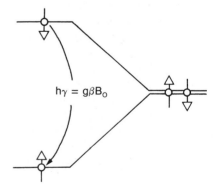

$$h\gamma = g\beta B_0$$

frequency ν is applied perpendicular to the external magnetic field B_0. The basic resonance condition in EPR is defined as

$$h\nu = g\beta B_0$$

where β is the Bohr magneton (0.92×10^{-23} J $\cdot$ T) and g is the electron g-factor, which for a free electron is 2.0023. We can also write (see Chapter 6) that the Larmor precession frequency ω is given by

$$\omega_e = \gamma_e B_0$$

where γ_e is the magnetogyric ratio of the electron (1.76×10^4 rad $\cdot$ s$^{-1} \cdot$ T^{-1}).

MEASUREMENT

For B_0 fields in the 0.3–1.2 T range, electron spin resonance occurs in the microwave range of frequencies (0.9×10^9–3.6×10^9 Hz, wavelength 3–0.8 cm).

Microwaves are readily conducted along hollow tubes whose dimensions are related to the microwave wavelengths. The most commonly used tube has a dimension of about 3 cm, corresponding to a frequency of about 9000 kHz. Such a system is referred to as an **X-band spectrometer**. The resonator in a microwave spectrometer is called a **cavity**. The cavity containing the sample is placed in a magnetic field that is "swept" through the resonance condition. Detection of the small absorption signal is improved by having the cavity in one arm of a balanced network (see Figure 7.2(a)). Detection is further improved by modulating the magnetic field. It can be seen from Figure 7.2(b) that if the modulation is less than the linewidth of the absorption signal, the signal detected appears as a *derivative* of the absorption line if the detector is sensitive to the phase of the signal.

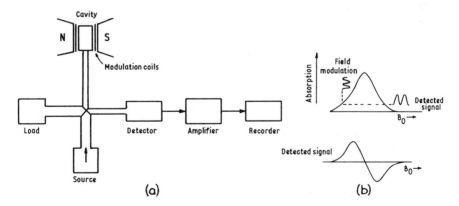

Figure 7.2 (a) The essentials of the EPR experiment. (b) Absorption curve and its first derivative obtained by field modulation and phase-sensitive detection at the modulation frequency.

Samples can be aqueous or solid and typically have volumes in the 10–500 μl range. The shape of the sample cells used varies depending on the nature of the sample. Flat quartz cells, for example, are often used for aqueous samples. The sensitivity of the method is high, and 1 μM can often be detected in a few minutes.

SPECTRAL PARAMETERS

Although EPR measures the absorption of energy, the spectrum is usually displayed as the first derivative of the absorption spectrum. An EPR signal is characterized by four main parameters: (1) intensity, (2) linewidth; (3) g-value, and (4) multiplet structure, or splitting.

THE INTENSITY

The integrated area under the signal is proportional to the concentration of the unpaired spins giving rise to the spectrum. A number of factors can alter the spectral intensity: (1) Quenching can arise if there is a high concentration of paramagnetic species in solutions. (2) Many resonances are easily saturated (that is, the populations of the energy levels are equalized during the measurement; see Chapter 2). (3) The presentation of the spectra as derivatives can make extrapolation to the integrated area difficult. However, in some cases (such as the measurement of free Mn(II) in a protein solution), EPR intensities are used quantitatively, often by measuring the peak-to-peak intensities of the derivative spectrum.

g-VALUE

The g-value is used to characterize the position of a resonance. It is a measure of the local magnetic field experienced by the electron (see chemical shift in NMR).* The g-value is defined as

$$g = \frac{h\nu}{\beta B_0}$$

where B_0 is the applied (or external) magnetic field at resonance. The g-value for a molecule containing unpaired electrons is regarded as a characteristic quantity of the electronic environment. Its measurement is an aid in the identification of an unknown signal. The g-value for a free electron (and also that found in a large number of free radicals) is 2.0023, while in transition metals, g-values as high as 10 have been observed.

*In NMR, it is assumed that the nuclear g-factor is constant and that the effective field at the nucleus is altered by screening effects. In EPR, it is usual to assume that the value of g alters in different environments.

In almost all cases the observed g-value is **anisotropic,** that is, its value varies according to the orientation of the molecule in the applied magnetic field. However, when a molecule is tumbling rapidly, so that the local fields are continuously altering, the g-value that is measured represents an average. As we shall see, anisotropy and its averaging by motion are extremely important in EPR.

LINEWIDTHS AND RELAXATION TIMES

The linewidth of an EPR line can be treated, rather simplistically, as comprising two kinds of contribution. The general principles of relaxation are similar to those in NMR, but in EPR are less well understood; this simplistic approach is sufficient for many purposes.

The first contribution to the linewidth is related to the **spin-lattice relaxation time** (T_1), which is characterized by a first-order rate constant $1/T_1$. The shorter the T_1, the broader the line. As with NMR, T_1 is a measure of the recovery rate of the spin populations after a perturbation. The basic mechanism that determines T_1 involves fluctuating magnetic fields experienced by the unpaired electron. Only those components of the fields that have the correct frequency can induce transitions between the energy levels.

The second main contribution to the linewidth arises because the local magnetic fields are slightly different at each molecule. This difference results in a spread in the values of the resonance frequencies and a corresponding broadening of the resonance, which is sometimes known as **inhomogeneous broadening** to distinguish it from pure relaxation broadening. This situation occurs when the molecules are either static or tumbling relatively slowly. If the molecules tumble more rapidly, this static broadening can be partially averaged as the local fields at the molecules change during the observation of the resonance. For convenience, we can also characterize these broadening processes by a lifetime T_2. It should be noted, however, that this is an approximation that assumes a certain kind of line shape (Lorentzian), which may not be observed in inhomogeneously broadened lines.

The observed linewidth is then defined in terms of a spin-spin relaxation time T_2 such that an expression of the type

$$\frac{1}{T_2} = \frac{1}{T_2'} + \frac{1}{T_1}$$

can be written. The term $1/T_2$ (per second) is related to the peak-to-peak separation $\Delta\nu$ (in millitesla) of a resonance by the equation

$$\frac{1}{T_2} = \pi\sqrt{3}\Delta\nu(2.8 \times 10^7)$$

In most transition metal complexes, the linewidth is determined by the T_1 contribution. Here T_1 is often so short that the resonance line is too broad to be detectable at room temperature. The lifetime T_1 often increases dramatically (sometimes exponentially) with decreasing temperature. Even so, very low temperature (1 K) may be required to observe an EPR signal at all. Conversely, in free radicals, the values of T_1 are usually long, so the line broadening arises mainly from T_2 contributions.

Finally, we note that a major contribution to the T_2' term with which we shall be concerned arises from dynamic (or exchange) processes in which the environment of the unpaired electron alters as a function of time.

MULTIPLET STRUCTURE IN EPR SPECTRA

In NMR we saw that resonances can be split into multiplets by interactions with other nuclei. In EPR there are two main types of interactions that lead to the observation of multiplets, namely, **hyperfine interactions** and **zero-field splitting**. We shall treat only hyperfine interactions here and postpone discussion of zero-field splitting.

HYPERFINE STRUCTURE

The hyperfine structure arises from the interaction of the electron spins with the nuclear spins. The interaction is of two types: (1) a contact interaction that is isotropic and that results from the delocalization of the unpaired electron onto the nucleus and (2) a dipolar interaction between spins of the electron and the nucleus. The dipolar interaction is directional (see Appendix IX) and is thus anisotropic (see next section). We shall see that the dipolar interaction is very important in many EPR applications, but for the rest of this section, we shall consider only the information available from the isotropic hyperfine interaction that is characteristic of free radicals tumbling rapidly in nonviscous solution. In such cases, the dipolar interaction is averaged to zero by the motion.

The result of the isotropic hyperfine interaction is that the nuclear spin produces a local field at the electron. In general, there will be $(2I + 1)$ orientations of the nuclear spin in a magnetic field, all of which correspond to a different energy. The unpaired electron spin "senses" these different orientations, so the EPR spectrum is split into $(2I + 1)$ lines of equal intensity. The magnitude of the splitting between the lines is called the **hyperfine splitting constant**. For n equivalent nuclei, the EPR spectrum consists of $(2nI + 1)$ lines whose relative intensities are given by the binomial coefficients obtained in the expansion of $(1 + x)n$. For example, if $n = 4$, then the relative intensities are 1:4:6:4:1. If there is more than one set of equivalent nuclei, *each* of the $(2nI + 1)$ lines is then further split according to the same rules.

Worked Example 7.1   Interpreting Hyperfine Splitting

Interpret the splitting patterns in the EPR spectra in Figure 7.3 (nuclear spins: I, N = 1, H = $\frac{1}{2}$, Mn = $\frac{5}{2}$).

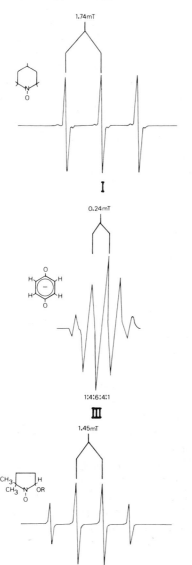

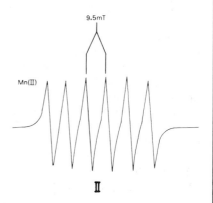

Figure 7.3   *(continued)*

Worked Example 7.1   (continued)

Solution

In (I), the three-line spectrum (which is characteristic of spin labels, as we shall see later) arises from interaction of the unpaired electron spin with the nuclear spin ($I = 1$) of the nitrogen. (Note the small satellite lines in this spectrum, which are caused by the ^{13}C nuclear spins ($I = \frac{1}{2}$, natural abundance 1.1%) of CH_3 groups adjacent to the ^{14}N of the nitroxide group). As a result, 1.1% of the radicals experience further superhyperfine splitting of each nitrogen hyperfine line into two). (II) The six lines arise from the nuclear spin of Mn($I = \frac{5}{2}$), giving ($2 \times \frac{5}{2} + 1$) lines. (III) In the benzoquinone radical, the electron is delocalized and interacts with four equivalent protons, giving rise to ($2 \times 4 \times \frac{1}{2} + 1$) = 5 lines of intensities given by the coefficients of $(1 + x)^4$, i.e., 1:4:6:4:1. (IV) The 1:3:3:1 splitting of the electron line comes from interaction of three equivalent protons from the CH_3. (*Note:*   The $-OH$ splitting is not resolved). (V) The four-line spectrum here arises from the splitting, by the nitrogen ($I = 1$), into three lines, which are further split by the proton, with the H and N splittings *equal*.

SPECTRAL ANISOTROPY

The orbitals containing the unpaired electrons will, in general, be anisotropic (i.e., different in different directions), not isotropic (spherical). Many of the parameters of EPR are anisotropic. For example, the *g*-value (position) and the *A*-value (hyperfine splitting) of a molecule often depend on the direction of the magnetic field relative to the molecular axes.

In many organic free radicals, the electrons are extremely delocalized, so the electron distribution is almost isotropic. However, for transition metals and spin labels, the electrons are primarily localized to *d*- or *p*-type orbitals and spectral anisotropy becomes very important. A fast rate of motion in all directions will average the anisotropy and give "isotropic" *A*- and *g*-values. Fast motion restricted to certain directions only partially averages the magnetic interactions. Spin label spectra can often be interpreted to give information about the amplitude and rate of motion. However, in this section we shall discuss the types of spectra that are obtained when the *g*- and *A*-values are anisotropic in the absence of motional averaging. The effects of motion will be discussed later.

Anisotropy of the g-Value

The *g*-value anisotropy is characterized by three principal *g*-values, g_{xx}, g_{yy}, and g_{zz}, which are those along the principal axes of the group containing the unpaired electron. The principal axes are fixed from symmetry considerations. For example, any twofold or greater axis of symmetry is a principal one.* A cube, a

---

*An *n*-fold axis of symmetry is one in which rotation by an angle $360°/n$ brings the molecule into itself again.

Figure 7.4 Variation of *g*-value with the
magnetic field orientation for an axially
symmetric system.

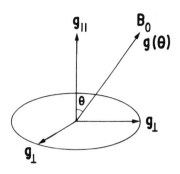

tetrahedron, and an octahedron have equivalent principal axes, and $g_{xx} = g_{xx} = g_{zz}$—that is, the *g*-value is **isotropic.** In the case of axial symmetry, as found for spin labels and many transition metals, $g_{xx} = g_{yy} \neq g_{zz}$. By convention, g_{zz} is defined as the *g*-value observed when the applied field is pointing along or parallel to the symmetry axis and it is designated $g_{\parallel}$. Similarly, for axial symmetry, g_{xx} and g_{yy} are designated $g_{\perp}$, which is the value when the applied field is perpendicular to the symmetry axis. By convention, g_{zz} is usually taken to be the value at lowest field in a spectrum and g_{xx} is that at highest field.

In a single crystal, the three principal *g*-values can be measured when the magnetic field is applied along the three principal axes of the molecule. At any intermediate orientation, the *g*-values depend on the angle θ that the magnetic field makes with the principal axes. For an axially symmetric system, as shown in Figure 7.4, the *g*-values will vary with magnetic field orientation as

$$g(\theta)^2 = g_{\parallel}^2 \cos^2\theta + g_{\perp}^2 \sin^2\theta$$

Thus, for single crystals, the angular variation of the *g*-values can give information on the orientation of the principal axes. A classic example by Ingram and colleagues is the determination of the orientation of the four heme groups in a single crystal of hemoglobin.*

In a liquid or a frozen sample, the molecules will be randomly oriented. There will be a spread of *g*-values comprising the individual absorptions of molecules, each with a different value of θ. The envelope of all the individual absorption curves and the corresponding derivative curve are shown in Figure 7.5 for cases with and without axial symmetry†. In these cases, the extreme *g*-values are sufficiently different that they are clearly "resolved." Note that the derivative spectra highlight edges of the absorption spectra. These edges usually correspond to the principal *g*-values.

*This is well and simply presented in Ingram's excellent book *Biological and Biochemical Applications of ESR* (London: Adam Hilger Ltd, 1969).
†Note that we have assumed that $g_{\parallel} > g_{\perp}$. Note also that the line shape for the axially symmetric shape has much more absorption concentrated near the $g_{\perp}$ than the $g_{\parallel}$ direction. This follows because there will be many more molecules distributed in the plane of directions represented by the $g_{\perp}$-value rather than the direction represented by $g_{\parallel}$. (A similar type of reasoning was used to discuss the angular dependence of fluorescence polarization.)

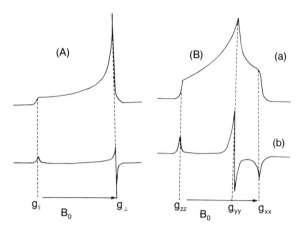

Figure 7.5   Idealized (a) absorption and (b) first-derivative EPR spectra for a system of randomly oriented molecules exhibiting g-anisotropy characteristic of (A) axial symmetry ($g_{zz} \neq g_{xx} = g_{yy}$) and (B) low symmetry $g_{xx} \neq g_{yy} \neq g_{zz}$.

Worked Example 7.2   Symmetry of Fe(III) Complexes

The 9.25-GHz EPR spectra of Fe(III) metmyoglobin and its azide complex are shown in Figure 7.6. Compare the symmetry around the metal ion.

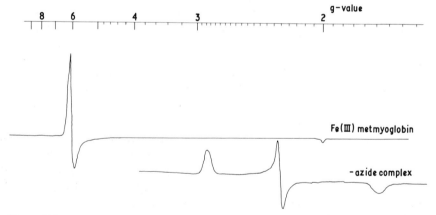

Figure 7.6

Solution

The Fe(III) metmyoglobin complex has g-values at 2 and 6, which suggest *axial* symmetry. By contrast, the azide complex has g-values of 2.8, 2.25, and 1.75, which reflect a lower symmetry around the metal ion (see also "Zero-Field Splitting").

Figure 7.7 Idealized EPR (a) absorption and (b) first-derivative spectra for a system of randomly oriented molecules exhibiting hyperfine anisotropy characteristic of a system with axial symmetry and for which $(A_\| > A_\perp)$.

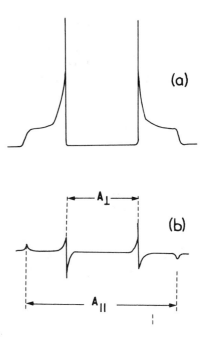

Anisotropy of the Nuclear Hyperfine Interaction A

The anisotropy of the nuclear hyperfine interaction A arises from the dipolar interaction of the magnetic field of the electron with the magnetic field of the nucleus. The considerations that apply to anisotropic A-values are the same as those for anisotropic g-values. Thus the A-value anisotropy is characterized by three principal A-values: A_{xx}, A_{yy}, and A_{zz}. In the case of axial symmetry, $A_\| = A_{zz}$ and $A_\perp = A_{xx} = A_{yy}$. Figure 7.7 shows the envelope of an idealized absorption spectrum and its first-derivative spectrum for a system of randomly oriented radicals that have axial symmetry and an anisotropic hyperfine interaction (with a single magnetic nucleus ($I = \frac{1}{2}$). (For simplicity, we have assumed that the g-factor is isotropic).

One very important example of hyperfine anisotropy is that involving stable nitroxide free radicals or spin labels (discussed later). The hyperfine interaction of the unpaired electron with the nitrogen nucleus ($I = 1$) results in three lines. The unpaired electron is primarily localized to a $2p$-electron orbital on the nitrogen. This orbital is anisotropic, therefore the hyperfine interactions and g-values are also anisotropic. The nitroxide radical can be oriented in a rigid matrix. When the magnetic field is then applied along the principal axes of the nitroxide (see Figure 7.8), the absorption spectra and corresponding first-derivative peaks obtained are shown in Figure 7.9.

Figure 7.8   The molecular coordinate system of the nitroxide spin label used to define the direction of the applied magnetic field. The z-axis is parallel to the nitrogen 2p-orbital associated with the unpaired electron.

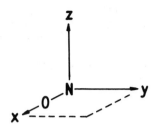

For a random orientation of rigid nitroxide radicals in the magnetic field, the spectrum in Figure 7.9(d) is obtained. It represents a sum of the spectra corresponding to all possible orientations of which Figure 7.9 (a)–(c) represents the extreme along each axis. This spectrum is called the **rigid-glass spectrum** and is observed for any randomly oriented spin label, assuming that there is no molecular motion.* Note that only the maximum splitting, designated $2A_{zz}$, can be resolved. The two smaller splittings are "lost" because of the broadening from spectral overlap in the center of the spectrum. In some circumstances, it may be possible to resolve the smaller splitting (see "Anisotropic Spin-Label Motion").

THE TIME SCALE FOR EPR

We have stated several times so far that various anisotropies are averaged if the molecular motions is rapid enough. In solutions of free radicals, for example, where the anisotropy of g is small and the tumbling rate is rapid ($\sim 10^{-10}$ s) all that can be measured is an average g-value, which is given by

$$g_0 = \tfrac{1}{3}(g_{xx} + g_{yy} + g_{zz})$$

Figure 7.9   (a)–(c) The 9.5-GHz spectra of a spin label in a rigid matrix (e.g., cyclobutane-1, 3-dione crystal). The host crystal orients the nitroxide molecules in a well-defined way. By suitable rotation of the crystal, the applied field can be made parallel to the principal axis. The dashed line indicates a reference marker g-2.0036. (d) The rigid-glass or powder spectrum.

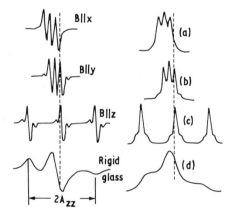

*Such spectra are also sometimes called **powder spectra.**

We can estimate the time scale required for averaging an EPR parameter as follows: Let $\Delta\omega$ (in $\text{rad} \cdot \text{s}^{-1}$) represent the difference between any two values of the parameters that are to be averaged. If the motion is characterized by a rate constant $1/\tau$, then the time scale is "fast" if $1/\tau >> \Delta\omega$ and "slow" if $1/\tau << \Delta\omega$. (This is formally similar to the situation in NMR of chemical exchange between two environments, separated by a shift $\Delta\omega$. As with NMR, intermediate cases on the time scale are also possible and are characterized by broadened lines).

For the g-value anisotropy, we can quantify the time scale that leads to complete averaging by assuming that the maximum anisotropy is given by $(\Delta g\beta B_0/\hbar)$ $\text{rad} \cdot \text{s}^{-1}$.

Worked Example 7.3   Averaging of g-Value Anisotropy in Spin Labels

A nitroxide spin label was oriented in a rigid host matrix. Its derivative EPR spectrum was similar to that obtained from the spin label oriented in a single crystal. Three g-values were obtained: $g_{zz} = 2.0027$, $g_{xx} = 2.0089$, and $g_{yy} = 2.0061$. At what rate would the spin label have to tumble to average out the g-value anisotropy in a magnetic field of 0.33 T? ($\beta = 0.927 \times 10^{-23}\,\text{J} \cdot \text{T}$; $\hbar = 1.05 \times 10^{-34}\,\text{J} \cdot \text{s}$).

Solution

The maximum g-value anisotropy, Δg, is given by $2.0089 - 2.0026 = 0.0062$. To average this out, the rate $1/\tau >> \Delta g\beta B_0/\hbar$ That is, $1/\tau >> 0.0062 \times 0.927 \times 10^{-23} \times 0.33/(1.05 \times 10^{-34}\,\text{s})$. Therefore, $1/\tau >> 1.8 \times 10^8\,\text{s}^{-1}$.

(*Note:*   More correctly, Δg is $g_{zz} - \frac{1}{2}(g_{xx} + g_{yy})$.

Different EPR Parameters May Have Different Time Scales

As well as having a spread of g-values, EPR lines may be broadened or split by zero-field splitting (discussed later) or by anisotropic hyperfine interactions. There may also be exchange between two environments (for example, a ligand binding to a protein), where there is a different relaxation time in each environment. In each case the relevant time scale is defined by the spread in values between the various directions or environments.

Worked Example 7.4   Averaging of Hyperfine Anisotropy in Spin-Label Spectra

A spin-label nitroxide has the following values of the hyperfine interaction: $A_{zz} = 87$ MHz, $A_{xx} = A_{yy} = 14$ MHz. If the label undergoes rapid motion, a three-line spectrum (such as that shown in Figure 7.10) is observed. Can you estimate very approximately a lower limit for the rate of this motion?

(continued)

Worked Example 7.4   (continued)

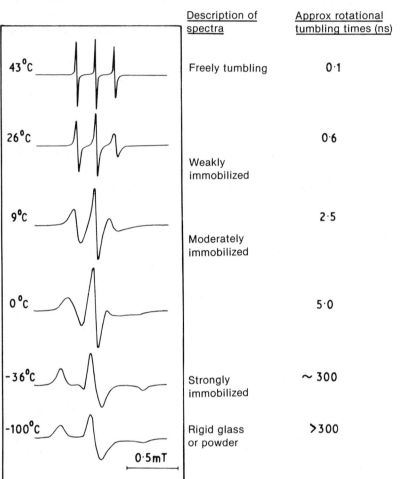

	Description of spectra	Approx rotational tumbling times (ns)
43°C	Freely tumbling	0·1
26°C		0·6
	Weakly immobilized	
9°C		2·5
	Moderately immobilized	
0°C		5·0
−36°C	Strongly immobilized	~ 300
−100°C	Rigid glass or powder	>300

0·5mT

Figure 7.10   The effects of the rate of motion on the EPR spectra of a spin label that is rotating isotropically. Rotational tumbling time is altered by changing the temperature.

Solution

The maximum hyperfine anisotropy is $(A_{zz} - \frac{1}{2}(A_{xx} + A_{yy}))$, or $(87 - 14) = 73$ MHz $= 73.10^6 \times 2\pi$ rad $\cdot$ s^{-1}. To average this out, the rate of tumbling, $1/\tau$, will have to be greater, that is, $1/\tau \gg 73 \times 10^6 \times 2\pi$s^{-1}. Therefore, $1/\tau \gg 4.6 \times 10^8$ s^{-1}.

Comparison of this result with that from the previous worked example shows that the anisotropy in g will be averaged out slightly more easily than that for A.

SPIN LABELS

The term **spin label** was introduced by the pioneer of this field, H. M. McConnell, to describe stable free radicals that are used as reporter groups or probes. The most commonly used spin labels contain the nitroxide moiety. The nitroxide moiety is not very reactive; it is stable up to about 80°C and over the pH range 3–10. It can be reduced to the N-hydroxyl-amine by mild reducing agents, such as dithionite.

By a suitable choice of X in molecules I and II, it is possible to meet the specific requirements of many different biological problems. For instance, spin labels may be either attached (via the group X) to specific functional groups on the molecule being investigated or intercalated into regions of interest. Additionally, it is possible to make spin-labeled substrates or ligands; thus, many different types of experiment can be designed. The major thrust of the spin-label method, however, has been to study motion—the time scale (10^7–10^{11} s^{-1}) is mainly determined by the anisotropy of magnetic interactions of the label.

Effect of Rate of Motion on Spin-Label Spectra

When the nitroxide group is tumbling rapidly and isotropically, the EPR spectrum consists of three narrow lines of almost equal heights. The motion is then sufficiently fast on the EPR time scale to average out the spectral anisotropies. This high degree of mobility is observed for small spin labels "freely tumbling" in nonviscous solutions.

As the rate of motion decreases, the EPR spectrum alters; Figure 7.10 illustrates this effect. Spectra of a spin label, obtained as the viscosity in the solution is increased by lowering the temperature, are shown. The spectra are described by qualitative terms such as "weakly immobilized"; the corresponding rotational tumbling times are also indicated.

The mobility of the label in the region between weakly and moderately immobilized labels in a magnetic field of 0.33 T is given approximately by the empirical relationship

$$\tau = 6.5\Delta\nu \sqrt{\frac{h(0)}{h(-1)} - 1}$$

where τ is in nanoseconds, $\Delta\nu$ is the peak-to-peak separation of the central line in milliTesla, and $h(-1)$ and $h(0)$ are the heights of the high-field and central lines, respectively. Rotational times slower than in the moderately immobilized region are usually obtained from comparison with model spectra or by a technique known as saturation transfer (for a review, see J. S. Hyde and L. R. Dalton in *Spin Labeling,* vol. 2, ed. L. J. Berliner (New York: Academic Press, 1979). The sensitivity of the label to motion makes it a useful probe for following environmental changes as a result of perturbations, such as the binding of ligands.

Worked Example 7.5   Spin-Labeled Phosphorylase b Gives a Weakly Immobilized EPR Spectrum That Changes on Activation of the Enzyme

The enzyme phosphorylase catalyzes the breakdown of glycogen. There are two forms of the enzyme: the "inactive" *b* form, which may be activated by addition of the ligand AMP, and the "active" *a* form (see Figure 7.11(a)). The essential difference between the *b* and *a* forms is that the *a* form is phosphorylated on one specific serine residue per subunit. Phosphorylase *a* or *b* can be specifically labeled on one —SH group with a nitroxide spin label of the type I (see section on spin labels), where X is an iodoacetamide group. The label does not affect the activity of the enzymes. The EPR spectrum of the labeled *b* form in the presence of AMP becomes identical with that of the *a* form alone (see Figure 7.11(b)). (a) What do you conclude from this? (b) Calculate the approximate tumbling times of the two forms. The peak-to-peak separation of the central peak is 3×10^{-4} T, and the ratios $h(0)/h(-1)$ are 4.8 and 5.6 for the *b* and *a* forms, respectively.

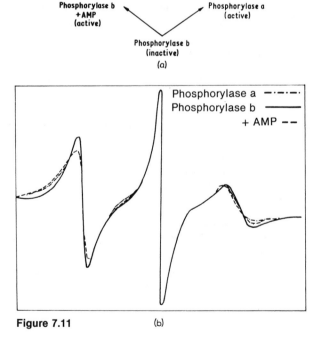

Figure 7.11   (b)

(continued)

Worked Example 7.5   (continued)

Solution

a. As judged by the *change* in the EPR spectrum, the two methods of activating phosphorylase *b* result in identical changes in conformation, which alter the tumbling time of the label.

b. Using the formula in the previous section, we obtain τ = 3.8 ns for phosphory-lase *b* and τ = 4.6 ns for phosphorylase *a*. (Note that this is much shorter than the overall tumbling time ($\sim 10^{-7}$ s) expected for the whole enzyme, which indicates that the label is probing a local environment.

Anisotropic Spin-Label Motion

Sometimes the spin label is not free to tumble in all directions, so it can perform only anisotropic motion. Motional averaging can then occur only in some directions. For example, consider a sample of lipid spin labels diluted into a bilayer membrane (see Figure 7.12). Most labels are designed so that the z-axis of the nitroxide is parallel to the long molecular axis of the lipid. If we assume that the chains can rotate rapidly only about the long axis, then this will average the A_{xx} and A_{yy} splittings. Since these splittings are almost equal (axial symmetry), this motion has little effect on the spectrum—the resultant spectrum is very similar to that of the no-motion, rigid-glass case (see Figure 7.9(d)). If the acyl chains can

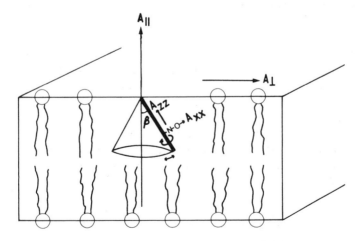

Figure 7.12   Anisotropic motion of a rigid spin label in a membrane bilayer. The axis of motional averaging is perpendicular to the membrane surface (‖ direction). The measured hyperfine splittings resulting from fast motional averaging of A_{zz} with A_{xx} and A_{yy} is termed $A_{\parallel}$. Similarly, $A_{\perp}$ reflects fast motional averaging of A_{xx} and A_{yy} with A_{zz}.

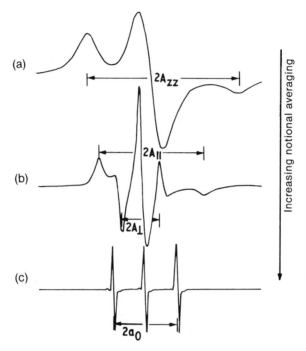

Figure 7.13   Increasing motional averaging of the hyperfine anisotropy of a nitroxide spin label.
(a) Rotation about the z-axis only (i.e., fast motion of restricted amplitude).
(b) "Wobbling" of the lipid chains in a bilayer increasing the amplitude of motion.
(c) Fast isotropic motion.

now also "wobble" (through fast *transgauche* rotational isomerism), the amplitude of the motion is increased (i.e., there is additional motion in other planes) and A_{zz} can be averaged with A_{xx} and A_{yy}. This reduces A_{zz} from its principal-axis value (~ 3.2 mT), but increases A_{xx} from its principal value (of about 0.6 mT), so a small splitting can often be resolved in the spectrum (see Figure 7.13(b)). The measured value of A reflecting fast-motional averaging of A_{zz} is termed $A_{\parallel}$. Similarly, $A_{\perp}$ reflects the fast-motional averaging of A_{xx} and A_{yy}. The narrow linewidths are also indicative of fast motion (see Figure 7.13). In the limit of fast isotropic motion, $A_{\parallel} = A_{\perp}$.

Worked Example 7.6   Lipid Motion in Membranes Is Anisotropic

Figure 7.14 shows the EPR spectra of a spin-label (SL) probe in a phosphatidyl-choline (PC) bilayer. The label has been attached to the carbon atoms on the lipid acyl chain on the probe molecules. The long axis of the lipid molecule is coincident with the z-axis of the nitroxide radical. How do you interpret the changes in the hyperfine splittings of the label at the different carbon positions?

(continued)

Worked Example 7.6   (continued)

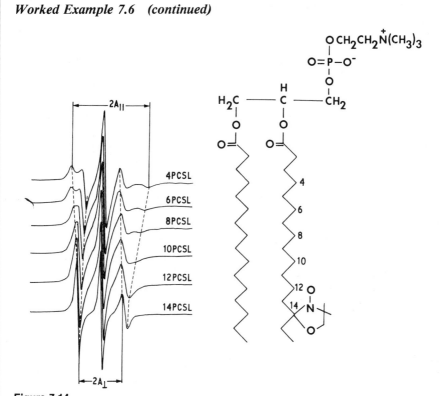

Figure 7.14

Solution

    The narrow linewidths confirm that the motion is fast. The minimum splitting $A_\perp$ does not change significantly in progressing down the chain. Therefore, there is not much amplitude of motion in the *x-y*–plane, and A_{xx} and A_{yy} are averaged through long-axis rotation. The amplitude of this motion is increased dramatically down the acyl chain as $A_\parallel$ becomes averaged with $A_\perp$, thus decreasing $A_\parallel$. At the end of the acyl chain (or in the center of the bilayer), $A_\parallel \simeq A_\perp$ and the motion is almost isotropic. (Figure by permission of A. Watts.)

Quantitation of Amplitude of Motion: The Order Parameter

The amplitude of motion of the spin label is governed by its environment. The extent of fast-motional averaging of the anisotropy can be defined by an order parameter (*S*), where

$$S = \frac{\text{observed anisotropy}}{\text{maximum anisotropy}}$$

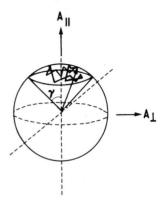

Figure 7.15 Models for a random motion of a spin label within a given cone of angle γ.

The observed anisotropy is $A_{\parallel} - A_{\perp}$. The maximum anisotropy is given by $A_{zz} - \frac{1}{2}(A_{xx} + A_{yy})$; these are the values of the hyperfine splittings obtained when the label is rigidly oriented in a single crystal. It follows that $|S| = 1$ for highly ordered systems and $S = 0$ for completely isotropic motion.

The order parameter is directly related to the angular amplitude of fast anisotropic motion.

$$S = \tfrac{1}{2}(3 < \cos^2\beta > - 1)$$

where the angle brackets indicate a time average over the instantaneous angular displacement β (see Figure 7.12). Note that to obtain a time average, the motion must be fast. There is no way of obtaining S for slow motion. To obtain values for β, the motional variations have to be modeled in some way. For instance, in the spin-labeled lipid example, rapid wobbling of the long axis of the label can be modeled by a random motion restricted within a cone of angle γ about an axis normal to the membrane surface (parallel direction) (see Figure 7.15). In this case,

$$S = \tfrac{1}{2}(\cos \gamma + \cos^2\gamma)$$

Note that when $S = 1$, $\gamma = 0°$, and when $S = 0$, $\gamma = 90°$.

Worked Example 7.7   Flexibility Gradients in Lipid Bilayers

The order parameters for the lipid spin labels in the previous worked example are shown in Figure 7.16. (a) Explain the change in the trend of S as the nitroxide label is positioned farther along the chain to the bilayer center. (b) Assume that the molecular motion can be described by a random motion restricted within a cone of angle γ. Calculate γ for the 4PCSL ($S = 0.5$) and for the 14PCSL ($S = 0.1$).

(continued)

Worked Example 7.7   (continued)

Solution

a. The gradual decrease of order parameter with label position reflects the increase in amplitude of chain motion along the length of the chain toward the center of the bilayer.

b. Using $S = \frac{1}{2}(\cos\gamma + \cos^2\gamma)$, we obtain that for $S = 0.5$, $\gamma = 52°$, while for $S = 0.1$, $\gamma = 85°$. This confirms the conclusion from (a): The increase in amplitude being experienced by the spin label on going down the fatty acid chain is reflected in a larger cone angle. (This arises from more cumulative *trans*-gauche rotational isomers in the chain).

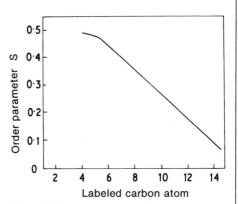

Figure 7.16

Lateral Diffusion in Membranes

As well as having intramolecular motions, lipids laterally diffuse within the membrane plane. Two experimental approaches utilizing spin labels are used to measure the rate of lateral diffusion.

In the first method, spin-labeled lipids are concentrated into one region of an *oriented* bilayer. When the spin labels are very close together, the electron spins of two labels can exchange. This exchange interaction can severely broaden the EPR spectrum. As the labels diffuse laterally and are diluted into the rest of the bilayer, the EPR spectra change with time, and the broadening is reduced. If the patch of spin-labeled lipids has a radius R_0, then after a time t, a typical (three-line) spectrum is obtained when the molecule has traveled about $2R_0$. The distance traveled in time t is given by $(4Dt)^{1/2}$. This relationship allows a crude estimate of the diffusion constant D to be made.

The second method does not require oriented bilayers. Relatively high levels of spin-labeled lipid (30%) are randomly incorporated into model membranes. Broadening occurs when the spin labels come into contact as a result of diffusion. If the rate of diffusion is sufficiently rapid, any dipole-dipole interactions will be averaged out and only the exchange process will contribute to the linewidth. For a mole fraction of spin-labeled lipid $P_M < 0.3$, the peak-to-peak linewidth of the central line in the spectrum is given by

$$\Delta \nu_{\text{exch}} = \frac{2}{\tau} P_M$$

Note that $\Delta \nu_{\text{exch}}$ must now be expressed in radians per seconds. The term $1/\tau$ is equal to the collision rate between the labels and can be related to D using various theoretical models.

Worked Example 7.8　Measurement of the Rate of Lateral Diffusion

The EPR spectrum of a highly concentrated region ($R_0 = 0.8$ mm) of spin-labeled phosphatidylcholine included into oriented bilayers of phosphatidylcholine changes with time, as shown in Figure 7.17. Estimate the lateral diffusion coefficient.

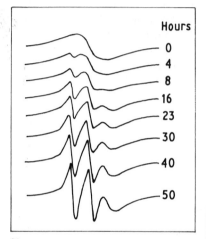

Solution

The main features of the three-line EPR spectrum of the label are seen after about 40–50 h. The equation $2R_0 = (4Dt)^{1/2}$ gives $D = 4 \times 10^{-12}$ m$^2 \cdot$s^{-1}.

Figure 7.17

Spin-Labeled Ligands Can Probe the Dimensions and Rigidity of Binding Sites

Spin-labeled ligands or substrates can be used to probe binding sites. The amplitude of motion of the bound label is limited by the steric effects of neighboring groups in the binding site, and the rate of motion is partially governed by the flexibility of these groups. The spin-label spectrum can give information on both dimensions and rigidity of binding sites.

Worked Example 7.9　Spin Labels as Molecular Dipsticks

Figure 7.18 shows the EPR spectra of four different spin-labeled Dnp ligands when bound to an anti-Dnp antibody. The maximum hyperfine anisotropies, termed $2A_{z'z'}$ are indicated. For (I), this value is essentially the maximum possible value. What can you conclude about the rigidity and depth of the combining site?

(continued)

Worked Example 7.9   (continued)

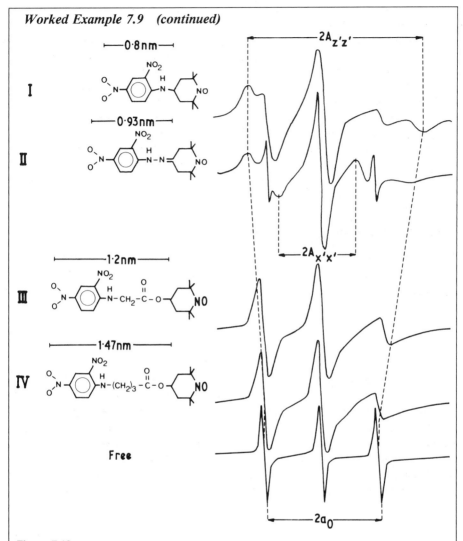

Figure 7.18

Solution

Comparison of the spectra shows that as the distance from the nitroxide to the Dnp ring increases, both the linewidths and anisotropy (spectral extent) of the spectra decrease. This corresponds to an increased amplitude and rate of motion of the nitroxide group, which partially averages out the anisotropy in the EPR spectrum. The shortest label has the largest value of $2A_{z'z'}$, which corresponds to the value expected in a rigid-glass spectrum. This, together with the broad lines, shows that the label is rigidly held in the combining site. The slightly longer spin-labeled hapten (II) has a somewhat smaller value of $2A_{z'z'}$ which indicates that it is more mobile. The

(continued)

Worked Example 7.9   (continued)

sharper lines here also enable a minimum splitting ($2A_{x'x'}$) to be measured. However, for haptens (III) and (IV), the motion is almost unlimited, since $A_{z'z'} = A_{x'x'}$, which indicates that the nitroxide ring is clear of the combining site. (Their motion is still somewhat hindered, however, since the spectral lines are considerably broader than those for the free label.) This places the depth of the site at 1.1–1.2 nm.

A more detailed study would involve comparison of the observed EPR hyperfine splittings with those expected as a result of different postulated motions for the spin label. This would lead to information on the dimensions of the site around the label (for further details see B. J. Sutton et al., *Biochem. J.,* 165(1977): 177–197).

Spin-Label Hyperfine Splittings Are Sensitive to Polarity

The hyperfine interaction depends in part on the unpaired-electron density at the nitrogen atom. We can consider the electronic structure of the nitroxide radical to be represented by two forms, namely,

$$\overset{\backslash}{\underset{/}{N}} - O \cdot \quad \longleftrightarrow \quad \overset{\backslash}{\underset{/}{N^+}} - O^-$$

(a)   (b)

The higher the polarity of the environment, the more structure (b) is favored because of hydrogen-bonding possibilities, and so forth. This structure increases the electron density at the nitrogen atom and therefore results in larger values of A.

The isotropic splitting constant (a_0) provides a relative index of polarity. It is calibrated by measuring its value for a given label in different solvents, e.g., decane and water. (The value changes by about 20% for these two solvents). Since a_0 is independent of the amplitude of motion of the label, we can use this parameter to probe polarity profiles in lipids, bilayers, and the like. In the general case,

$$a_0 = \tfrac{1}{3}(A_{xx} + A_{yy} + A_{zz})$$

(Note that a_0 is also the splitting measured for small molecules tumbling isotropically in solution.)

Estimation of the Separation Between Two Paramagnetic Centers

We have met quenching in the discussion of lateral diffusion, where spin-labeled lipids in close proximity resulted in undetectably broad EPR lines. In certain cases, interactions can be quantified to yield information about the separation of,

Figure 7.19 Variation of the relative amplitude of the peak height of a nitroxide radical with the dipolar interaction coefficient C.

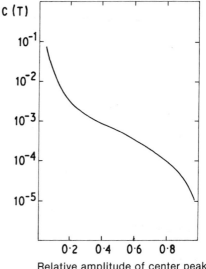

Relative amplitude of center peak

for example, a paramagnetic transition metal ion and a spin label. The basis of this is the fact that the interaction is a dipolar one, involving a $1/r^6$ dependence on the separation r between the two centers. We will not go into the theory here, but merely quote the semiempirical formula

$$C = 1.5 \times 10^{13}\frac{\mu_M^2 \tau_s}{r^6}$$

where C (milliTesla) is a **dipolar interaction coefficient** relating the observed quenching (see Figure 7.19) of the spin-label signal to r (nanometers), μ is the magnetic moment of the metal ion (Bohr magnetons), and τ_S (seconds) is the electron-spin relaxation time of the metal.

Worked Example 7.10   Distance between MnADP Site and Spin-Label Site in Creatine Kinase

Creatine kinase can be specifically labeled on one SH group. The EPR spectrum of spin-labeled creatine kinase is quenched on addition of MnADP (see Figure 7.20). Compared with the diamagnetic MgADP control, the quenching of the spectrum is about 0.9 (i.e., the relative amplitude is 0.1). Calculate the distance between the spin-label and MnADP sites. (Assume μ_{MnADP} = 5.9 Bohr magnetons and τ_S = 1 ns.)

(continued)

Worked Example 7.10   (continued)

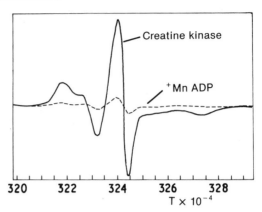

Figure 7.20

Solution

From Figure 7.19, a relative amplitude of 0.1 gives a value of C of about 5×10^{-2}. Substituting the other values into the formula for C, we have $5 \times 10^{-2} = 1.5 \times 10^{13} \times (5.9)^2 \times 10^{-9}/r^6$; therefore, $r = 0.69$ nm (for further details, see J. S. Taylor, J. S. Leigh, Jr., and M. Cohn, *Proc. Nat. Acad. Sci. U.S.A.,* 64(1969):219, and for a discussion of the theory, see R. A. Dwek, *N.M.R. in Biochemistry,* Oxford: Clarendon Press, 1973.)

TRANSITION METAL IONS

The transition metals that have been most frequently studied in biological systems are Mn, Fe, Co, Cu, Mo, and V. In transition metals, the unpaired electrons are primarily localized to d-orbitals, and the number of electron-spin energy levels is given by $2S + 1$, where S represents the *total* spin of the unpaired electrons. There are several characteristic features of the EPR spectra of transition metal ions.

1. The EPR spectrum depends on the number of unpaired electrons and on the arrangement (symmetry) and nature of the coordinated ligands.

2. The g-values vary over a wide range. (Values of 1.4 to 10 have been observed.)

3. Very low temperature (1 K) are often required to observe an EPR spectrum.

In the brief discussion on ligand field theory in Appendix VI, we note that the nature of the ligand can determine the number of unpaired electrons in the complex. This leads to concepts of high-spin and low-spin complexes (see Figure 7.21). For instance, Fe(III) has five d-electrons, which can all be unpaired ($S = \frac{5}{2}$) in a high-spin complex. Alternatively, they can be arranged to give only one unpaired electron ($S = \frac{1}{2}$) in a low-spin complex. These complexes have different

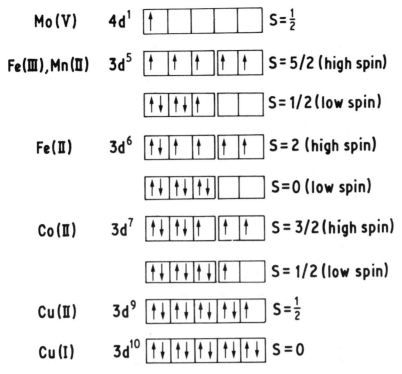

Figure 7.21 Arrangement of d-electrons in some transition metal complexes (see also Appendix VI). The nature of the ligand determines whether the high- or low-spin arrangement is favored. The net spin is given by S. (Note that Cu(I) will have no ESR signal.)

EPR spectra. The dependence of the EPR spectrum on the spin state and on the nature and arrangement of the ligands provides a "fingerprint" of the metal site.

As with electronic spectra of transition metal ion complexes, the identification of a metal-binding site in a biological compound is, in the first instance, essentially made by reference to well-characterized metallocomplexes. This section highlights some of the concepts that are important in understanding the nature of the EPR spectra of the transition metal ions and, in particular, the interactions that influence the electron-spin energy levels.

Spin-Orbit Interaction: g-Values and Low Temperature

The magnetic dipole of an electron arises because the electron is a charged particle that generates a magnetic field by moving. The electron both spins about its axis and orbits the nucleus. The spin magnetic dipole and orbital magnetic dipole can interact. This dipolar interaction produces local magnetic fields at the electron and is anisotropic. The extent of the anisotropy depends on that of the unpaired electron distribution, which for some transitions metal complexes can be very large. This distribution mainly accounts for the observed variation in g-values (which are a measure of the local fields).

If the spin-orbit interaction is large, the EPR line can often be observed only at very low temperatures. Some caution is therefore needed in extrapolating from conclusions made under these conditions to physiological conditions—one should be aware that artifacts can arise from pH changes in freezing the samples.

Worked Example 7.11   Some EPR Signals from Transition Metals Can Be Observed at Room Temperature

$[Mn(H_2O)_6]^{2+}$ has five unpaired electrons ($S = \frac{5}{2}$) and gives an observable EPR spectrum at 300 K. $[Mn(CN)_6]^{4-}$ has one unpaired electron ($S = \frac{1}{2}$), but gives an observable EPR spectrum only at 20 K. Why?

Solution

    Spin-orbit interaction is important in the $[Mn(CN)_6]^{4-}$ complex, but not in $[Mn(H_2O)_6]^{2+}$. The interaction so broadens the signal in $[Mn(CN)_6]^{4-}$, that it is unobservable at normal temperature.

Hyperfine Structure

Many of the transition metal ions have a nuclear spin, which gives rise to hyperfine structure. Complexes with anisotropic g-values usually also have anisotropic hyperfine interactions.

If the ligands have nuclear spin, these spins can also interact with the unpaired electrons. The resultant interaction is manifest in the spectrum as a super-hyperfine (s.h.f.) splitting, which is usually considerably smaller than the nuclear hyperfine splitting and is not always resolved.

Worked Example 7.12   Cu(II) Nuclear Hyperfine Interaction

Cu(II) has one unpaired electron and a nuclear spin $I = \frac{1}{2}$. In superoxide dismutase, its environment is axially symmetric and at 9.25 GHz it has the EPR spectrum at $-150°C$ shown in Figure 7.22. Interpret this information.

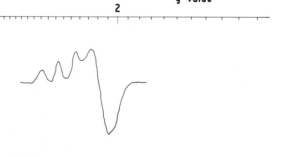

Figure 7.22

(continued)

Worked Example 7.12   (continued)

Solution

One would expect the EPR spectra for an axially symmetric Cu(II) to consist of two peaks, corresponding to $g_{\parallel}$ and $g_{\perp}$ (see Figure 7.5). Interaction of the electron with the nuclear spin will result in spitting each of these into $(2I + 1) = 4$ peaks. These are apparent for the $g_{\parallel}$ peak but not for the $g_{\perp}$ peak. because the hyperfine interaction is anisotropic. Thus, while $A_{\parallel}$ can be resolved (equal to about 0.015 T), $A_{\perp}$ is too small to be resolved.

Zero-Field Splitting

We have seen that the effect of a magnetic field on the spin system is to remove the degeneracy of the electron-spin energy levels. However, there may be various magnetic interactions in the molecule (such as the dipolar interaction between spins or spin-orbit coupling) that create local magnetic fields. The local fields can remove the degeneracy and cause a splitting of the electron-spin energy levels even in the absence of an applied magnetic field. These local fields are termed **zero fields,** and the resulting splitting is called **zero-field splitting** (ZFS).

The presence of ZFS can result in further fine structure in the EPR spectrum. Sometimes, though, the ZFS splitting of the energy levels may be so large compared with the microwave irradiation quanta that it may be impossible to observe any EPR transitions at all. This is often the case when the number of unpaired electrons is *even* (for instance, for high-spin Fe(II) complexes, where there are unpaired electrons and $S = 2$; see Figure 7.21). When the number of unpaired electrons is *odd,* the ZFS results in double-degenerate energy levels. The applied field then splits these levels in the usual way, and EPR transitions are allowed, subject to the expected selection rule of $\Delta M_s = \pm 1$. Thus, for Mn(II), the effect of ZFS should further split each of the six ^{55}Mn nuclear ($I = \frac{5}{2}$) hyperfine lines into five (see Figure 7.23).

The ZFS is anisotropic, so as with A and g (see section on anisotropy), a random orientation of the molecules in the magnetic field will result in a spread of resonances (inhomogeneous broadening). In such cases, the transitions may become too broad to observe (except possibly the one from $\frac{1}{2}$ to $-\frac{1}{2}$, which is almost independent of the ZFS). In some cases where the ZFS is small, such as in aqueous solutions of Mn(II), the tumbling of the molecule may be sufficiently fast to average out the ZFS. However, on binding to a macromolecule, the ZFS could increase and the tumbling rate might not be fast enough to average out the anisotropy of the resonances. The spectrum could then become so broad as to be undetectable. This is the basis for monitoring the binding of Mn(II) to proteins by measuring the decrease in peak height of the free Mn(II) on addition of protein.

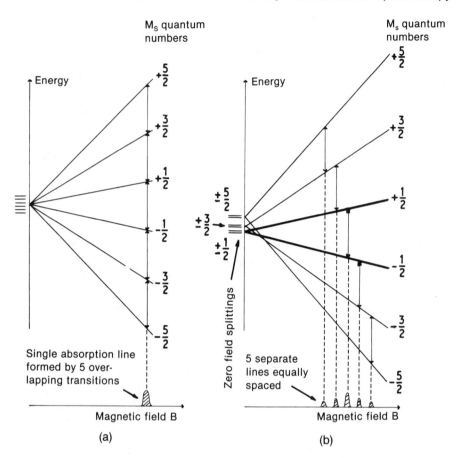

Figure 7.23   Energy levels for high-spin Mn(II) or Fe(III) (d^5) in a magnetic field. (The energy depends on M_s; see Chapters 2 and 6.) (a) In the absence of ZFS, all six levels are assumed to have the same energy in zero field. (b) The presence of ZFS results in three pairs of doubly degenerate levels. (Note that all transitions ($M_s = \pm 1$) except for the $\frac{1}{2}$ to $-\frac{1}{2}$ reflect the ZFS.)

An interesting case arises when the ZFS becomes very large (see Figure 7.24). Only the transition indicated will then be observed; the system behaves as if it had a spin of $\frac{1}{2}$ with only two magnetic levels present. This phenomenon is sometimes known as a "fictitious" spin $\frac{1}{2}$. It is, however, distinguished from a genuine spin $\frac{1}{2}$ case (for example, low-spin Fe(III) with one unpaired electron) by the very much larger g-value anisotropy associated with the high-spin Fe(III) case. For instance, high-spin Fe(III) metmyoglobin has $g_\perp = 2$ and $g_\parallel = 6$ (axial symmetry). In contrast, the low-spin azide complex of metmyoglobin has a lower symmetry and has g values at 2.8, 2.25, and 1.75 (see Worked Example 7.2). The differences can be even more dramatic if the symmetry of the high-spin form is low. In this event, the g-values can range from about 1.8 to 10.

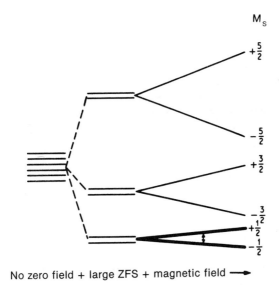

$$M_S$$

$+\frac{5}{2}$

$-\frac{5}{2}$

$+\frac{3}{2}$

$-\frac{3}{2}$

$+\frac{1}{2}$

$-\frac{1}{2}$

No zero field + large ZFS + magnetic field ⟶

Figure 7.24   Energy levels for high-spin Mn(II) and Fe(III) with a large ZFS. The arrow indicates the relative size of the microwave quantum, so the only possible transition is between $-\frac{1}{2}$ and $\frac{1}{2}$. (This is the "fictitious" spin $= \frac{1}{2}$ case).

Worked Example 7.13   Heme Environments in Cytochrome Oxidase

Cytochrome P450 is an oxidase containing a heme group. At 9.25 GHz and 4 K it has an EPR spectrum typical of low-spin ($S = \frac{1}{2}$) ferric heme in a low-symmetry (rhombic) environment. The addition of a substrate leads to a change in the spectrum (see Figure 7.25). Comment on the nature of the change.

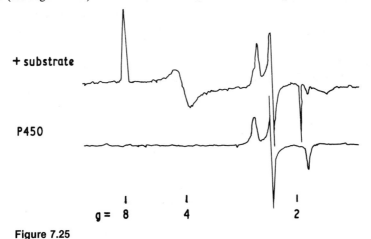

+ substrate

P450

$g =$ 8   4   2

Figure 7.25

(continued)

Worked Example 7.13   *(continued)*

Solution

       Substrate addition results in three new peaks in the spectrum at about $g = 8.0$, 4.0, and 1.8. Such a spectrum indicates a system with $S = \frac{1}{2}$. However, the very large anisotropy is more suggestive of a system with a fictitious $S = \frac{1}{2}$ caused by a large ZFS, and for which there is a very low symmetry. The addition of substrate ($S = \frac{1}{2}$) therefore must result in a change of the Fe(III) spin state from low to high spin ($S = \frac{5}{2}$).

OTHER APPLICATIONS

So far, we have discussed the biological applications of EPR only in terms of transition metals and spin labels (see Chapter 6). Other applications include studies of molecules in their triplet (phosphorescent) state (see Chapter 13) and work with short-lived free radicals and flavins. One important recent extension of this work is spin trapping.

Spin Trapping

With existing intrumentation, a practical limit of detection of free radicals is about $10^{-8}M$. To resolve hyperfine splitting constants, however, concentrations of about $10^{-6}M$ are often required. Many free radicals of biological interest never reach a concentration high enough to be detected by EPR. For instance, the hydroxyl radical reacts with various species at rates controlled only by diffusion. Some short-lived radicals of lesser reactivity can sometimes be detected by continuous-flow methods, but large quantities of starting materials may then be required.

       In the method of **spin trapping,** a compound such as a nitrone (see figure) forms a stable free radical by reacting covalently with an unstable free radical. The unstable free radical is "trapped" in a form that often has a relatively long lifetime. It can then be observed at normal temperatures and pH and in aqueous solution. For instance, in the case of the nitrone 5,5-dimethylpyrroline-1-oxide (DMPO), the trapping reaction can be written as follows:

DIAMAGNETIC   PARAMAGNETIC

Table 7.1 Splitting constants of DMPO—R adducts

Radical	a_N (mT)	$a_{H\beta}$ (mT)	$a_{H\gamma}$ (mT)
OH$^\bullet$	1.49	1.47	—
O$_2^{\bullet-}$/OOH$^\bullet$	1.43	1.17	1.25
Ph$^\bullet$	1.59	2.48	
Ph$^\bullet$SO$_3$H	1.58	2.44	
Vitamin K$_1$ semiquinone	1.45	1.45	

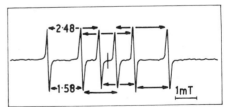

The hyperfine splittings of the adducts, notably those involving the β-proton, depend on the nature of the radical R$^\bullet$ (see Table 7.1). In general, if R$^\bullet$ has the unpaired electron localized mainly on a carbon atom, the EPR spectrum of the adduct consists of six lines from a triplet splitting (from the N, $I = 1$) of a doublet (from the β-proton). The splitting by the β-proton is generally greater than that by the nitrogen atom, except when R$^\bullet$ is an oxyradical (RO$^\bullet$), in which case an additional splitting from one of the γ-protons can sometimes be resolved.

Although this method allows the detection of free radicals, the identification of a particular radical depends on knowing the expected hyperfine splitting pattern for that radical. Very often this may involve quite complicated chemistry in order to synthesize the suspected free radical and obtain its EPR fingerprint. Several different spin traps may be necessary for an unambiguous assignment of the free radical.

Worked Example 7.14   Phenylhydrazine-Induced Hemolysis Involves Free Radical Intermediates

The hemolytic action of hydrazine derivatives was originally thought to involve the radicals O$_2^-$ and OOH$^\bullet$ (depending on pH) and OH$^\bullet$. When a 1-*mM* phenylhydrazine solution was used in a 1% red cell suspension containing 100-*mM* DMPO as a spin trap, at pH 7 and room temperature, an adduct was formed that had the EPR spectrum (at 9.256 GHz) with

Figure 7.26

the two sets of splittings marked as shown in Figure 7.26. Does this support the original hypothesis?

(continued)

Worked Example 7.14   (continued)

Solution

The EPR spectrum can be analyzed in terms of the six lines that arise from the triplet splitting ($a_N = 1.58$ mT) of a doublet ($a_H = 2.44$ mT). This is the characteristic pattern for radicals with the unpaired electron localized mainly on carbon atoms and not for oxyradicals, so the original hypothesis is not supported. In fact, comparison of the hyperfine constants with those in Table 7.1 shows them to be characteristic of the Ph$^\bullet$ radical adduct (for further details, see H. A. O. Hill, and P. J. Thornalley, *FEBS Lett.*, 125(1981):235–238).

PROBLEMS

1. Use Table 7.1 to interpret the splitting pattern in Figure 7.27 of the nitrone spin trap

$$\begin{array}{c} CH_3 \\ CH_3 \end{array}\!\!\diagdown\!\!\underset{\underset{O}{|}}{\overset{}{N}}\!\!\diagup R'$$

when R$'$ = Ph$^\bullet$ in (I), OH$^\bullet$ in (II), and OOH$^\bullet$ in (III).

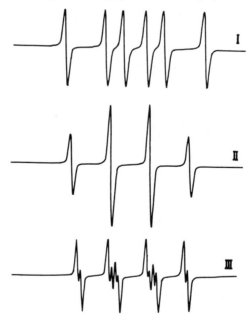

Figure 7.27

2. Predict the EPR spectrum of the benzene negative ion:

3. The EPR spectra at 9.25 GHz of three spin-labeled proteins are shown in Figure 7.28. How would you describe these spectra in terms of immobilization of the label? All the proteins are uniquely labeled on one SH group, yet the spectrum of labeled phosphofructokinase seems to show two sets of spectra. Can you suggest a reason?

4. Some representative 9.25-GHz EPR spectra of transition metal ions that have been discussed in this chapter and that are commonly encountered in biological systems are shown in Figure 7.29. Can you deduce the symmetry of the metal-ion site in each case and identify any hyperfine coupling? In rubredoxin, the Fe(III) atom is surrounded by four S atoms. Can you identify the spin state of the Fe(III)?

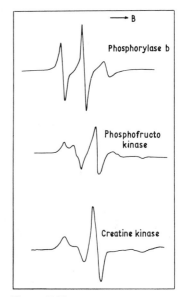

Figure 7.28

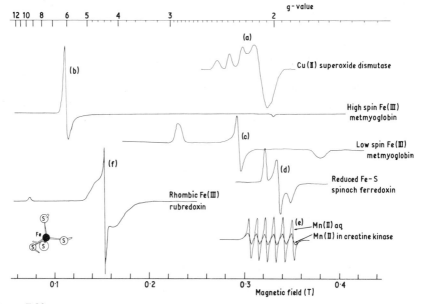

Figure 7.29

5. The structures of two basic types of 2- and 4-Fe sulfur centers are shown in Figure 7.30(a). The 8-Fe centers consist of two 4-Fe S clusters. Some representative EPR spectra of Fe–S proteins are shown in Figure 7.30(b). Comment on the feasibility of using EPR as a diagnostic technique for the different types of protein centers. Why do these spectra have to be obtained at low temperatures?

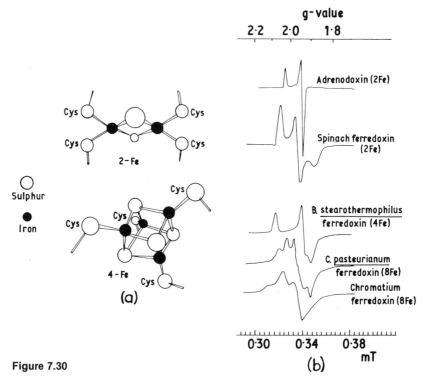

(a)

Figure 7.30

(b)

6. Putidaredoxin is involved in electron transport. It undergoes one electron reduction, and an EPR signal is observed only in the reduced state. The molecule has two 2-Fe labile S centers (see Figure 7.30(a)). When (a) ^{57}Fe ($I = \frac{1}{2}$) is substituted for ^{56}Fe ($I = 0$) or (b) ^{32}S ($I = 0$) is replaced by ^{33}S ($I = \frac{3}{2}$) in the native protein, the EPR spectrum changes. (See Figure 7.31(a) & (b)). Interpret these experiments.

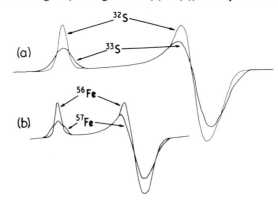

Figure 7.31

7. A series of phosphatidylcholine spin labels (PCSL) was introduced into a PC bilayer. The label could be attached to various carbon atoms down the lipid acyl chain. The values of $A_{\parallel}$ and $A_{\perp}$ listed below were obtained from the spectra. Calculate the polarity profile, a_0, of the bilayer and comment on any changes that are observed. (Assume that there is fast axial motion.)

	4PCSL	6PCSL	8PCSL	10PCSL	12PCSL	14PCSL
$A_{\parallel}$ (mT):	2.44	2.32	2.14	1.99	1.71	1.62
$A_{\perp}$ (mT):	1.09	1.14	1.13	1.19	1.29	1.33

8. The EPR spectra of a series of Dnp-spin-labeled haptens when bound to two different Dnp-binding antibodies gave the following values of A_{zz} (millitesla):

Hapten		Antibody	
		XRPC	MOPC 560
(I)	Dnp—NH— ($\leftarrow$0.8 nm$\rightarrow$) (N—O)	3.18	2.45
(II)	Dnp—NH—N= ($\leftarrow$0.93 nm$\rightarrow$) (N—O)	2.93	2.39
(III)	Dnp—NH—CH$_2$—CO—O— ($\leftarrow$1.2 nm$\rightarrow$) (N—O)	1.59	1.60
(IV)	Dnp—NH(CH$_2$)$_3$—CO—O— ($\leftarrow$1.47 nm$\rightarrow$) (N—O)	1.59	1.60

Estimate the depths of the site using the given distances between the nitroxide label and the Dnp group. (Assume $A_{zz} = 3.5$ mT and $A_{xx} = 0.7$ mT for each label.)

9. Mark out the expected splittings of the resonances for a rigid nitroxide label with the following parameters (all in milliTesla): $a_{\parallel} = 3$; $a_{\perp} = 0.6$; $h\nu/\beta(g_{\perp} - g_{\parallel}) = 1$. In the limit of fast motion, a three-line spectrum is observed. Use the splitting pattern to explain why the three lines may be broadened differentially as the rate becomes slower. (*Note:* 1 mT $\simeq$ 1.76 rad$\cdot$s^{-1}.)

CHAPTER 8

Scattering

OVERVIEW

1. Scattering is a result of oscillations induced in a particle by applied radiation. The size of the oscillations induced depends on the properties of the molecule (its polarizability) and the wavelength of the incident radiation.

2. The angular dependence of scattering can give information on the size and shape of a particle, provided that the dimensions of the scatterer (usually defined by the radius of gyration, R_G) are comparable to or greater than the wavelength of the incident radiation.

3. X-rays and neutrons are very useful because of their short wavelength ($\lambda \sim 0.1$ nm). X-rays are scattered from atoms by the electrons; the scattering increases linearly with atomic number. Neutrons are scattered by the nuclei. Since the scattering from hydrogen and deuterium is very different, the scattering from solvent water can be matched to different components of a heterogeneous biological sample, for example, to protein or to DNA.

4. Transmitted light is affected by scattering in two main ways: (a) It is reduced (turbidity) and (b) its apparent velocity is changed (refraction).

5. Motion of the scatter can be detected by analysis of the scattering. This can be done by measuring either the frequency dependence of the scattering or the fluctuations in intensity.

6. In biological applications, scattering techniques have been used primarily to gain information about the shapes and movements of macromolecules and cells in solution.

INTRODUCTION

In general, when electromagnetic radiation interacts with a particle, the radiation is scattered. Detailed analysis of this scattering can give information about the shape and motion of the particle in solution. In addition, scattering causes the phenomena of refraction, diffraction (see Chapter 12) and turbidity.

217

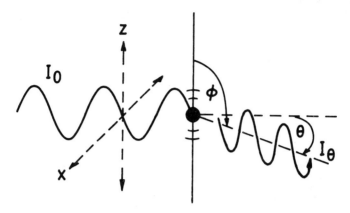

Figure 8.1 An incident wave of intensity I_0 and polarized along the z-direction causes a particle to oscillate and emit radiation in all directions. Note the definition of the angles ϕ and θ.

THE OBSERVED SCATTERING FROM AN ISOLATED PARTICLE THAT IS SMALL COMPARED WITH THE WAVELENGTH

As shown schematically in Figure 8.1, the applied radiation of intensity I_0 and frequency ν causes the particle to oscillate. These oscillations produce secondary radiation in all directions.

Consider the case of an electron in an atom. The applied radiation, amplitude $E_0\cos(2\pi\nu t)$, will induce a displacement of the electron with respect to the nucleus. (The nucleus is essentially unaffected by most electromagnetic radiation because its mass is much greater than that of the electron.) The magnitude of this displacement will depend on how easy it is to displace the electron from the nucleus. A parameter used to define this is the **molecular polarizability** (α) of the system; it is a function of the shape of the electron cloud and the frequency of the applied radiation. The electron displacement that oscillates with the applied radiation produces a dipole moment proportional to $\alpha E_0\cos(2\pi\nu t)$. This oscillating dipole transmits electromagnetic radiation like an aerial. The amplitude (E_ϕ) of the transmitted wave at a distance r and angle ϕ to the direction of the dipole (see Figure 8.1) is given by the equation

$$E_\phi = \frac{1}{c^2}\frac{d}{dt^2}(\alpha E_0\cos 2\pi\nu t)\sin \phi r$$

where c is the velocity of light. This equation is a result of electromagnetic wave theory derived by Maxwell in the nineteenth century. On differentiation, this equation becomes

$$E_\phi = \frac{\alpha E_0\, 4\pi^2\nu^2}{c^2}\cos 2\pi\nu t(\sin \phi r)$$

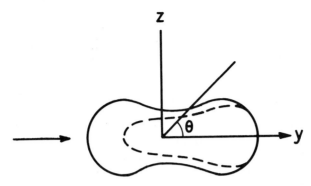

Figure 8.2   An illustration of the expected distribution of intensity of scattering in the zy-plane for unpolarized radiation incident along the y-direction (a) for a particle that is much smaller than the wavelength (solid line) and (b) for a particle that is larger (dotted line).

In fact, the observed intensity is the square of this quantity. It is common to use unpolarized incident radiation and the angle θ rather than ϕ (see Figure 8.1). In this case, the $\sin^2\phi$ term becomes $(1 + \cos^2\theta)/2$ (see problem 8.1) and the scattering is given by (see also Figure 8.2)

$$I_\theta = I_0 \frac{8\pi^4\alpha^2}{r^2\lambda^4} (1 + \cos^2\theta) \qquad (8.1)$$

since $I_0 = E_0^2\cos^2(2\pi\nu t)$ and $\nu/c = 1/\lambda$.

Another useful expression of this equation is

$$R_\theta = \frac{I_\theta}{I_0} \frac{1 + \cos^2\theta}{r^2} = \frac{8\pi^4\alpha^2}{\lambda^4} \qquad (8.2)$$

The preceding is known as the Rayleigh ratio, after Lord Rayleigh (1842–1919). Note that in this derivation, which applies to an isolated particle that is small compared with λ, R_θ is independent of r and θ.

MOLECULAR POLARIZABILITY: THE ANALOGY WITH A WEIGHT ON A SPRING

In the preceding derivation, α was assumed to be frequency independent, which is not, in general, true. Some insight into the properties of α can be deduced because the situation of an electron oscillating with the applied electromagnetic radiation is analogous to the driven oscillations of a weight on a spring. It can be shown from the equations in Appendix II that the displacement of a mass m by a driving force $F_0\cos 2\pi\nu t$ is

$$x = \frac{F_0\cos 2\pi\nu t}{4\pi^2 m (\nu_0^2 - \nu^2)}$$

where ν_0 is the resonance, or natural, frequency of the system. This displacement is proportional to the dipole moment of the oscillation, since it is a measure of the induced charge separation between electron and nucleus. The amplitude of the wave produced by this oscillating dipole is proportional to

$$\frac{d^2x}{dt^2} = \frac{F_0\cos 2\pi\nu t \, (2\pi\nu)^2}{4\pi^2m \, (\nu_0^2 - \nu^2)} \tag{8.3}$$

Comparing this with the equation for E_ϕ it is clear that the frequency dependence of α is given by $1/(\nu_0^2 - \nu^2)$. Thus, α becomes very large near resonance. (This equation was derived for the case where there is no damping. There would normally be a damping term in the denominator.)

THE WAVELENGTH DEPENDENCE OF SCATTERING

From equation 8.1 and the frequency dependence of α, we can predict that scattering will have a wavelength dependence of the form $\lambda_0^4/(\lambda^2 - \lambda_0^2)^2$. If $\lambda \gg \lambda_0$, then the wavelength dependence is $1/\lambda^4$. This is the situation commonly encountered with light scattering and was predicted by Lord Rayleigh. The wavelength dependence of scattering is also the explanation for the blue color of the sky, since blue light is scattered more strongly than red. If $\lambda \ll \lambda_0$, then there is no predicted wavelength dependence. This situation is known as **Thomson scattering** and is encountered with X-rays, where the wavelength is very short.

In addition to the two limits just described, other situations are possible. For example, in X-ray diffraction work, **anomalous scattering** is sometimes observed and put to good use. In this situation, the scattering is wavelength dependent because the incident radiation has a wavelength near a resonance frequency of the scatterer.

Another interesting prediction from the weight-on-the-spring analogy is that there is a phase shift of 180° between the two frequency limits $\lambda \gg \lambda_0$ and $\lambda \ll \lambda_0$ (see Appendix II). This phenomenon is also indicated by the change in sign of the scattering amplitude between the two extremes.

THE SCATTERING FROM MANY PARTICLES WHOSE DIMENSIONS ARE SMALL COMPARED WITH λ

Scattering from a Rigid Array

In an infinite array of small molecules, it is possible to choose pairs of particles, at any angle other than $\theta = 0$, such that the scattered light is canceled out because the phase difference between the two members of the pair is 180° (see Chapter 2). The net result is that there is no observed scattering from such an array.

When $\theta = 0$, the phase difference between scattering from different particles is zero. The effect of the superposition of unscattered waves and the forward-scattered waves is to change the apparent velocity of the transmitted wave. This is the phenomenon known as **refraction** (discussed later). Note that the extent of scattering and thus refraction will depend on α and λ.

Solution Scattering and Information on Molecular Weight

In a solution, motion of the particles causes the concentrations to vary, or fluctuate, as a function of time. Canceling pairs of scattering centers are no longer always found and scattering is observed, in general. The fluctuations in a solution are caused by the Brownian, or thermal, motion of the particles.

We have stated that n is related to α. In fact, the relationship is

$$\alpha = \frac{n_s^2 - n_0^2}{4\pi N}$$

where n_0 is the refractive index of the pure solvent, n_s is the refractive index of the solution, and N is the number of particles per unit volume. $N = c\mathfrak{N}/M$ where c is the concentration (grams per volume), $\mathfrak{N}$ is Avogadro's number (particles per mole), and M is the molecular weight of the solute (grams per mole).

For a dilute solution, $n_s \simeq n_0$, so the term $(n_s^2 - n_0^2) = (n_s + n_0)(n_s - n_0)$ becomes approximately $2n_0\Delta n$, where $\Delta n = n_s - n_0$. Thus,

$$\alpha = \frac{2n_0\Delta nM}{4\pi c\mathfrak{N}} \simeq \frac{n_s M}{2\pi \mathfrak{N}} \cdot \frac{dn}{dc}$$

Here dn/dc is the change in refractive index with change in solute concentration, which can be measured separately and is a constant for the solution. Thus $\alpha =$ constant $\times M/\mathfrak{N}$ and $R_0 =$ constant $\times M^2/\mathfrak{N}^2$. To obtain the total solution scattering, we must multiply the latter by $N = c\mathfrak{N}/M$. Thus

$$R_\theta = KcM \tag{8.4}$$

where K is a constant $[(2\pi^2 n_0^2 (dn/dc)^2/\mathfrak{N}\lambda^4]$. Equation 8.4 shows that the observed scattering from a solution is related to the molecular weight of the solute.

If the solution is not homogeneous (for example, if it contains macromolecules of different molecular weight), then the observed R_θ will be the sum of the scatterers, that is,

$$R_\theta = \sum_i R_{\theta i} = \sum_i Kc_i M_i$$

Since

$$c = \sum_i c_i \quad \text{and} \quad c_i = \frac{N_i M_i}{\mathfrak{N}}$$

then,

$$R_\theta = \frac{Kc \sum_i c_i M_i}{\sum_i c_i}$$

$$= \frac{Kc \sum_i N_i M_i^2}{\sum_i N_i M_i}$$

$$= Kc\overline{M}_w \tag{8.5}$$

where $\overline{M}_w$ is defined as the weight-average molecular weight. Note that high molecular weight components (for example, dust particles) in the solution have a relatively strong effect.

SCATTERING FROM LARGER PARTICLES

So far, we have considered the scattering from "small" particles, by which we mean small compared with the wavelength (see further for definition of particle size in terms of R_G). If λ is of the same order as the particle size, then the scattered radiation produced from different points on the same particle can interfere. Consider the observed effect at P_1 and P_2 when elements A and B of an object scatter an incident beam, as shown in Figure 8.3. The bigger the phase difference between the beams, the stronger the effects of interference will be. It is clear from the diagram that while $S_1AP_1 \simeq S_2BP_1$, $S_1AP_2 < S_2BP_2$. Thus, finite particles scatter less at large angles than small ones (see also Figure 8.2).

The angular dependence of scattering predicted by equation 8.1 is thus incorrect for other than very small angles. It is convenient to define a parameter to describe the angular dependence of the scattering from a large particle. The function used is $P(\theta)$, which is the ratio of the scattering from a finite particle at an angle θ to the scattering from an infinitely small particle of the same mass. $P(\theta) \to 1$ as $\theta \to 0$. Moreover, $P(\theta)$ can be calculated for various molecular shapes and the calculated curves can be compared with observed scattering curves (discussed later). At small θ, a convenient but very approximate form of $P(\theta)$ for scattering from a random array of scatterers in solution is

$$P(\theta) \simeq \exp\left(\frac{-Q^2 R_G^2}{3}\right) \tag{8.6}$$

where $Q = (4\pi/\lambda)\sin(\theta/2)$ and R_G is the **radius of gyration** of the particle. (*note:* Q is often called h in some texts, and θ is often defined as the angle 2θ.)

Figure 8.3   Illustration of the effect of scattering from a particle that is large compared with the wavelength of the incident radiation. S_1 and S_2 represent the source.

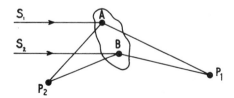

The term R_G is used as a measure of the dimensions of the particle. For a polymer with i segments of mass m_i a distance r_i from the molecular center of mass, it is defined by

$$R_G^2 = \frac{\sum\limits_i m_i r^2}{\sum\limits_i m_i}$$

This equation can be evaluated for various shapes. For example, for a sphere of radius r, $R_G = \sqrt{\frac{3}{5}}\, r$; for a long rod of length L, $R_G = L/\sqrt{12}$.

A VECTORIAL DESCRIPTION OF SCATTERING

It is often convenient to describe scattering in terms of vectors. Consider the situation where an electromagnetic wave is incident on a particle and there is a scattered wave at an angle θ to the incident beam (see Figure 8.4). The incident wave is usually defined as the vector $\mathbf{k}_0$ with amplitude $2\pi/\lambda$ (the units are merely a convenient form).

If the scattering is elastic (energy conserving) the amplitudes of the scattered vectors are equal, that is, $|\mathbf{k}'| = |-\mathbf{k}_0|$. We can define a vector $\mathbf{Q}$ such that $\mathbf{Q} = \mathbf{k}_0 - \mathbf{k}'$ which, from simple geometry, has amplitude $(4\pi/\lambda)\sin(\theta/2)$.

Consider now a beam incident on an object of length r. There will be a path difference between radiation scattered at an angle θ from the two ends of this object, which is $p + q$, as shown in Figure 8.5. If β is the angle between $\mathbf{k}_0$ and $\mathbf{r}$, the angle between $\mathbf{k}'$ and $\mathbf{r}$ is $180° - (\theta + \beta)$. We also have

$$p = r \cos \beta \quad \text{and} \quad q = r \cos[180° - (\theta + \beta)]$$

A property of vectors is that $\mathbf{a} \cdot \mathbf{b} = |\mathbf{a}||\mathbf{b}| \cos \phi$, where ϕ is the angle between $\mathbf{a}$ and $\mathbf{b}$, thus

$$\mathbf{p} = \mathbf{k}_0 \mathbf{r} \left(\frac{\lambda}{2\pi} \right) \quad \text{and} \quad \mathbf{q} = -\mathbf{k}' \mathbf{r} \left(\frac{\lambda}{2\pi} \right)$$

So

$$\mathbf{p} + \mathbf{q} = \left(\frac{\lambda}{2\pi} \right) \mathbf{r} \, (\mathbf{k}_0 - \mathbf{k}')$$

Figure 8.4   A vectorial description of scattering.

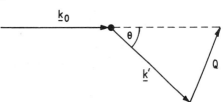

Figure 8.5   A wave incident on an object of length r. The path difference between the scattered waves, at angle θ, from the two ends of the object is $p + q$.

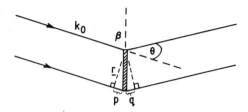

A path difference can be converted into a phase difference by multiplying by $2\pi/\lambda$, since unit wavelength corresponds to a phase shift of 2π. The phase difference induced by scattering from the object represented by $\mathbf{r}$ is thus simply the product $\mathbf{r} \cdot \mathbf{Q}$. This phase difference is one of the most important factors determining the properties of the scattered wave.

PLOTS TO DETERMINE R_G and $\overline{M}_w$

The Guinier Plot

The approximate relationship given by equation 8.6 immediately suggests that a plot of $\ln(I_\theta)$ versus Q^2 would give a straight line with slope $R_G^2/3$). This is known as a **Guinier plot**.

Worked Example 8.1 $\mathbf{R_G}$ *of C1*

C1, macromolecular assembly of proteins, is an important component of the mammalian immune defense system. A Guinier plot for activated C1, from a neutron scattering experiment in 100% D_2O, is shown in Figure 8.6. The value of $I(Q)/C$ is 2.9 for $Q = O$ nm^{-1} and 0.6 for $Q = 0.17$ nm^{-1}. If $I(Q)/C$ for hemoglobin is 0.22 under the same experimental conditions, calculate the molecular weight and R_G for C1 if the molecular weight of hemoglobin is 62 k daltons.

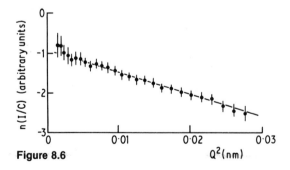

Figure 8.6

Solution

The molecular weight is $62,000 \times 2.9/0.22 \simeq 820,000$ daltons. The slope of the graph is $(\ln(2.9) - \ln(0.6))/(0.17)^2 = 54.5 = R_G^2/3$. Thus, $R_G = 12.8$ nm.

Figure 8.7 A Zimm plot of a DNA sample. The slope at zero angle gives B and the slope at zero concentration gives R_G. The results of this experiment gave $M_w = 3.7 \times 10^6$ daltons, $R_G = 213$ nm, and $B = 5.3 \times 10^4$ mL $\cdot$ mol $\cdot$ g^2 (from D. Jolly and H. Eisenberg, *Biopolymers*, 15(1976):61).

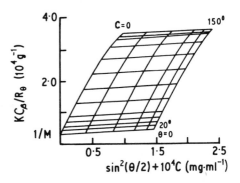

The Zimm Plot

It was shown in equation 8.5 that $R_\theta = KC\,\overline{M}_w$ for cases where $R_G \ll \lambda$. This condition can be relaxed by writing $R_\theta = KCP(\theta)\overline{M}_w$. We have also assumed that the solution is "ideal", that is, that there are no interactions between different particles. This problem can be overcome in the same way as for other solution properties, such as osmotic pressure, by using a second virial coefficient B. Then we can write

$$\frac{KC}{R_\theta} = \frac{1}{P(\theta)}\left(\frac{1}{\overline{M}_w} + 2BC\right)$$

This equation will yield the correct value for $\overline{M}_w$ when $P(\theta) \to 1$ (when $\theta \to 0$) and when $C \to 0$.

These two limiting conditions cannot be achieved experimentally, of course, because the transmitted beam interferes with the measurement at $\theta = 0$ and because there is no scattering when $C = 0$. However, Zimm showed that a double extrapolation can be used in a plot of KC/R_θ versus $\sin^2(\theta/2)$. The intercept on the vertical axis is $1/\overline{M}_w$ and the slope of the plot at $\theta = 0$ gives the second virial coefficient B. The slope at zero concentration yields R_G from an expansion of equation 8.6, as follows:

$$P(\theta) = 1 - \frac{Q^2 R_G^2}{3} + \cdots$$

If the Q^2 term is small, then $1/P(\theta) \simeq (1 + Q^2 R_G^2/3)$. Thus, the slope of the extrapolation line to zero concentration is $R_G^2/3$. Note that in a Zimm plot (see Figure 8.7), some arbitrary factor (e.g., $10^4 C$) is often added to the horizontal axis for convenience.

THE DETERMINATION OF MOLECULAR SHAPE

We have indicated how R_G and $\overline{M}_w$ can be determined from scattering measurements at small angles using Guinier or Zimm plots. A detailed study of the angular dependence of scattering at larger angles can give more information, especially if short-wavelength radiation, such as X-rays or neutrons, is used.

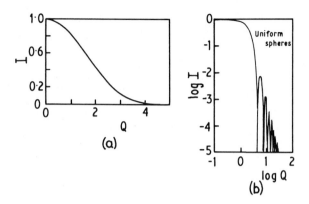

Figure 8.8   The calculated variation in scattering density with Q for uniform spheres (a) in a direct I versus Q plot and (b) in a log I versus log Q plot (adapted from O. Kratky, *Prog. Biophys. Mol. Biol.* 13 (1963):105.

The term $P(\theta)$ can be calculated for different shapes of particles that are randomly oriented in solution. For a sphere of radius r, $P(\theta)$ is of the form

$$P(\theta) = \left(\frac{3\,(\sin Qr - Qr \cos Qr)}{(Qr)^3} \right)^2$$

The form of this function leads to a predicted angular dependence of the scattered intensity, as shown in Figure 8.8.

Example 1   The Shape of t-RNA in Solution.   Small-angle X-ray scattering measurements on yeast phenylalanine t-RNA gave $R_G = 2.44$ nm. More detailed comparisons of the angular dependence of the scattering with theoretical models showed that the molecule must be elongated with two different cross sections. In a trial-and-error procedure, the model shown in Figure 8.9 was found to give the

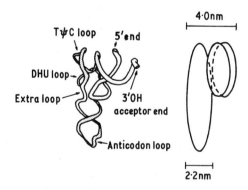

Figure 8.9   The model for phenylalanine t-RNA from X-ray crystallography and from small-angle X-ray solution scattering (from O. Kratky and I. Pilz, *Quart. Rev. Biophys.*, 11(1978):39).

best fit to the experimental data. This model, which consists of three ellipsoids, is shown on the same scale as the model predicted from X-ray diffraction studies.

SCATTERING USING X-RAYS AND NEUTRONS

X-rays and neutrons have proved very valuable in experiments to determine the shapes of molecules in solution. The wavelength of commonly used X-rays is around 0.1 nm, while neutrons are in the range 0.1–1 nm. When discussing light, we use the molecular polarizability to describe the relative degree of scattering produced by different species. In dealing with neutrons, we use instead a factor b with dimensions of length. The greater the b, the greater the scattering. For X-rays, an atomic scattering factor is generally used. The scattering of the elements increases linearly with atomic number in a systematic way for X-rays (see Figure 8.10). However, for neutrons the variation with atomic number is unsystematic. Moreover, for some atoms, such as hydrogen, the scattering length is negative. This variability in the neutron b-values results from resonance effects, and the negative values correspond to a situation where $\lambda_0 < \lambda$. In such a case, the scattered wave is 180° out of phase with respect to cases where $\lambda_0 > \lambda$, when b is positive. (Remember that neutrons are scattered by the nuclei of atoms, whereas X-rays are scattered by the electrons.)

In neutron scattering from molecules and other objects, the b-factors of the constituent atoms can be averaged to yield an **average scattering length.** Alternatively, a **scattering-length density** can be defined, where the b-values are summed over a volume element; thus, density is usually expressed in units of $cm \cdot cm^{-3} = cm^{-2}$.

Nuclei with the property of spin scatter neutrons differently for different spin orientations. This kind of scattering is termed **incoherent** and will be ignored in this book.

Figure 8.10   The visibilities of some atoms for X-rays and neutrons. The crosshatched atoms have negative amplitudes in neutron scattering (from G. E. Bacon, *Endeavour*, 25(1969):129).

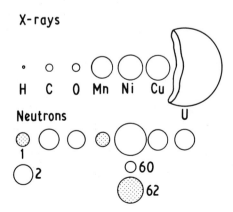

Figure 8.11   The variation in scattering
density of water and various solutes as a
function of D_2O/H_2O ratio (adapted from H.
Stuhrmann and A. Millar, *J. Appl. Cryst.*,
11(1978):325).

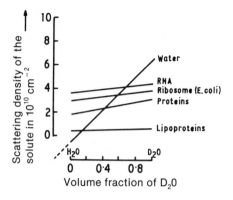

CONTRAST VARIATION IN NEUTRON SCATTERING

The main advantage of neutrons over X-rays as an irradiation source in biological
samples is the large but different scattering between hydrogen and deuterium.
The scattering-length density (ρ) from solvent water is -0.0562 cm^{-2} for H_2O
and 0.6404 cm^{-2} for D_2O. It is a simple matter to vary the ratio of H_2O to D_2O in
a sample and thus to vary the scattering density of the solvent over a wide range
(see Figure 8.11). The scattering density of macromolecules also changes with
percentage D_2O because some of the hydrogens on $-OH$ and $-NH$ groups
exchange with solvent. However, this effect is small compared to the solvent
effect. Note also in Figure 8.11 that macromolecules of different composition
(RNA, protein and so forth) behave differently.

The point when $\rho_{sol} = \rho_{particle}$ (where the lines intersect in Figure 8.11) is
known as ρ_{cm}, the **matchpoint, or contrast matching scattering density.** At the
matchpoint, the particle becomes indistinguishable from the solvent, on average.
When a particle comprises different materials—for example, a ribosome
(RNA + protein), chromatin (DNA + protein), or membranes (lipid, protein,
carbohydrate)—each component has a different matchpoint and the various parts
can be studied selectively by adjusting ρ_{sol}.

It is useful to consider the special case of an object made from material with
two scattering densities ρ_1 and ρ_2 with a mean scattering density of $\bar{\rho}$. If the difference in solvent contrast $\Delta\bar{\rho}$, defined as $\bar{\rho} - \rho_{sol}$, is varied, then the effective scattering density will also vary (see Figure 8.12). It is clear that the radius of gyration
R_G determined from a Guinier plot will vary with $\Delta\bar{\rho}$, since the apparent dimensions of the sample change (compare the visible shapes in Figure 8.12 when
$\rho_{sol} = \rho_1$ and when $\rho_{sol} = \rho_2$.

H. Stuhrmann showed that it is possible to extract structural information
from contrast variation experiments. Variation in R_G as a function of $\Delta\bar{\rho}$ is given
by

$$R_G^2 = R_c^2 + \frac{\alpha}{\Delta\bar{\rho}} - \frac{\beta}{\Delta\bar{\rho}^2}$$

Figure 8.12 Illustration of the nomenclature used in neutron scattering. A scatterer is composed of two components of density ρ_1 and ρ_2. ρ_{sol} is the solvent scattering density, which can be varied by changing the D_2O/H_2O ratio. $\bar{\rho}$ is the mean density of the solute and $\Delta\bar{\rho} = \bar{\rho} - \rho_{sol}$.

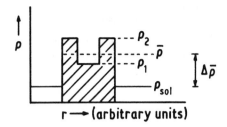

where R_c is the radius of gyration measured at infinite contrast. A plot of R_G^2 versus $1/\Delta\bar{\rho}$ yields estimates of α (the slope) and β (the curvature). If $\beta = 0$, then the centers of mass of the two components (that is, ρ_1 and ρ_2) coincide. The slope α is the radial variation in scattering density within the particle. It is a spatial measure of the difference in scattering density of the two components within the particle.

Worked Example 8.2   Neutron Scattering from the Ribosome

The ribosome from *E. coli* consists of 55 proteins and three molecules of RNA. The ribosome has a sedimentation coefficient of 70S and it dissociates into two subunits with sedimentation coefficients 50S and 30S. These three components are labeled according to their S-values. The following values of R_G were obtained for these particles. (For proteins, $\rho_{cm} \simeq 42\%$ D_2O; for RNA, $\rho_{cm} \simeq 68\%$ D_2O; for ribosomes, $\rho_{cm} \simeq 60\%$.)

	70S	50S	30S	
$R_G = R_c$ (infinite contrast)	8.1	7.0	6.8	(nm)
R_G at 68% D_2O	9.8	9.0	7.8	
R_G at 42% D_2O	7.0	5.9	6.1	

What can you conclude from these data?

Solution

The three contrast points have $\Delta\bar{\rho} = \rho_{cm} - \rho_{sol} = \infty$, -8% and 18% D_2O. It is then simple to construct a plot of R_G^2 versus $1/\Delta\rho$, as shown in Figure 8.13. We can conclude that $\beta \simeq 0$; thus, the centers of mass of the protein and RNA are about equal (note that β is never very well determined and is prone to large errors.) The value of α implies that the component with the high scattering density (RNA) has a lower radius of gyration than the component with the lower scattering density. In other words, the particle has an RNA-rich core surrounded by a protein-rich shell. This can also be noted directly from the table, since in 42% D_2O the scattering is

(continued)

Worked Example 8.2   (continued)

effectively only from RNA because the protein is matched out by the solvent, while in 68% D_2O it is dominated by the protein because the RNA is matched out by the solvent.

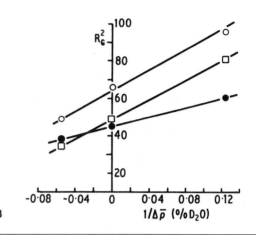

Figure 8.13

Example: Selective Deuteration of Ribosomal Proteins Leads to a Simple Scattering Pattern.   The 30S subunit of the *E. Coli* ribosome contains 21 proteins of molecular weights between 10,000 and 65,000 daltons, plus one RNA molecule of molecular weight 560,000. In a series of experiments, Engelman, Moore, and coworkers (see, for example, *Scientific American,* October 1976) assembled the 30S subunit from its constituent parts. The only change made was that 2 of the 21 proteins had been prepared from bacteria grown in D_2O. The scattering density of the solvent was adjusted to equal that of the small ribosome in its native state. Hence, scattering was observed only from the two deuterated proteins. The scattering from a pair of spherical (or ellipsoidal) particles gives a relatively simple $P(\theta)$ function that depends on the separation between them. This experiment allowed a series of distances to be calculated for different protein pairs in the ribosome.

THE EFFECT OF SCATTERING
ON TRANSMITTED LIGHT

Scattering produces two effects on the transmitted beam, namely, (1) a loss of transmitted intensity due to scattering of the radiation to angles other than $\theta = 0$ (**turbidity**) and (2) an apparent change in velocity of the transmitted beam (**refraction**).

Figure 8.14 Turbidity: If a solution scatters light the transmitted intensity is reduced.

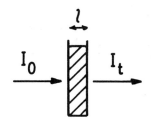

Turbidity

Consider an incident beam of intensity I_0 after its passage through a cuvette of thickness 1, as shown in Figure 8.14. It is possible to define a turbidity parameter τ exactly analogous to the extinction coefficient ϵ (see Chapter 4) that describes the reduction in intensity, namely,

$$I_t = I_0 \exp(-\tau c l)$$

It is possible to show that τ is related to the observed scattering by the relationship

$$\tau = \frac{16\pi}{3} R_{90}$$

where R_{90} is the Rayleigh ratio when $\theta = 90°$.

Worked Example 8.3   Absorbance Measurements in Suspensions

In suspensions of bacteriophage, the nucleic acid content of the sample can be determined by measuring the absorbance at 260 nm in the absence of scattering. Such suspensions scatter significantly as well as absorb. Can you suggest methods to correct for the scattering contribution?

Solution

The 260-nm absorbance band has a finite width and the absorbance drops rapidly to zero above 300 nm. The scattering contribution, however, appears as an apparent absorbance with a $1/\lambda^4$ dependence at wavelengths above 300 nm. Thus, an extrapolation can be made. Enzymatic cleavage of the bacteriophage would also reduce the scattering, since the average R_G would then be very small.

Example: Bacterial Cell Counts.   Bacterial cell counts can be made by measuring the turbidity at wavelengths far from absorbance bands, e.g., at 640 nm because τ is proportional to R_{90}, which depends on the number of scatterers in solution.

Example: Membrane Transport.   Membrane transport can be observed using turbidity. The volume of cells and organelles depends on the osmotic forces in the solution. When small molecules, such as sugars or amino acids, are added to a suspension of cells, there is an osmotic imbalance across the cell membrane. This

Figure 8.15 The transport of threonine into mitochondria, as measured by changes in absorbance. Solutions of D, L, and DL mixtures were added to a suspension of mitochondria at $t = 0$ (adapted from R. L. Cybulski and R. R. Fisher, *Biochemistry*, 16(1977):5116).

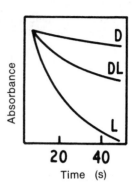

results in the rapid flow of water from the cell, causing cell shrinkage. As the small molecule equilibrates across the membrane, the imbalance is removed and water flows back into the cell with the small molecule. These cell-volume changes can be measured by the turbidity of the solution (see Figure 8.15).

Example: Muscle Action.   Muscle action is a result of interactions between the macromolecules actin and mysin, fueled by ATP. Szent Gyorgyi noted that when actin and myosin were mixed together, the solution became turbid. When ATP was added to the solution, however, the solution cleared as a result of ATP-induced dissociation of myosin and actin. The observation has been studied in detail to yield useful kinetic information about the macromolecular complex. For example, it has been shown that actin and myosin dissociate before the ATP is cleaved to ADP and P_i (R. W. Taylor and A. G. Weeds, *Biochem. J.,* 159 (1976):301).

Note that all three examples could also be carried out in a normal scattering experiment with $\theta \neq 0°$. Of course, such equipment is less readily available than the normal spectrophotometer, which is all that is required to measure turbidity.

Refractive Index

If a wave interacts with an array of scatterers, e.g., a slab of glass, a forward-scattered beam is produced by a large number of scatterers. It can be shown that an infinite array of scatterers produces a wave that is 90° out of phase with respect to the incident beam. This scattered wave combines with the transmitted wave to produce a new wave that is phase shifted with respect to the incident beam (see Figure 2.13). This phase shift is equivalent to the wave's having a different velocity in the slab than in the external medium. The **refractive index** is defined as

$$n = \frac{\text{velocity in vacuum}}{\text{velocity in material}} = \frac{c_0}{c_s}$$

The extent of scattering, and therefore n, depends on $\sqrt{\alpha}$, λ, and the number and size of the scatterers.

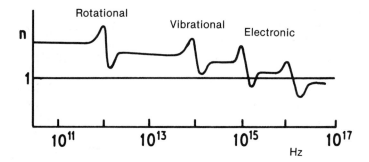

Figure 8.16 Variation in refractive index with frequency. The different frequencies and dispersions correspond to different molecular processes (see Chapter 2).

The Frequency Dependence of the Refractive Index: Dispersion. We have said that n is proportional to $\sqrt{\alpha}$. Thus, the wavelength dependence of n will have the form $\lambda_0^2/(\lambda^2 - \lambda_0^2)$. When $\lambda_0 < \lambda$, the scatterers oscillate 180° out of phase with respect to the case where $\lambda_0 > \lambda$. Arrays of scatterers can thus produce a forward-scattered beam either 90° ahead or 90° behind the transmitted beam. In addition, the scattering becomes large when $\lambda_0 \simeq \lambda$. The net result is that the wavelength dependence of n has the form shown in Figure A.5.

In a real molecule, there are several "λ_0" values corresponding to electronic transitions, molecular vibrations, and so on; thus, the frequency-dependence of n has several "dispersion" regions of the kind shown in Figure 8.16.

In the case of visible light, $\lambda_0 < \lambda$ and the refractive index increases with decreasing λ. Typical plots of the variation of refractive index are shown for different kinds of glass in Figure 8.17. Note that dispersion curves have much broader "wings" than absorption curves (see Chapter 2 and Appendix II). It is thus sometimes possible to distinguish materials on the basis of their refractive index even if their absorptions are essentially equal. This is the basis of the **phase contrast microscope** (see Chapter 11). In this mode of operation, the microscope is made sensitive to refractive index differences between materials.

Figure 8.17 Typical dispersion curves for flint and crown glass.

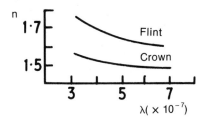

Figure 8.18 A rotating cylinder device that orients long molecules so that birefringence can be detected.

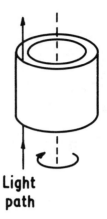

Light path

The Directional Properties of the Refractive Index: Birefringence.   If a sample is anisotropic—for example, an oriented fibre—the refractive index may be different for incident light polarized in two perpendicular directions. This phenomenon, known as **birefringence,** is related to linear dichroism (see Chapter 2) in the same way that optical rotatory dispersion is related to circular dichroism.

It is necessary to orient the molecules in order to detect birefringence. One suitable method is to use flow. A flow apparatus is shown in Figure 8.18. The refractive index depends on the asymmetry induced by the flow, which depends on the dimensions of the particles. For example, long molecules, such as DNA, tend to line up along the direction of flow. The birefringence, Δn, can be detected by measuring the optical rotation induced (see Chapter 10).

Other methods that can be used to induce orientation include the application of electric and magnetic fields. Molecules tend to orient in such fields because of their dipolar nature. The rate of response, as well as the degree of response, of these molecules to a field can be measured. A typical response to an electric field would be as shown in Figure 8.19. Both the relaxation time and the steady-state value of Δn are related to the shape of macromolecules and their dipolar character.

DYNAMIC OR QUASI-ELASTIC LIGHT SCATTERING

So far, we have analyzed solution scattering in terms of the angular dependence of the scattered intensity. We have indicated that small-particle solution scattering arises because of local fluctuations in concentration. These fluctuations can

Figure 8.19   The birefringence induced by an applied voltage as a function of time.

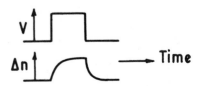

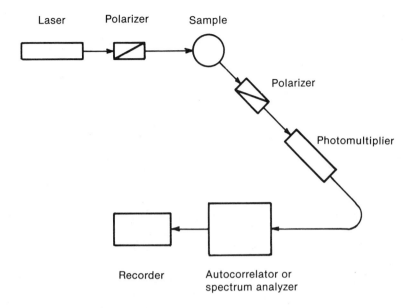

Figure 8.20 Typical experimental layout for a dynamic light-scattering experiment.

also be detected directly as noise in the scattered radiation. Methods that do this are usually called **dynamic** or **quasi-elastic** light-scattering techniques. They have advanced rapidly in recent years because of improvements in laser technology and are much more versatile and convenient than older light-scattering methods.

The noise in the scattered radiation is usually analyzed in one of two ways: (1) by a spectrum analyzer or (2) by an autocorrelator. These two methods are often known as **optical mixing** and **intensity of fluctuation spectroscopy,** respectively. The two methods of analysis give equivalent information. The first is a **frequency-domain** experiment, while the second is a **time-domain** experiment. The experiments are related by a Fourier transformation in exactly the same way as are NMR experiments that are carried out in these two domains (see Chapter 6).

A typical experimental set up is shown in Figure 8.20. The laser produces an intense monochromatic light beam; the sample scatters the light and this is detected by the photomultiplier, whose output is analyzed either by an auto-correlator or by a spectrum analyzer.

Spectrum Analyzer or Optical Mixing Spectroscopy

The frequency of the scattered light is best analyzed using a method that mixes the light from the laser source with the output from the photomultiplier (optical mixing). This gives an apparent frequency spread around zero frequency,

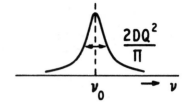

Figure 8.21 The linewidth at half-height of the spectrum of frequencies observed in an optical-mixing experiment is related to the diffusion coefficient of the scatterer.

although it is really about the source frequency. The deviations in frequency arise from the Brownian motion of the scattering particles. At any instant, some of the particles will be moving away from the detector, some toward it, and some will have no net velocity with respect to the detector. This motion results in a spread of frequencies because such movements result in Doppler shifts in the scattered light. (If light is scattered at the same frequency, energy is conserved and the scattering is said to be elastic. In dynamic light scattering, energy is conserved but there are shifts of frequency because of the motion, so the term *quasi-elastic* is used.)

The resulting spectrum of frequencies is related to the motion of the scatterers, which depends on their diffusion coefficient (*D*) (see Figure 8.21). The spread of frequencies is usually defined by a Lorentzian line shape (see Figure A.5) with a width at half-height ($\Delta\nu_{1/2}$) that is related to the product DQ^2, where Q is the scattering vector:

$$\Delta\omega_{1/2} = 2\pi\Delta\nu_{1/2} = 4DQ^2$$

Note that $\Delta\nu_{1/2}$ increases with increasing θ and decreasing incident wavelength (λ).

Worked Example 8.4   Diffusion Coefficient of Lysozyme

An optical-mixing spectrum of a 1% solution of lysozyme was obtained using a laser with a wavelength of 632.8 nm. The linewidth at half-height was 13,754 Hz when the scattering angle was 61.2°. Calculate the translational coefficient of lysozome in this solution.

Solution

$\Delta\omega_{1/2} = 2\pi\Delta\nu_{1/2} = 2\pi \times 13{,}754 = 4D[(4\pi/632.8)\sin(61.2/2)]^2$; therefore, $D = 10.8 \times 10^{-7} \text{cm}^2 \cdot \text{s}^{-1}$.

Example: The Electrophoretic Mobility of Red Cells.   In optical-mixing experiments, the random motion of particles is detected as a spread of frequencies around zero frequency. If there is net movement of particles in one direction caused by, for example, a voltage applied to the scattering cell or by bacteria swimming toward a chemical source (chemotaxis), then there is a net frequency shift as well as a frequency spead in the detected spectrum.

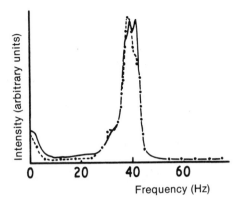

Figure 8.22 Electrophoretic light-scattering spectra for two sets of red cells with different ages. Note the width of the spectra and the position around 40Hz (from S. J. Luner et al., *Nature,* 269(1977):719).

In Figure 8.22, the light-scattering spectra of two populations of human red cells are shown superimposed. There is a mean Doppler shift of 39.5 Hz for both sets of cells. This shift arises from the voltage of $16.2V \cdot cm^{-1}$ applied to the solution. It had been suggested from previous electrophoretic measurements that there was an age-related change in the net charge on the cell surface of erythrocytes. This was a possible way in which cells could be recognized and removed from circulation after reaching an age of about 120 d. However, the results shown in Figure 8.22 do not support this hypothesis, since this type of electrophoretic experiment, which takes place in free solution, is extremely sensitive to small changes in charge.

Intensity Fluctuation or Correlation Spectroscopy

In the correlation spectroscopy method of analyzing scattering data, the photomultiplier output is fed into a device that compares the output at some time t with the output at time $(t + \tau)$. If the outputs are similar at the two times, they are said to be **correlated**. If the motion of the scattering particle is rapid compared to τ, then there will be no average correlation between the two output levels. If the motion is slow compared to τ there will be, on average, a correlation between the output levels at times t and $(t + \tau)$. This correlation is defined by a function $g(\tau)$. It is common for the correlation function $g(\tau)$ to decrease exponentially with increasing τ. The time constant of the decay is related to the product of the scattering vector Q and the diffusion coefficient D. Thus,

$$g(\tau) \simeq e^{-DQ^2}$$

A plot of $\ln[g(\tau)]$ versus τ then gives a straight line with slope $-DQ^2$. Note that this experiment, carried out using an autocorrelator, gives the same information as the spectrum analyzer experiment.

PROBLEMS

1. In the text it was stated that the angular dependence of the scattering intensity from an isolated particle is $\sin^2\phi$ for a driving wave polarized in the zy-plane. Prove that the angular dependence is $1 + \cos^2\theta$ for unpolarized incident radiation (see Figure 8.1 for a definition of θ and ϕ).

2. What happens to a Zimm plot when the solution conditions are changed so that a protein dimerizes?

3. Derive Snell's law, which states that $\sin\theta_1/\sin\theta_2 = c_1/c_2 = n_2/n_1$, where the angles θ, velocities c, and refractive indexes n refer to two media through which a light beam passes. What happens when the condition $\sin\theta = n_1/n_2$ holds? Can you suggest a method for the accurate measurement of n.

4. What is the apparent molecular weight of a molecule in a light-scattering experiment when its true molecular weight is 40 kilo-daltons but there is a 0.1% contamination with particles of molecular mass 10^3 k daltons.

5. A certain DNA molecule can exist either in a circular form or in a rigid double-stranded form 200 nm long. In an experiment with $\lambda = 500$ nm, calculate the ratio of the scattering produced by each of these forms when (a) $\theta = 20°$, (b) when $\theta = 90°$ (R_G for a rod is $L/12$, while for a circle it is equal to the radius).

6. In an experiment on calf thymus DNA ($M \simeq 15 \times 10^6$ daltons) using correlation spectroscopy, a time constant of 10 μs was observed together with a spectrum of slower time constants in the range 0.5–18 ms. Can you interpret these data? (At room temperature, the rotational diffusion coefficient of a spherical molecule is given approximately by $10^{13}/5Ms^{-1}$.)

7. Tomato bushy stunt virus is an RNA-protein virus with a neutron scattering matchpoint of 44.1% D_2O. The matchpoints of the RNA and the protein are 70% D_2O and 41% D_2O, respectively. Calculate the volume ratio of RNA to protein on the assumption that the H–D exchange properties of the RNA and the protein are the same.

8. In Worked Example 8.1, estimate whether C1 is compact or elongated, given the additional information that R_G for hemoglobin is 2.5 nm.

CHAPTER 9

Raman Scattering

OVERVIEW

1. Raman spectra give information on molecular vibrations and are obtained from changes in the frequency of light observed in a scattering experiment (inelastic scattering).

2. The physical picture arises from considering changes in polarizability (induced dipole moment) that arise if a vibration occurs during the time the electrons are oscillating in response to the applied radiation.

3. The gross selection rule is that the vibrational motion must produce a change in the polarizability of the molecule.

4. The anisotropy of the polarization of the scattering can be measured. Comparison of the spectra polarized perpendicular and parallel to the incident radiation gives information on the symmetry of the vibrational motions.

5. Raman spectra can be obtained in water. This is a major advantage over infrared spectra.

6. Resonance Raman spectra result when the wavelength of the exciting light falls within an electronic absorption band of a chromophore in the molecule. Some vibrations associated with such a chromophore may be enhanced by factors of 1000 or more.

7. The experimental parameters of a band in a spectrum are its position ($\Delta\nu$) (which is independent of the frequency of the exciting light), its intensity (which is directly poportional to concentration), and its polarization.

8. The main biological applications of conventional Raman are very similar to those for infrared. Resonance Raman affords a means of probing selective sites in molecules. For example, in metalloproteins, Raman can give information on the nature of the ligand directly attached to the metal.

RAMAN AND RESONANCE SCATTERING

Raman spectroscopy gives information on molecular vibrations. In infrared spectroscopy, this information is obtained from the absorption of the incident electromagnetic waves. In contrast, Raman spectroscopy involves scattering of the incident light and the information is obtained from the resulting changes in the frequency of the light. For biological samples, Raman spectroscopy has a distinct advantage compared with conventional infrared in that it can be performed easily in water. Furthermore, the Raman-scattered light occurs in the visible and ultraviolet regions of the electromagnetic spectrum and the detection systems in this range are far superior to those in the infrared region. However, the intensity of a Raman spectrum is usually quite weak—it is about one-thousandth the intensity of the light that is scattered at the same frequency as the incident beam (Rayleigh scattering). If, however, the frequency of the exciting light used has a value that is within an electronic absorption band of a chromophore in the molecule, then some vibrations associated with that chromophore can be dramatically enhanced. If the exciting frequency matches the frequency at the absorption maximum of the chromophore, the technique is known as **resonance Raman spectroscopy**. This is becoming increasingly important in studies of biological systems.

PHYSICAL PICTURE

Usually, scattering arises when electrons in a molecule oscillate in sympathy with an applied electromagnetic wave. The extent of the scattering depends on the polarizability of the electrons in the molecule (i.e., the dipole moment induced by the electric field). Most of these oscillating electrons scatter light at the same frequency as the incident beam (Rayleigh scattering). However, during the time that the electrons oscillate, a vibration may occur that causes some of the energy of the incident light to be transferred to the molecule (or vice versa). This leads to a change in frequency of the scattered light and the phenomenon is known as the **Raman effect.**

Energy is quantized, so the transfer or acceptance of energy by the molecule must correspond to a difference (ΔE) between the energy levels of the molecule doing the scattering. It also follows that the consequent change or displacement in frequency of the scattered photon reflects only this value of ΔE and must be independent of the frequency of the exciting light.

A schematic representation of the processes involved in Raman scattering experiments is shown in Figure 9.1. For the case of resonance Raman, the energies to which the molecules are initially raised by interaction with the photons correspond to the excited electronic state of the chromophore. In ordinary Raman spectroscopy, however, the energy level to which the oscillating molecule is raised is sometimes termed a **virtual energy level**. From such a level, the molecule can lose its energy elastically by a direct scattering process, that is, it can emit light at

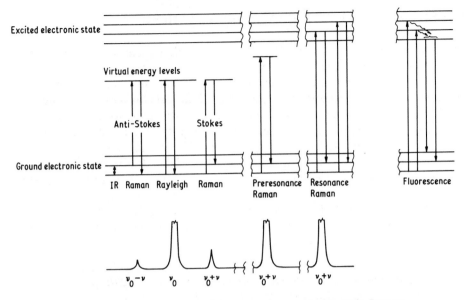

Figure 9.1 Processes leading to normal, preresonance, and resonance Raman scattering. (For comparison, the processes involved in IR and fluorescence are shown.) The horizontal lines represent different vibrational energy levels in the two electronic states. The Raman scattering spectrum is also indicated. Note that the intensity of the Stokes lines is greater than that of the anti-Stokes.

the same frequency as the irradiation (Rayleigh scattering). Alternatively, if the molecule changes vibrational state, the scattering is inelastic. The frequency of the scattering is lower than the irradiation frequency if the molecule goes to a higher vibrational state during scattering. This process gives Stokes lines in the Raman spectrum. If the molecule goes to a lower vibrational state during scattering, the frequency of scattering is higher and anti-Stokes lines are produced.

Worked Example 9.1   Only Stokes Lines Are Usually Measured Experimentally

Use Figure 9.1 to help explain the statement in the title of the problem.

Solution

Anti-Stokes lines involve transitions from excited vibrational states of the samples. The populations of these states are very much less than the population of the ground state (see Chapter 2). Fewer molecules are therefore available to give up a quantum of vibrational energy; thus, it is customary to measure the Stokes lines in Raman experiments.

SELECTION RULE FOR RAMAN SCATTERING

Raman scattering occurs when the vibrational motion produces a change in the polarizability of the molecule. The polarizability is a measure of the dipole moment *induced* in a molecule by an external electric field (see Chapter 2), which in this case is the incident radiation. This radiation distorts the molecule because of the forces that act on the electrons and the nuclei. A high polarizability therefore indicates that the molecule can be distorted quite easily. A molecule may be easier to distort in some directions rather than others. We then say that the polarizability is **anisotropic**. If it can be distorted equally well in all directions, the polarizability is **isotropic**. For example, a diatomic molecule can usually be distorted more easily when the field is applied along the bond direction than when it is applied perpendicular to it—the molecular polarizability is anisotropic.

The selection rule in Raman scattering requires that the vibration result in a change in the molecular polarizability.

Worked Example 9.2   Diatomic Molecules Comprising Identical Atoms Give Raman Spectra But Not Infrared Spectra

Explain the statement in the title of the problem using the selection rules for the two techniques.

Solution

The selection rule for observation of an infrared spectrum requires the vibration to cause a change in dipole moment. This clearly cannot happen for a diatomic molecule, which does not have a dipole moment—the vibration is infrared inactive or forbidden. However, during the vibration, the molecule distorts, thus changing its electronic distribution with respect to the nucleus. Accordingly, the molecular polarizability that reflects this distortion varies, so the vibration will be Raman active.

Worked Example 9.3   The Raman Spectrum of CO_2

CO_2 has four fundamental modes of vibration (see Figure 9.2) and yet only one of these is Raman active. Why?

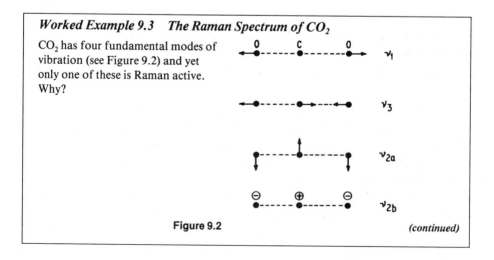

Figure 9.2

(continued)

> **Worked Example 9.3** *(continued)*
>
> *Solution*
>
>   The symmetric stretch of CO_2 alters the relative position of the nuclei and electrons, thereby causing a change in molecular polarizability and Raman activity. In the antisymmetric stretch, the effect of the expansion of one bond is compensated for by the contraction of the other, leaving the net polarizability unchanged. The same must also be true for the other modes. Note that the Raman-active vibration is infrared inactive (see Chapter 3). In fact, in a molecule with a center of symmetry, those vibrations that are infrared active are Raman inactive, and vice versa.

RESONANCE RAMAN SCATTERING

Scattering results when the incident light causes a charge displacement in the molecule. The charge displacement is reflected in the molecular polarizability. The scattering intensity depends on the square of the polarizability (see Chapter 8). When the frequency of the exciting light is near (preresonance) or matches (resonance) the frequency difference between two electronic energy states, then there is a high probability of a significant electronic charge displacement (or distortion of the molecule). This can result in absorption, but it also means that for some vibrations the polarizability becomes relatively large at or near the absorption transition ($\nu \approx \nu_0$ in equation 8.3 of Chapter 8). Such vibrations are said to be **resonance enhanced.** The enhancement depends on the strength of the electronic transition and the symmetry of the vibration. If there is only a single electronic state to consider, then the enhanced vibrations have to be symmetric, that is, they do not alter the symmetry of the molecule. However, if there is more than one electronic transition of a particular chromophore that is stimulated, then the symmetry of the vibration is less important.

   The enhancement in the resonance Raman (RR) spectrum is typically between 10^2 and 10^3. Resonance Raman can be obtained at concentrations of 10^{-4} mol $\cdot$ dm^{-3} or less. At these concentrations normal Raman spectra are usually undetectable. Thus, resonance Raman provides a means of selectively probing vibrational frequencies of a chromophore with a sensitivity approaching that of ultraviolet spectroscopy.

EXPERIMENTAL PARAMETERS

Position

The peak position is a property of the vibrational states of the ground electronic state of the molecule. The frequency shifts are independent of the exciting frequency (compare this with fluorescence, where the absolute frequency remains constant).

Intensity

The Raman scattering intensity is directly proportional to the concentration of species giving rise to the line. (Unlike infrared, it is not exponentially related to the concentration). The intensity generally increases as the molecular weight of the sample increases, since Rayleigh scattering increases (see Chapter 8). For a given sample, the intensity (I) is given by

$$I = CI_0\nu_s^4\,\alpha^2$$

where C is a constant, I_0 is the intensity of the incident light, ν_s is the frequency of the scattered light, and α is the molecular polarizability. In resonance Raman, it is the increase in α that results in the enhancement compared with the normal spectrum.

Polarization

The intensity of the scattered light depends on the molecular polarizability. If this is isotropic, the intensity of the scattered light will be the same in all directions. However, if the molecular polarizability is anisotropic, the scattered light will be anisotropic and consequently depolarized with respect to the incident light.

It is usual to define a depolarization ratio (ρ) as

$$\rho = \frac{I_\perp}{I_\parallel}$$

The depolarization ratio is useful in assigning the symmetry properties of a vibration and also in fingerprinting different bonds. For a totally symmetric vibration, ρ can have values between 0 and $\frac{3}{4}$, depending on the polarizability changes and symmetry of the bonds making up the molecule. However, nonsymmetric vibrations all have the value $\rho = \frac{3}{4}$.*

Worked Example 9.4   Polarization Measurements Probe Symmetry of Vibrations

Figure 9.3(a) shows the Raman scattering spectrum of CCl_4 between 500 and 150 cm^{-1} in two directions of polarization obtained with 488-nm excitation. Work out the approximate depolarization ratios of each band. What conclusions can you make about the symmetry of the vibrations of CCl_4 giving rise to each band? Can you assign any of the vibrations (I), (II), or (III) in Figure 9.3(b) to particular bands?

(continued)

---

*In resonance Raman, when excitation involves two electronic states—as is often the case in the hemes—ρ can approach∞ (i.e., $I_\parallel \simeq 0$). This is known as **inverse polarization.**

Worked Example 9.4   (continued)

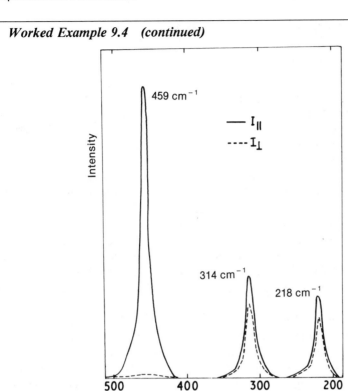

(a)

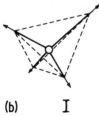 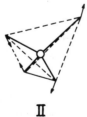 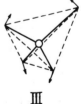

(b)   I   II   III

Figure 9.3

Solution

 The values of ρ are approximately 0.02 (459 cm^{-1}), 0.75 (314 cm^{-1}), and 0.75 (218 cm^{-1}). The 459-cm^{-1} band must arise from a totally symmetric vibration, in contrast to the other two, which arise from nontotally symmetric vibrations. The vibrations shown in (I), which is symmetric, is assigned to the 495 cm^{-1} band; (II) and (III) could be assigned to either of the other two bands.

APPLICATIONS OF RAMAN SPECTROSCOPY

Conventional Raman Scattering

Conventional Raman scattering has the same general application as infrared spectroscopy. For biological applications, the main advantage of Raman spectroscopy is that water scatters so weakly that it no longer obscures much of the spectral range, as it does in the infrared. In particular, the 200–2000-cm^{-1} region is completely accessible in Raman in either D_2O or H_2O. In this range, there are many low-frequency vibrations that are sensitive to conformational changes. These include the S—S stretching vibration in cystine bridges, which gives a strong Raman line at about 675 cm^{-1}, the C—S vibration at about 510 cm^{-1}, and several vibrations from aromatic side chains and nucleic acids. All these groups are highly polarizable and often give rise to intense lines.

In proteins, the amide I band can be studied either in the infrared or Raman. However, the amide II band, which is intense in the infrared, is very weak in the Raman. There is, however, another band associated with a coupled C—N stretching and N—H in-plane bending of the peptide group: the amide III band. This band, which is sensitive to conformation, occurs between 1230 and 1295 cm^{-1}. In aqueous solutions, it can be effectively studied only by Raman scattering. The band is intense at around 1235 cm^{-1} for the α-helix conformation, weak at 1260–1295 cm^{-1} for the β-conformation, while the random-coil conformation has a frequency of around 1245 cm^{-1}. Similar conformationally dependent bands can be found in the Raman scattering spectra of nucleic acid–protein complexes.

Worked Example 9.5   The Raman Scattering Spectrum of Lysozyme

The laser-excited Raman spectrum of lysozyme in water at 28°C and pH 5.2 is shown in Figure 9.4 (a). Assignments in terms of types of amino acid residues are shown in Figure 9.4 (b). These mainly follow from comparison of the spectrum at pH 1.0 of the constituent amino acids.

(a) What can you conclude about the ionization state of the protein?

(b) Why do you think the lines from the aromatic residues phenylalanine tryptophan, and tyrosine are so intense?

(c) These aromatic peaks seem to be insensitive to changes in conformation of the protein. What peaks might you expect to be sensitive to changes?

Solution

(a) The lack of a COOH peak in the protein spectrum at pH 5.2 suggests that these groups are fully ionized at this pH under these conditions.

(b) The mobility of electrons associated with aromatic rings makes these residues highly polarizable and hence good Raman scatterers.

(continued)

Worked Example 9.5 *(continued)*

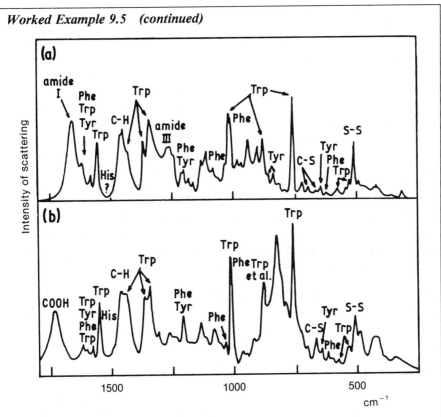

Figure 9.4

(c) The amide I, II, and III bands should be sensitive to conformational changes. However, since the chance of assigning these vibrations to individual bonds of a particular residue is unlikely, the information so obtained would relate mainly to macroscopic conformation (In lysozyme, the amide III peak is split into three: at 1240, 1262, and 1274 cm^{-1}. It has been suggested that these *may* reflect three distinct features: (1) the α-helix associated with residues 5–15, 24–34, and 88–96, (2) the antiparallel pleated sheet, and (3) a sequence folded in irregular ways; for more details, see R. C. Lord, and N. T. Yu, *Mol. Biol.,* 50 (1970): 509–524).

Resonance Raman Scattering

Resonance Raman spectroscopy affords the possibility of probing vibrations of selective sites that are associated directly with a chromophore. Only those vibrations that distort the molecule in the same direction as the allowed electronic transitions of the chromophore are efficiently enhanced. Chromophores to which the technique can be readily applied include hemes, certain metal ions, carotenoids, chlorophylls and the like. In principle, the applications to other chromophores will be limited only by the availability of suitable light sources. (A laser source is required because the applied radiation must be monochromatic.)

The two examples that follow illustrate the potential of the resonance Raman technique. The first is representative of the application to metalloproteins, in which, as is quite common with metalloproteins, the excitation frequency corresponds to a ligand–metal charge-transfer band ($d \rightarrow d$ transitions do not lead to significant enhancements because they are forbidden). The resonance Raman spectrum is used to probe the oxidation state of a ligand directly attached to the metal. In the second example, the heme group is probed by exciting the electronic transitions associated with the porphyrin ring. Although there are relatively few vibrational modes in the resonance Raman spectrum compared with a conventional Raman spectrum, these can rarely be fully assigned to particular vibrations, although methods of assignment such as isotropic substitution and the use of model compounds can often help. At worst, however, resonance Raman spectra provide an elegant fingerprint of some of the vibrational modes of the chromophore. Interesting and important biological information can often result from comparative studies or from the use of specific perturbations.

Worked Example 9.6   Does Oxygen Bind in the Peroxide Form, O_2^{2-}, to Hemocyanin?

Hemocyanin is an oxygen carrier in some invertebrates. It is a metalloprotein containing two Cu atoms that bind the oxygen. (It does not have a porphyrin-like heme group). It is thought that the oxygen becomes O_2^{2-} on binding to the metal. The O—O stretching modes for peroxides occur at about 742 cm^{-1}. There is a band at 570 nm tentatively assigned to a $O_2^{2-} \rightarrow Cu^{2+}$ charge-transfer band. Irradiation at 570 nm results in an enhanced intensity of the band at 742 cm^{-1}. What does this suggest? How would you confirm your conclusion?

Solution

(a) This suggests that oxygen does indeed bind as the peroxide.

(b) We must prove that the band at 742 cm^{-1} arises from an O—O stretch. Isotopic substitution with $^{18}O_2$ should cause the band to shift to 704 cm^{-1}. It does (see J. S. Loehr, T. B. Freedman, and T. M. Loehr, *Biochem. Biophys. Res. Commun.*, 56, (1974): 510–55).

Worked Example 9.7   Resonance Raman Scattering Can Be Used to Probe Heme Groups

The near-ultraviolet (Soret band) and visible absorption bands of a typical heme are shown in Figure 9.5(a). The resonance Raman spectra of oxyhemoglobin and deoxyhemoglobin, with an excitation frequency of 514 nm, are shown in Fig. 9.5(b).

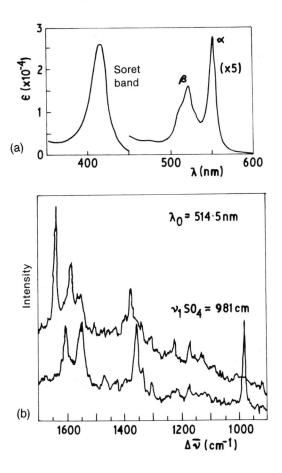

Figure 9.5

The solutions are ~ 0.4-mM in heme and $0.4M$ ammonium sulfate is also present in the deoxy sample. (SO_4^{2-} has a Raman scattering band at 981 cm^{-1}.)

(a) What do the resonance Raman spectra probe and what do the differences mean?

(b) Estimate the approximate increase in intensity of the larger band over a conventional Raman spectrum in the deoxy spectra.

(c) Would any other irradiation frequency result in a resonance Raman spectrum?

(continued)

Worked Example 9.7   (continued)

Solution

(a) The ultraviolet and visible spectra arise from $\pi \rightarrow \pi^*$ transition in the porphyrin ring (see Chapter 4). The stimulation of the 514-nm electronic transitions results in a significant increase in molecular polarizability, which is reflected in an enhancement of vibrations associated with the heme chromophore. Not all the vibrations associated with the ring are identical in the oxy and deoxy cases; hence, resonance Raman is a good probe of the heme group.

(b) The sulfate band is not resonance enhanced. It is approximately the same size as the largest peak that is resonance enhanced. The sulfate concentration is 10^3 times that of the heme which suggests that the enhancement of the Raman spectrum is of this order.

(c) Irradiation in the Soret band would also result in a resonance Raman spectrum.

Pre-Resonance Raman Scattering

Often an experimentalist does not have a suitable light source to irradiate at the maximum of an absorption band of a chromophore. The enhancement in the scattering is maximal when the irradiation frequency corresponds to the resonance condition. As the irradiation frequency deviates from this frequency, the enhancement rapidly decreases. For instance, 4-nitroaniline has an absorption band centered at 350 nm. Irradiation at 457.9 nm still produces a relative enhancement of about 20 in the band at 1334 cm^{-1} assigned to the nitro group compared with irradiation at about 633 nm (which essentially produces the conventional Raman spectrum). It is thus possible to use a range of wavelengths depending on the available laser light sources and still obtain a reasonable enhancement. It is clear, though, that while this so-called pre-resonance Raman scattering can lead to enhancements, the nearer the irradiation frequency to the actual maximum in the absorption peak, the larger the enhancement.

Worked Example 9.8   Pre-Resonance Raman Spectra Probe the Binding of Dinitrophenyl Haptens to Antibodies

Figure 9.6(a) shows an absorption spectrum of a 0.1 mM solution of a typical dinitrophenyl (Dnp) compound. The arrow indicates the laser Raman excitation wavelength. The pre-resonance laser Raman spectra of three Dnp derivatives was studied in the region of NO_2 group frequencies. The positions of the spectral lines alone ($-$) or in the presence of a Dnp-binding antibody ($+$) are presented in Figure 9.6(b). The bands at 1337 cm^{-1} for the free hapten arise from the vibrations of the 4-NO_2 group. All others are associated with the vibrations of the NHR $\cdots\cdots$ 2-NO_2 moiety. What can you conclude about the mode of binding of the Dnp ring?

(continued)

Worked Example 9.8   (continued)

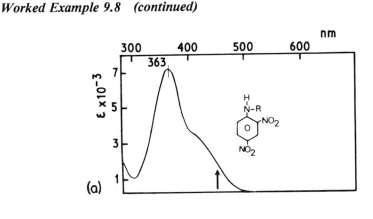

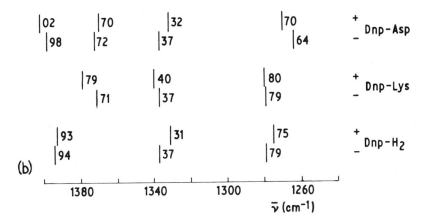

Figure 9.6

Solution

Binding to the antibody results in differential shifts of the bands for all three haptens. Considering the 4-NO$_2$ group vibration first, we see that, Dnp-Asp and Dnp-NH$_2$ both shift by about -5 cm^{-1}, but Dnp-Lys shifts $+3$ cm^{-1}. The 4-NO$_2$ group appears to bind in a different manner, depending on the Dnp derivative. This conclusion is reinforced from a comparison of the shifts in the other vibrations on binding. We conclude that at the atomic level, the exact mode of binding in the antibody site is influenced by the side chain on the Dnp and not simply by the Dnp itself (for further details, see P. Gettins, R. Perutz, and R. A. Dwek, *Biochem. J.* 197 (1981):119–125).

PROBLEMS

1. It has been suggested that the conversion of interchain —SH groups of lens proteins to S—S bonds leads to less soluble, higher molecular weight products whose presence may help to account for some of the properties of aging and senile cataractous lens. Laser Raman scattering spectroscopy provides a unique advantage in being able to study the intact living lens. The Raman scattering spectra in the —SH and S—S vibrational regions of an intact lens on a rat at ages 25 d and 7 mo are shown in Figure 9.7. The signal at 2580 cm^{-1} arises from the —SH stretching vibration, and the band at 508 cm^{-1} from the S—S stretching frequency. Interpret the changes in these bands. Would any other information help your conclusion?

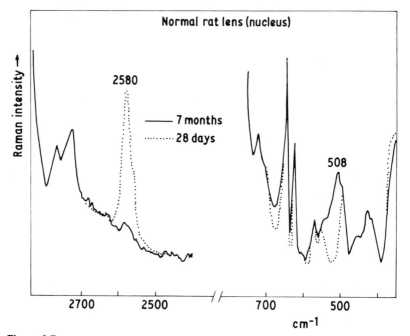

Figure 9.7

2. The development of resonance Raman in biochemistry has been held up because many of the chromophores are fluorescent. Explain this.

3. Hemerythrin is an oxygen carrier and binds one O_2 molecule per two Fe atoms. The two Fe atoms exist as Fe(II). Excitation of an absorption feature at 500 nm produces an intensity-enhanced O—O stretch at 844 cm^{-1}, which is characteristic of the peroxide form. (a) Calculate the shift expected when $^{18}O_2$ is used instead of $^{16}O_2$. (b) The use of a mixed isotope, $^{16}O-^{18}O$, in the experiments resulted in two vibrational O—O frequencies. Does this help differentiate between the oxygen-metal coordination schemes shown in Figure 9.8?

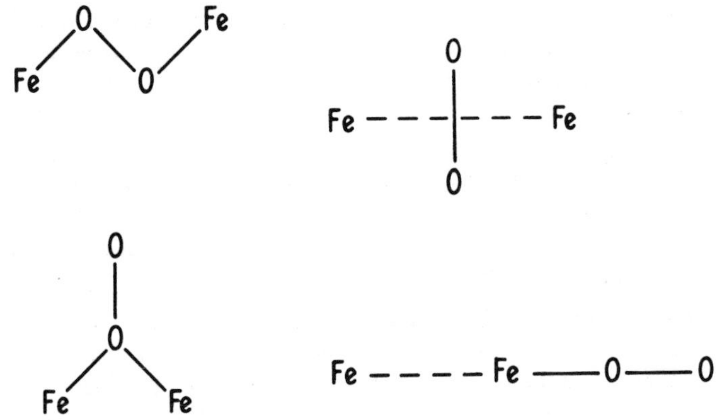

Figure 9.8

4. Transferrin is used in the transport of iron. It has a strong absorbance at 475 nm, which is assigned to Tyr(0) → Fe(III) charge transfer. The iron atom probably has three tyrosines and two histidines coordinated to it. How might you use resonance Raman to attempt to probe this metal-binding site?

5. Carbonic anhydrase is a zinc metalloprotein. It can be inhibited by sulfonamides. The aromatic sulfonamide $^-OPh-N=N-Ph\ SO_2NH_2$ has broad absorption bands centered at about 420 nm. Irradiation of these transitions produced a resonance Raman spectrum of the inhibitor. The resonance Raman spectrum of the free inhibitor changes when the SO_2NH_2 group ionized at pH 13. However, the latter spectrum was very similar to that of the inhibitor when bound to the protein complex at pH 9. Suggest an interpretation of the result.

6. The binding of a Dnp hapten to three anti-Dnp antibodies with differences in their amino acid sequences was studied by preresonance Raman scattering in which the Dnp absorption spectrum was irradiated. The shifts of the hapten on binding were very similar in each case. What can you conclude about the binding sites?

7. Carboxypeptidase was labeled at Tyr-248 with diazotized p-arsanilic acid to give the product shown in Figure 9.9.

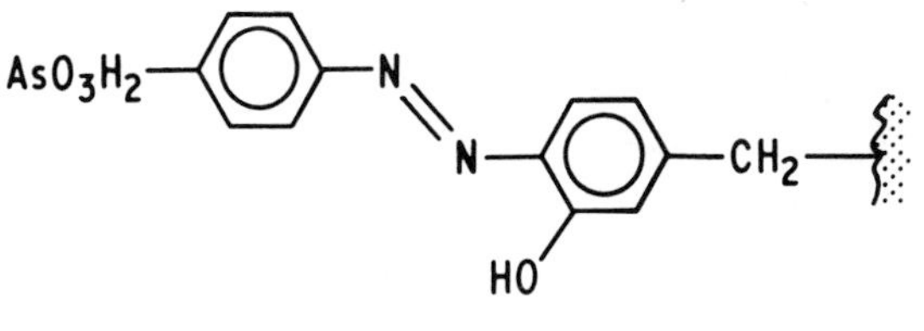

Figure 9.9

This label does not alter the activity of the enzyme. The enzyme contains a zinc atom at the active site. Excitation of the absorption band of the label results in the resonance Raman spectrum associated with the label. A frequency of 1440 cm^{-1} for the N=N stretching frequency is characteristic of an azo group in the

trans conformation. At pH 8.5, the resonance Raman spectrum has bands at
1338 cm^{-1} and 1539 cm^{-1} associated with the phenol-azophenolate oxygen binding to a zinc atom. Suggest a structure for the azotyrosyl-248 zinc complex in the
enzyme.

8. An enzyme is discovered that contains Mn(III) and has a charge-transfer band at
   515 nm. Irradiation of this band produces resonance Raman bands at 1230,
   1298, 1508, and 1620 cm^{-1}. In transferrin, Fe(III) is coordinated to tyrosines and
   has bands at 1174, 1288, 1508, and 1613 cm^{-1}. Suggest an assignment for the
   resonance Raman bands in the Mn(III) enzyme.

9. Figure 9.10 shows the resonance Raman scattering spectra of oxyhemoglobin
   when the heme chromophore is irradiated both in the near-ultraviolet (Soret
   band) and in the visible (α-β) absorption spectrum (see Figure 9.5(a)). Since both
   these electronic transitions arise from the heme, suggest an explanation for the
   differences in the two spectra.

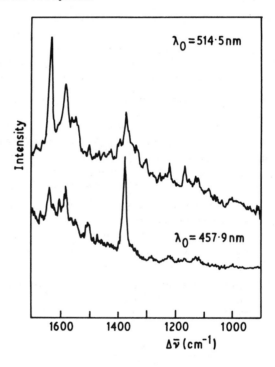

Figure 9.10

CHAPTER 10

Optical Activity

OVERVIEW

1. A molecule is optically active if it interacts differently with left- and right-circularly polarized light. This interaction can be detected either as a differential change in velocity of the two beams through the sample—**optical rotatory dispersion** (ORD)—or as a differential absorption of each beam—**circular dichroism** (CD).

2. ORD spectra are characterized by $[\alpha]_\lambda$, which is the specific rotation at a given wavelength, or the molar rotation $[\phi]_\lambda$. Both have units of degree $\cdot cm^2 \cdot dmol^{-1}$. CD spectra are characterized by ΔA (the differential absorption of the two beams) or the molar ellipticity $[\theta]_m$, which at a given wavelength is related to A. CD or ORD bands are often referred to as **Cotton effects**. These can be positive or negative.

3. CD is more frequently used than ORD because of superior instrumentation and the shapes of the CD curves.

4. Very few chromophores are intrinsically optically active; those that are active include the amides and disulfide cystine in proteins. Most optical activity of chromophores arises from optical activity induced by interactions with asymmetrically placed neighboring groups.

5. One of the main applications of CD spectra is based on their sensitivity to the secondary structure of proteins. Other uses include detection of conformational changes and measurement of ligand binding.

6. Optical activity can also be induced by the application of a magnetic field, which perturbs the energy levels of the system. This is the basis of magnetic circular dichroism (MCD). Unlike CD, MCD is largely insensitive to molecular conformation, but it is sensitive to the total concentration of MCD-active chromophores and their local environment.

OPTICAL ACTIVITY

A molecule is said to be optically active if it interacts differently with left- and right-circularly polarized light. This differential interaction gives rise to two separate but related phenomena known as **optical rotatory dispersion** (ORD) and **circular dichroism** (CD). In practice, these phenomena are usually observed for electronic transitions. Optical activity measurements in the 200–700-nm wavelength range are sensitive probes of molecular conformation and in some instances are related to the secondary structure of macromolecules.

OPTICAL ACTIVITY AND CIRCULARLY POLARIZED LIGHT

We saw in Chapter 2 that an electromagnetic wave can be plane or circularly polarized and that plane-polarized light is equivalent to two circularly polarized beams rotating in opposite directions.

An understanding of the phenomena of CD and ORD arises from considering the differential effects a sample can have on two circularly polarized beams **L** and **R** (see Figure 10.1). These effects are caused by (1) absorption by the sample, which is characterized by the extinction coefficient; and (2) the velocity of the beam through the sample, which is characterized by n, the refractive index. Optical activity can be detected either as the differential absorption (dichroism) of **L** and **R** ($\epsilon_L \neq \epsilon_R$) or as the differential refractive index of **L** and **R** ($n_L \neq n_R$).

If $\epsilon_L \neq \epsilon_R$, then after beams **L** and **R** have passed through the sample, their sum is no longer a plane-polarized beam but is elliptically polarized (see Figure 10.2(a)). In this case, the phenomenon of circular dichroism (CD) is observed.

If $n_L \neq n_R$, then the rotation velocity of **L** and **R** in the sample will be different, and the resultant of **L** and **R** will be rotated with respect to the situation in the absence of the sample (see Figure 10.2(b)). This is the phenomenon of optical rotatory dispersion (ORD). (Note that the term *dispersion* is used to denote the frequency (or wavelength) dependence of refractive index.) ORD and CD are related, and in practice both occur together. They are related by a Fourier transform relationship (known as the Kronig-Kramer transformation).

Figure 10.1   Two circularly polarized beams of equal amplitude, where the **E** vector is rotating in a right-handed (R) and left-handed (L) sense. The resultant of these two components is a plane-polarized wave (see Chapter 2).

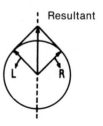

Resultant

Figure 10.2 (a) CD results in the L and R circularly polarized light beams having different amplitude after passing through the sample. Two circularly polarized beams combine to give an elliptically polarized beam. (b) Optical rotation arises from different relative values for L and R polarized beams. If only the L component is retarded with respect to R, the resultant is plane polarized, like the incident light, but the plane is rotated by α.

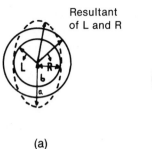

Resultant
of L and R

(a)   (b)

PARAMETERS FOR OPTICAL ACTIVITY

In ORD the **rotation** is defined as

$$[\alpha]_\lambda = \frac{\alpha_{obs}}{cL}$$

where $[\alpha]_\lambda$ is the specific rotation at wavelength λ, α_{obs} is the observed rotation measured in degrees, c is the sample concentration in grams per milliliter, and L is the path length in decimeters (1 cm = 0.1 dm).

The **molar rotation** $[\phi]_\lambda$ is defined by

$$[\phi]_\lambda = \frac{[\alpha]_\lambda M_w}{100} = \frac{(\alpha_{obs} M_w)}{100Lc}$$

where M_w is the molecular weight. The units of $[\phi]_\lambda$ are degree·cm²·dmol⁻¹.

Another commonly used parameter is the **mean residue rotation** $[m]_\lambda$, which is defined as

$$[m]_\lambda = \frac{[\alpha]_\lambda MRW}{100}$$

where MRW is the mean residue weight of the macromolecule monomers (e.g., for a protein it is the molecular weight divided by the number of amino acids). The units of $[m]$ are again degree·cm²·dmol⁻¹.

In CD the experimentally observed parameter is ΔA, which is defined as

$$\Delta A = \Delta\epsilon cL = A_L - A_R$$

where A_L and A_R are the absorbances of the **L** and **R** beams. (Note that $\Delta\epsilon = \epsilon_L - \epsilon_R$, which has typical values of 10^{-3} to 10^{-6} (mol·dm⁻³)⁻¹·cm⁻¹, is much smaller than ϵ.) Although ΔA is measured, the parameter $[\theta]_\lambda$, the **molar ellipticity**, is often used. These parameters can be related by the expression

$$[\theta]_\lambda = 3300\Delta\epsilon \text{ degree·cm}^2\text{·dmol}^{-1}$$

As with ORD, the **molar ellipticity** is defined as

$$[\theta]_\lambda = \frac{(\theta_{obs} M_w)}{100Lc}$$

where θ_{obs} is the observed ellipticity in degrees. (For the mean residue ellipticity, M_w is replaced by MRW.) To convert the units of c to mole residue per liter (dm^3) (c'), the preceding equation has to be multiplied by $10^3/M_w$, lhich gives

$$[\theta]_\lambda = \frac{(\theta_{obs} \times 10)}{Lc'}$$

Sometimes the subscript λ is omitted in reporting CD data.

Note that the ellipticity θ_{obs} is defined by the relationship

$$\tan^{-1}\theta_{obs} = \frac{b}{a}$$

where b/a is the ratio of the minor to major axes of the ellipse in Figure 10.2(a).

Worked Example 10.1   Prove the Relationship $[\theta] = 3300(\epsilon_L - \epsilon_R$ Where $[\theta]$ is in degree $\cdot$ cm^2 $\cdot$ dmol^{-1}

Solution

From Figure 10.2(a) the ratio of the ellipse axes (R) is $(\epsilon_L - \epsilon_R)/(\epsilon_L + \epsilon_R)$ where E_L and E_R are the electric vectors of the transmitted L and R waves. These electric vectors are differentially attenuated by the sample. The absorption is given by the Beer-Lambert law (see Chapter 4). Thus,

$$\log_{10}\left(\frac{I_R}{I_0}\right) = \epsilon_R cl$$

where c is in mol $\cdot$ dm^{-3}, l is in cm, and ϵ_R is in (mol $\cdot$ dm^{-3})$^{-1} \cdot$ cm^{-1}. Using the relationships $\ln x = 2.303 \log_{10}x$ and $E = \sqrt{I}$, we can deduce that

$$E_R = \sqrt{I_0} \exp(-2.303\epsilon_R cl)$$
$$= \sqrt{I_0}\exp\left(\frac{-2.303\epsilon_R cl}{2}\right)$$

Similarly

$$E_L = \sqrt{I_0}\exp\left(\frac{-2.303\epsilon_L cl}{2}\right)$$

These values for E_L and E_R can be substituted into the equation for R. Multiplying top and bottom lines of this equation by $\exp(2.303\epsilon_L cl/2)$, we obtain

$$R = \frac{e^0 - \exp(-2.303cl(\epsilon_R - \epsilon_L)/2)}{e^0 + \exp(-2.303cl(\epsilon_R - \epsilon_L)/2)}$$

Since $\epsilon_R - \epsilon_L$ is always a small quantity, we can expand the exponential terms using the relationship $e^{-x} \approx 1 - x$. Thus,

$$R \approx \frac{1 - 1 - (\epsilon_R - \epsilon_L)(2.303/2)cl}{1 + 1 - (\epsilon_R - \epsilon_L)(2.303/2)cl}$$
$$\approx 2.303(\epsilon_L - \epsilon_R)cl/4 \qquad \text{(since } \epsilon_R - \epsilon_L \text{ is small)}$$

(continued)

Worked Example 10.1   (continued)

Now $\theta_{obs} = \tan^{-1}R$, but since θ is small, $\theta \simeq R$. We must also convert from radians to degrees by multiplying by $360/2\pi$. Thus, $\theta_{obs} \simeq 2.303 / 4 \times (360/2\pi)$ $(\epsilon_L - \epsilon_R)cl \simeq 33(\Delta\epsilon)cl$. To convert the units of θ from $(mol \cdot dm^{-3})^{-1} \cdot cm^{-1}$ to $cm^2 \cdot mol^{-1}$ we have to multiply by 10^3 and those of l (cm) to L (dm), we have to divide by 10, so that

$$\theta_{obs} = 3300 \, \Delta\epsilon \, cL$$

where $\Delta\epsilon$ is now expressed as $cm^2 \cdot mol^{-1}$.

Using $[\theta] = \theta_{obs} \, 10/Lc$

we obtain $[\theta] = 3300 \, \Delta\epsilon$ degree $\cdot cm^2 \cdot dmol^{-1}$

MEASUREMENT OF ORD

Optical rotation arises from differential refractive indexes in a sample for the **L** and **R** circularly polarized light beams. An incident plane-polarized wave is rotated by the sample by some angle α, as shown in Figure 10.3. In modern instruments, the angle is determined by using a detection system sensitive to the direction of the plane of polarization of the incident beam. Rotation of the beam thus leads to a reduction in the signal detected. The instrument varies the wavelength and produces a plot of $[m]$ versus wavelength.

MEASUREMENT OF CD

CD arises from the differential absorption of **L** and **R** polarized light. Thus, it could be detected by a double beam instrument that puts **L** in one beam and **R** in the other. In practice, it is possible to use a crystal that can be made to pass either

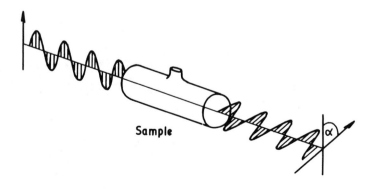

Figure 10.3 Rotation of plane-polarized light by an optically active sample.

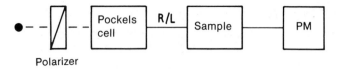

Polarizer

Figure 10.4   Detection of CD using a Pockels cell.

R or **L** depending on the applied voltage of an electric field. This device, known as a **Pockels cell**, can thus produce a beam that is alternately switched between **L** and **R** (see Figure 10.4). The beam then passes through the sample to a photomultiplier. The detected signal can then be processed to give a plot of ΔA versus λ.

THE FREQUENCY DEPENDENCE OF ORD AND CD

A molecule may have several absorption bands, which may be optically active or inactive. The active bands may be such that $\epsilon_L - \epsilon_R$ is positive or negative. A schematic frequency dependence for a molecule with three resolved bands, one with $\epsilon_L - \epsilon_R$ positive, one with $\epsilon_L - \epsilon_R$ negative, and one that is optically inactive, is shown in Figure 10.5.

Note that:

1. The CD bands are better resolved than the ORD bands. However, the wings of ORD bands may be detected in regions inaccessible to CD because of other absorptions or instrumental limitations.

Figure 10.5   Representation of three absorption bands. The first band is optically active and has a positive CD band. The second band has a negative CD band. The third absorption band is optically inactive.

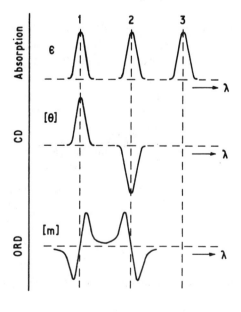

2. The ORD bands are *dispersion curves* because they depend on the sign change of the refractive index at the absorption maxima (see Appendix II).

Worked Example 10.2   CD versus ORD

Most recent investigations of proteins use CD rather than ORD. Can you suggest reasons for this?

Solution

The main advantages of CD over ORD measurements for studying proteins are the following:

(a) CD bands are more easily assigned because each optically active electronic transition gives rise to a CD band of only one sign (i.e., positive or negative) instead of the dispersion curves of ORD.

(b) The dispersion curves of ORD are spread over a wider range of wavelengths than those of CD curves. This means CD bands are better resolved.

(c) The CD method is technically superior to the ORD detection method because of the alternating nature of the CD detection system.

THE PHYSICAL BASIS OF OPTICAL ACTIVITY

As we discussed in Chapter 2, transitions from the ground state to the excited state involve a displacement of charge. A linear displacement with an electric transition dipole moment is denoted by μ. The transition can also have a circular component; this (rotating current) then generates a magnetic dipole moment **m** perpendicular to the plane of the circular motion (see Figure 10.6(a)).

Optical activity requires a finite μ and a finite **m**. This corresponds to a helical displacement of charge. The result is that left- or right-circularly polarized beams then interact differently with the molecules in solution. These interactions do *not* average to zero even in molecules randomly oriented in solution (compare "Linear Dichroism"). (The magnitude of the transition is proportional to the vector product of μ and **m**).

The required helical displacement of charge may arise from the intrinsic nature of the chemical bonds, for example, in an $n \rightarrow \pi^*$ transition (see Figure 10.6(b)). Chromophores that have intrinsic optical activity are often termed **asymmetric**. The required helical displacement of charge may also arise from an induced effect because of the environment of the molecule undergoing the transition. The resulting optical activity is then often referred to as an **extrinsic Cotton effect**.

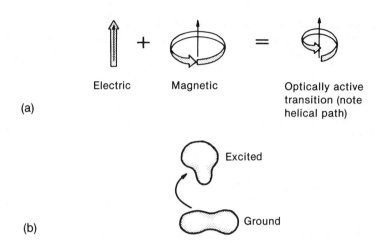

(a)

(b)

Figure 10.6 (a) A helical displacement of charge can be considered to comprise a linear displacement plus a circular displacement. (b) Rotation and translation of charge in an orbital.

OPTICALLY ACTIVE CHROMOPHORES

Optical activity is observed only when the environment in which a transition occurs is asymmetric. An inherently asymmetric chromophore, such as the peptide bond, is always optically active irrespective of the surrounding groups—which can, however, modify its optical properties. In contrast, any optical activity from inherently symmetric chromophores is induced and results from interactions with asymmetrically placed neighboring groups. The sign and magnitude of the induced optical activity (Cotton effects) of a particular chromophore depend on the local environment.

Very few chromophores are inherently optically active. In proteins, apart from the amide chromophores (< 240 nm), only the disulfide cystine is intrinsically optically active. The positions of the CD bands from the disulfide are found in the range 240–360 nm. Their CD spectra are complex to interpret and depend on the $-S-S-$ dihedral angle and the interactions with neighboring groups. They are not usually resolvable from the induced CD bands of the normally optically inactive aromatic side chains, which occur in the range 250–310 nm. Induced optical activity is also observed in the heme absorption transitions of most hemeproteins and for the metal absorption bands in most metalloproteins.

In nucleic acids, the individual bases of DNA and RNA are not optically active, but their incorporation into nucleosides or nucleotides results in induced optical activity. These effects depend on the nature of the base and the glycosidic linkage.

Carbohydrates generally absorb only in the far ultraviolet, but as instrumentation is improved, there will be more applications in this field. In polysaccharides, both the nature of the individual sugars and the linkages between them are important in determining the optical activity.

Worked Example 10.3   Use of CD to Monitor Ionization of Catalytic Carboxyl Groups, Asp-52 and Glu-35, in Lysozyme

The absorption spectrum of the model compound N-acetyl-L-tryptophanamide has a band around 300 nm. The CD spectrum of hen lysozyme is shown in Figure 10.7(a). It has a small negative band at 305 nm. The CD band at 305 nm changes with pH and reflects several pK$_a$ values; in particular, a pK$_a$ value of 6 is observed. Chemical modifications and kinetic studies together with a knowledge of the crystal structure have allowed this abnormal pK$_a$ value to be associated with Glu-35. The crystal structure also shows that Glu-35 and Asp-52 are close to Trp-108. Note that a pK$_a$ of 6 is abnormally high for a carboxyl group.

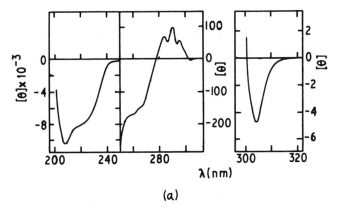

(a)

The pH dependence of the CD band at 305 nm as shown in Figure 10.7(b) is 4. When Asp-52 is esterified, the titration curve alters and the change from pH 7 to 3 can be analyzed in terms of a single pK$_a$ transition. However, the CD spectrum between 200 and 250 nm of the Asp-52–esterified lysozyme is the same as that of native lysozyme.

Using the given data can you suggest (a) an assignment of the CD band at 305 nm, (b) why its ellipticity is so small, (c) what is the effect of esterification of Asp-52 on the protein structure, and (d) why the pH dependence of the CD band at 305 nm alters after esterification?

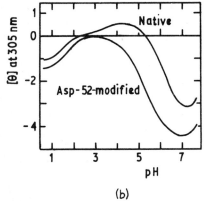

(b)

Figure 10.7

(continued)

Worked Example 10.3   (continued)

Solution

(a) Based on the absorption of the model compounds and the proximity of Glu-35 to Trp-108, the band at 305 nm can be assigned mainly to Trp-108.

(b) The CD is small because it is *induced* through interactions with neighboring groups and it arises from only one amino acid side chain.

(c) The CD spectrum between 200 and 250 nm arises from the conformation of the polypeptide backbone. This is unaltered on esterification.

(d) The CD band at 305 nm reflects the *ionization* of both Glu-35 and Asp-52. After esterification, the pH dependent contribution from Asp-52 is removed.

(In fact, the pH titration behavior of the active enzyme has been analyzed in detail to give two macroscopic pK_a values of 3.4 (Asp-52) and 5.9 (Glu-35) at 25°C and 0.1 ionic strength (for further details, see S. Kuramitsu et al., *J. Biochem.*, 76 (1974): 671–683).

CD SPECTRA OF INTERACTING CHROMOPHORES

If chromophores are near in space, then the transition dipole moments interact, which causes a splitting of the absorption bands (see Chapters 3 & 4 and Appendix X). This effect is especially important in CD, since with a pair chromophores the two absorption bands have CD spectra of opposite sense (see Figure 10.8).

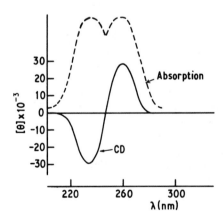

Figure 10.8   Typical appearance of two CD bands forming a "couple" with equal and opposite spectra.

Worked Example 10.4   CD Spectra of DNA

Nucleic acid bases are intrinsically optically inactive, but they are induced to have activity in polynucleotides. Such spectra normally show both a positive band around 270 nm and a negative band around 240 nm because of stacking interactions among the bases, which cause a splitting of the $n \rightarrow \pi^*$ transitions. Figure 10.9 shows the CD spectra of the A and B forms of DNA, which are obtained by changing the humidity of the sample from 75% to 92%. Why do these changes occur?

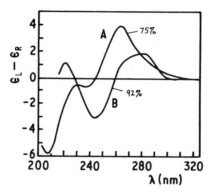

Figure 10.9

Solution

The CD spectra of DNA arise from the helical structure, which provides an asymmetric environment for the bases, resulting in optical activity. The rotational strength of a transition depends on the interaction of transition dipoles between adjacent groups and therefore on the geometrical relationship between them (see Appendix V). Any changes in the angle between adjacent chromophores alters the transition probability, the position of the absorption bands, and also the rotational strengths. The CD spectrum is therefore very sensitive to local environment. The changes in the spectrum with hydration imply that the structure of the DNA has been altered. (In the A form, there are about 11 residues per turn in a double helix, with the base pairs tilted 20° away from the perpendicular to the helix axis. In the B form, the double helix has about 10 residues per turn, with the bases perpendicular to the helix axis).

THE USE OF CD TO DETERMINE SECONDARY STRUCTURE

It was widely believed in the late 1960s that CD (and even ORD) could be used to determine the secondary structure of a protein. The approach was largely empirical and was based on the CD spectra of the peptide backbone (190–250 nm) for model polypeptides. The structure of these models was known from X-ray data

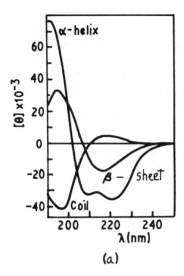
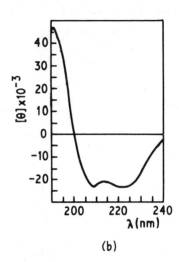

(a)   (b)

Figure 10.10   CD spectra of the α-helix, β-sheet, and random-coil forms of a model polypeptide poly-L-lysine (after N. Greenfield and G. D. Fasmin, *Biochemistry*, 8(1969):4108).

and it was assumed that in solution these polypeptides had only a single conformation. For proteins, the main standards were the three forms of poly-*L*-lysine: α-helix, β-sheet, and random coil (R), the spectra of which are shown in Figure 10.10(a). The CD spectrum of the protein was then simulated by a suitable graphical addition of the three curves as in Figure 10.10(b).

Worked Example 10.5   The Secondary Structure of Myoglobin

The CD spectrum of myoglobin is shown in Figure 10.10(b). Using the data in Figure 10.10(a), what qualitative deductions can you make about the protein conformation?

Solution

From a comparison of the two figures, it is clear that myoglobin is mainly made up of α-helical structures. (In fact this curve can be exactly matched using the X-ray structure of myoglobin, in which the protein has the following content: 68.3% α-helix, 4.7% β-sheet, and 27.0% random coil—see next section.)

MULTICOMPONENT ANALYSIS

The procedure of matching the observed and calculated CD spectra is known as **multicomponent analysis**. The curve is simulated from equations of the form

$$S = a[\alpha] + b[\beta] + r[R]$$

The variation of α, β, and R with wavelength is taken from suitable standards; then a, b, and r varied so as to obtain a fit to the experimental curve over the entire wavelength region.

For proteins whose structure is unknown, such graphical simulations must be used with caution because the reference spectra are not universally valid. CD spectra of the peptide bond may also be affected by the presence of aromatic side chains, disulfide bridges, and prosthetic groups. When it is also realized that the peptide bond may exist in conformations other than the α, β, or random coil forms (for example, the β-bend), some of the difficulties of using CD in this way become apparent.

Nevertheless, there are highly sophisticated graphical curve-fitting procedures that make allowances for these factors by using empirical corrections, which are often based on the results of X-ray structures of a variety of proteins. Estimates of secondary structures from CD spectra (especially of α-helix and collagen-like structures) are of value in initial assessments of X-ray diffraction data, in model building from calculations based on amino acid sequences, and, importantly, in cases where it is impossible otherwise to obtain any structural data.

Worked Example 10.6   CD Studies of the Structure of Human C1q: Does the Use of CD Confirm the Presence of Collagen-like Structure in Human Subcomponent C1q?

C1q has a molecular weight of 400,000 and is one of the first components in the complement cascade of proteins that controls the secondary immunological response following combination of antibody with antigen. Electron microscopy studies of C1q show that it has a structure that looks like "a bunch of tulips" (see Figure 10.11(a)). From sequence studies, C1q is known to have 18 polypeptide chains arranged in 3s. 40% of its amino acid sequence is in the form of collagenlike regions composed of the repeating triplet Gly—X—Y, and X is often proline and Y is often hydroxyproline or lysine. The molecule can be cleaved by pepsin at pH 4.4 (see Figure 10.11(a)). Sequence studies then show that about 84% of the residual tails have the collagen-like sequence.

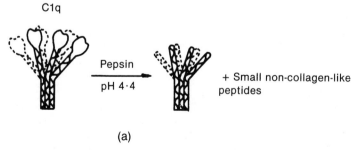

(a)

Figure 10.11

(continued)

Worked Example 10.6   (continued)

Using the data in Table 10.1 and the spectra in Figure 10.11(b) of C1q, its pepsin fragment, and lathyritic rat skin collagen, comment on the use of CD in probing the collagen-like nature of C1q.

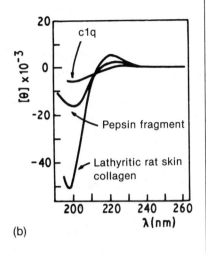

(b)

Figure 10.11

Table 10.1   CD features obtained for the different types of conformations found in proteins and polypeptides in solution

	CD features	
Conformation	nm	Mean residue ellipticity
α-Helix	193	+ 73,000
	208	− 35,000
	222	− 38,000
β-Structure	198	+ 50,000
	217	− 8,000
Collagen	198	− 50,000
	220	+ 6,000
"Random" form heat-denatured collagen	200	− 15,000
	198	− 20,000
Charged polypeptides: Extended		to − 40,000
Random form	211	+ 50,000

Solution

From Table 10.1, we see that CD spectra should give characteristic bands for a triple-helical (collagen-type) structure. The CD spectrum in Figure 10.11(b) of the whole C1q shows a small positive peak at 230 nm, which suggests the presence of some collagen-like structure. The only noncollagen conformation that has a positive band in the 215–230-nm region is that found for charged polypeptides, such as poly-lysine at low pH. It seems unlikely, however, that the C1q positive band arises from charged polypeptide effects, since after digestion of the noncollageneous regions of the molecule with pepsin, the positive band is shifted to 223 nm, which is more typical of collagen (see Figure 10.11(b)).

(continued)

Worked Example 10.6   (continued)

For the pepsin fragment, the magnitude of the CD is much less than that expected for collagen molecules of 84% triple-helical structure. One obvious explanation is that the 16% of the noncollageneous region in the fragment are in α-helical and β conformation, since these residues show high negative ellipticities at this wavelength, but it is difficult to be more precise than this (for further details, see B. Brodsky-Doyle et al., *Biochem. J.,* 159 (1976):279–286).

EMPIRICAL USES OF CD

That there are limitations in attempting to use CD spectra of the peptide backbone to determine secondary structures in proteins is obvious, since many of the difficulties arise from the sensitivity of the CD spectra to different environments. However, this sensitivity can be used in the detection of conformational changes in macromolecules resulting from the ionization of groups, the binding of ligands, variations in temperature, and so on. Used in an empirical way, CD is an extremely sensitive technique to report on conformational changes—but, of course, it does not give information on the extent and nature of the changes. Occasionally, however, the CD band may be assigned to a particular residue (see Problem 2), but as is usual in spectroscopy, the assignment procedure is far from trivial and is not always unambiguous.

Worked Example 10.7   The Binding of NAD⁺ to Glutamate Dehydrogenase

The binding of NAD^+ to glutamate dehydrogenase, which has six subunits, shows "negative cooperativity" in the presence of glutarate. The CD spectrum of the enzyme, shown in Figure 10.12(a), is unchanged in the presence of excess NAD^+. For the ternary complex with glutarate, the ellipticity also remains essentially unchanged until 50% of the NAD^+ is bound, when there is a marked change (see Figure 10.12(b)). Can you suggest an explanation for this?

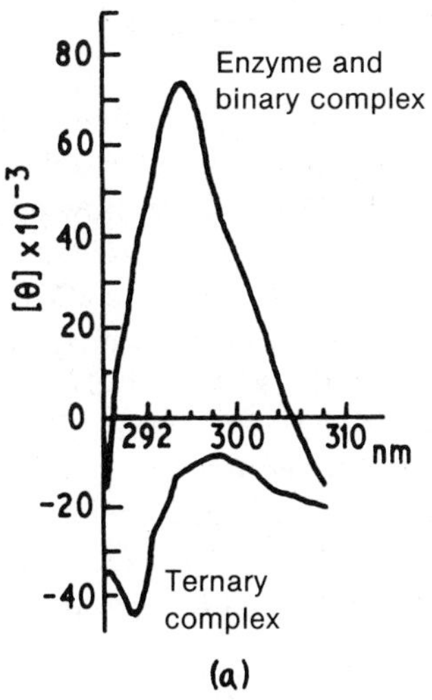

Figure 10.12   (a)

(continued)

Worked Example 10.7   (continued)

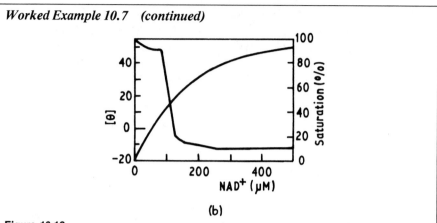

(b)

Figure 10.12

Solution

The CD studies provide evidence for a conformational change in the protein accompanying the binding of NAD^+, but only in the ternary complex. The conformational transition occurs when 50% of the coenzyme is bound. This suggests that the enzyme consists of two sets of three subunits between which interactions can occur. That is, the binding of NAD^+, as a ternary complex with glutarate, to three of the six subunits causes a conformational transition in the enzyme resulting in reduced affinities to the remaining three subunits of the enzyme (for further details, see J. E. Bell, and K. Dalziel, *Biochim. Biophys. Acta,* 309 (1973):237–242).

EFFECTS OF MAGNETIC FIELDS ON OPTICAL ACTIVITY

In 1845, Faraday observed that optical activity could be induced by the application of a magnetic field. This activity can be observed as magnetic ORD or magnetic CD (MCD). MCD has now been applied to a variety of biological problems and is the only technique discussed here. MCD arises if the magnetic field influences the electronic states of a chromophore in such a way that right- and left-circularly polarized beams are differentially absorbed.

There are three distinct effects a magnetic field can induce on transitions; these are designated as the *A, B,* and *C* terms. The *A* term arises in situations where there is a degenerate excited state (see Figure 10.13(a)) in the absence of an applied field. However, in the presence of the magnetic field B_0, the energy levels split and two effects occur: The absorption band is split into two components and each component is optically active (see Figure 10.13(b)). The shape of the curve

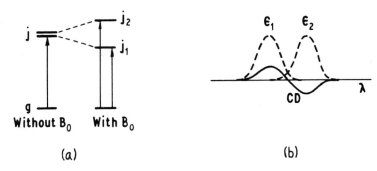

Figure 10.13 Effect of an applied magnetic field on a degenerate excited state. The energy level is split and each of the two resulting absorption bands is optically active.

arises because the two states absorb **L** and **R** with equal intensity but opposite sign. The C term arises when the ground state is initially degenerate and is split by the applied field. In such a case, the differential populations of the ground state are temperature dependent because the levels are relatively close in energy. The C-term MCD spectra are thus temperature dependent. The B term arises from mixing of states; it is not so temperature dependent as the C term.

It is important to emphasize the difference between CD and MCD. CD arises from an asymmetry in a chromophore or its environment—it is very sensitive to environment or molecular conformation. MCD arises from field-induced perturbations in the energy levels of the system. As a result, MCD is insensitive to molecular conformation. It is, however, sensitive to total concentrations of MCD-active chromophores (e.g., tryptophan), to distinctions between chromophores (e.g., purines and pyrimidines), and to the precise ligand geometry around a chromophore (e.g., a metal ion).

Experimentally, MCD is very similar to CD, except that in MCD a strong magnetic field is applied to the sample, usually with the field in the same direction as the light propagation. A CD measurement in the presence of a magnetic field has the form

$$\Delta A = A_{\mathbf{L}} - A_{\mathbf{R}} = \Delta A_{\mathrm{CD}} + B_0 \Delta A_{\mathrm{MCD}}$$

The simplest way to measure ΔA_{MCD} is to measure ΔA in the absence and presence of the applied field. The usual units of MCD are given by

$$\Delta \epsilon_M = \frac{\Delta \epsilon}{B_0} = \frac{\Delta A}{bcB_0} = \frac{\theta}{33bcB_0}$$

where $\Delta \epsilon$, θ, and ΔA are values induced by the magnetic field B_0 (in tesla), b is the path length in centimeter, and c is the molar concentration. The units of $\Delta \epsilon_M$ are $(\text{mole dm}^3)^{-1} \cdot \text{cm}^{-1} \cdot \text{T}^{-1}$.

Worked Example 10.8   Tryptophan Content of Protein

Tryptophan is the only amino acid that gives a strong positive MCD band at 290 nm; the intensity of this band is proportional to the tryptophan concentration. $\Delta\epsilon_M$ for N-acetyltryptophanamide is 2.35 (mole dm^3)$^{-1}$·cm^{-1}·T^{-1}. Assess the tryptophan content of a lysozyme molecule if A for a sample at a concentration of 10 μM in a 1-cm sample at 4 T is 4.8×10^{-4}.

Solution

Using the equation $\Delta\epsilon_M = \Delta A/bcB_0$, we have $\Delta\epsilon_M = 1.2 \times 10^{-4}/10 \times 10^{-6} =$ 12. The number of tryptophan residues is then $12/2.35 \simeq 5$.

Worked Example 10.9   Environment of Co(II) in Enzymes

The MCD spectra of model compounds of Co(II) that are tetracoordinate and pentacoordinate are shown in Figure 10.14(a). For carbonic anhydrase, the MCD spectra of the Co(II) enzyme are shown at high pH in Figure 10.14 (b) and at low pH in Figure 10.14(c). In the presence of inhibitor the spectrum in (d) is obtained, which is independent of pH. Comment on the results.

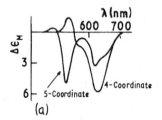

(a)

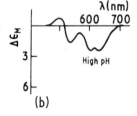

(b)

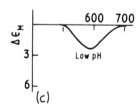

(c)

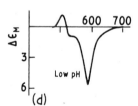

(d)

Figure 10.14

Solution

The spectrum of the inhibitor complex is characteristic of tetrahedral coordination. The spectrum at high pH is suggestive of pentacoordination (although distorted). The low-pH form is complex and is either pentacoordinate or some form of distorted tetrahedral symmetry (see B. Holmquist, B. L. Vallee, *Methods in Enzymology,* 49 (1978):149–179.

PROBLEMS

1. The functional groups involved in catalysis by serine proteases and their parent zymogens are the same. The inferior catalytic activity of zymogens is thought to arise from an underdeveloped binding site. The hydrolysis of NPGB (see Figure 10.15(a)) by trypsin and trypsinogen proceeds via the formation of an acyl intermediate, which can be isolated. Figure 10.15(b) compares CD spectra of trypsin and trypsinogen with those of their GB derivatives (dotted curves). The difference CD spectrum between trypsin and GB-trypsin has a broad bell-shaped curve with a maximum ellipticity at 270 nm identical with the difference absorption spectrum.

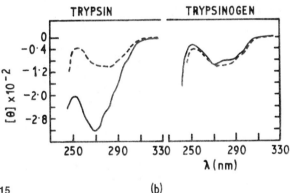

(a)

Figure 10.15   (b)

   (a) What does this suggest about the origin of the ellipticity changes on acylation?
   (b) Can you suggest why there are changes in the CD spectrum of trypsin but not trypsinogen on acylation?

2. (a) Dinitrophenyl (Dnp) ligands absorb in the range 280–470 nm. Dnp–lysine is optically inactive, yet on binding to a Dnp antibody M315, extrinsic Cotton effects appear in the range 280–470 nm. Can you suggest a reason for this?
   (b) The CD difference spectra shown in Figure 10.16(a) for the intact M315, the Fab fragment, and the Fv fragment (see Figure 10.16(b)) were obtained by subtracting the CD spectra of the proteins from those of the respective protein-Dnp complexes.

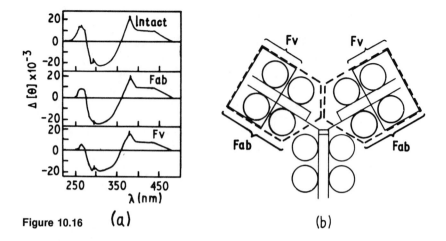

Figure 10.16   (a)   (b)

What do these spectra tell us about the nature of the Dnp binding sites in the intact antibody and its two fragments?

3. The antibody molecule consists of two polypeptide chains—the heavy and the light. In Figure 10.17, the CD difference spectrum resulting from the binding of a trinitrophenyl (Tnp) ligand to the intact antibody M315 is compared with those obtained when the Tnp is bound to either the light or the heavy chain. What conclusions can you make about the residues contributing to the binding site?

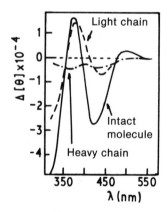

Figure 10.17

4. Glycophorin A spans the red blood cell membrane; its CD spectrum is shown in Figure 10.18(a). Its primary structure is known and is shown schematically in the Figure 10.18(b). There are 15 carbohydrate units attached to serine or threonine on the 1–50 part of the chain, which is known to be extracellular. The 70–92 region is very hydrophobic, while the 92–131 region is hydrophilic and intracellular. The CD spectra of these fragments are shown in Figure 10.18(c). The CD spectrum of the hydrophobic peptide 60–90 has to be measured in 1% detergent to make it soluble. (All the other CD spectra here are essentially unchanged in the presence of this level of detergent.)

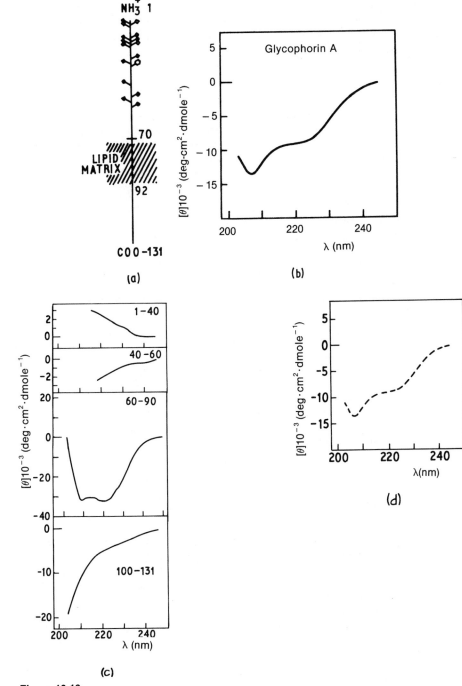

(a)

(b)

(c)

(d)

Figure 10.18

(a) What can you deduce about the tertiary structure of the peptides.

(b) When the individual spectra A–D are summed, the composite spectrum in Figure 10.18(d) is obtained. What conclusion can you draw from this?

5. (a) DNA can exist in either double- or single-stranded forms,

$$(DNA)_2 \rightleftharpoons DNA$$

The essential activity of any DNA-unwinding protein is to displace this equilibrium to the right. All DNA and RNA polymerases must have unwinding activity, since they need to unwind DNA to create a template. The protein made by the small bacterial virus fd, coded by its gene 5, is the simplest DNA-unwinding protein known. A study of the interaction of this protein with DNA provides a model system for protein-DNA interactions in general.

The CD spectrum of fd DNA shows a maximum at 270 nm, which decreases on the addition of gene-5 protein. The change on DNA ellipticity, $[\theta]_{270}$, was measured as a function of the molar ratio (R) of gene-5 protein to DNA base by titrating gene-5 protein solution into fd-DNA solution

R:	0.03	0.04	0.10	0.17	0.24	0.32	0.50	0.64
$-\Delta[\theta]_{270}$:	0.20	0.17	0.35	0.75	1.05	1.30	1.35	1.40

Using these data, suggest how the gene-5 protein binds to single-stranded DNA.

(b) The following data were also obtained in this system:

(i)   The addition of $0.1 mMg^{2+}$ induces a complete reversal of the change in $[\theta]_{270}$ discussed in (a).

(ii)   Acetylation of all six lysyl residues of gene-5 protein with N-acetyl inidazole inhibits any change in $[\theta]_{270}$ on addition of the modified protein to fd DNA.

(iii) Tetranitromethane nitrates three of the five tyrosines, almost completely inhibiting the change in $[\theta]_{270}$. Nitration is inhibited by the presence of an excess of fd DNA.

(iv)   The CD spectra of the acetylated and nitrated gene-5 proteins were measured in the range 200–250 nm. They were not significantly different from the native protein.

Suggest a possible molecular mechanism for the binding of gene-5 protein to fd DNA.

6. The optical rotation of oligosaccharides is very sensitive to the anomeric linkage. Hudson's rules for the additivity of chiral centers can successfully be applied to oligosaccharide sequences; that is, the molar rotation $[\phi]_\lambda$ of the whole molecule can be predicted from the sum of the values of the individual sugars. From the data for constituents of potato lectin, calculate the $[\phi]_\lambda$ values of the hydroxyprolyl diarabinofuranoside, triarabinofuranoside, and tetraarabinofuranoside. Using the values given for the model compounds, make a table of theoretical values of all the possible permutations of the α- and β-anomeric linkages that could be present in these molecules. Can an anomeric composition for these oligosaccharides be deduced from these results? Suggest reasons for any discrepancy.

Model compounds	$[\phi]_\lambda$	
	365 nm	589 nm
Methyl-α-L-arabinofuranoside	− 591	− 208
Methyl-β-L-arabinofuranoside	+ 555	+ 192
Hydroxyproline[a]	− 54	− 21

Contituents of potato lectin	$[\alpha]_\lambda$		
	365 nm	589 nm	Mol. wt.
Hydroxyprolyl[a] diarabinofuranoside (HA$_2$)	+ 242.5	+ 86.7	395
Hydroxyprolyl[a] triarabinofuranoside (HA$_3$)	+ 436.4	+ 174.6	527
Hydroxyprolyl[a] tetraarabinofuranoside (HA$_4$)	+ 185.1	+ 60.7	659

[a]Equimolar mixture of *L*-hydroxyproline and allo-*D*-hydroxyproline. This is produced by the alkaline conditions used for isolating the hydroxyprolylglycosides.

CHAPTER 11

Microscopy

OVERVIEW

1. Microscopes magnify an object using a system of lenses. The usual illumination sources are light and electrons.

2. The resolution of light microscopes is limited by diffraction phenomena to around 1 μm, whereas the resolution of an electron microscope is usually limited by sample preparation procedures and radiation damage to around 2 nm.

3. With the light microscope, contrast in the object can be enhanced by staining with dyes or utilizing refractive index differences (phase contrast).

4. In the electron microscope, the sample must be prepared on some sort of grid, fixed, stained, or shadowed with metals. Improvements in the electron images can be obtained using techniques such as underfocusing, optical filtering, image reconstruction, and electron diffraction.

5. The widespread uses of the light microscope in biology are familiar to almost everyone. The electron microscope has also given beautiful pictures of tissues, cells, macromolecular assemblies, and macromolecules. The fluorescence microscope is particularly useful in detecting specifically labeled regions of a cell and for the technique known as photobleaching recovery.

INTRODUCTION

The purpose of a microscope is to *magnify* an object and to *resolve* as much detail as possible in the magnified image. It is important to realize, however, that magnification alone does not necessarily produce better resolution.

The two forms of radiation most used in microscopes are light (λ in the range 200–600 nm) and electrons (λ in the range 0.001–0.01 nm).

Figure 11.1   The appearance of the image
of a slit.

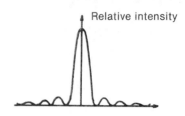

The light microscope has been of enormous value in giving detailed information about the structure of cells and tissues for hundreds of years. More recently,
it has also become possible to use microscopes to obtain information about macromolecules and macromolecular assemblies.

FACTORS THAT AFFECT RESOLUTION

Apart from considerations such as the mechanical construction of the microscope
and the lenses (see next section), there are some fundamental physical principles
that are important in determining the **resolving power** of a microscope. One of
these is diffraction. When radiation strikes an object, the light is diffracted (scattered). Light incident on a slit, for example, produces a pattern as shown in Figure 11.1. The reason for the pattern is that different diffracted rays interfere with
each other and, in some cases, cause cancellation. This causes a fundamental lack
of definition in the image formed by a lens.

---

*Worked Example 11.1   The First Minimum in the Diffraction Pattern
of a Slit*

Show that the first minimum in a slit diffraction pattern is $0.5\lambda \sin U$, where U, the
angle defined in Figure 11.2(a), is the angle subtended by the slit (width a) and the
plane in which the diffraction pattern is formed. Assume that U is small, so that
$\sin U \simeq a/2R$.

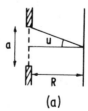
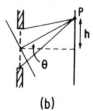

Figure 11.2   (a)   (b)

Solution

   Consider the position of the first minimum at a point P (see Figure 11.2(b)).
The path difference between the two extreme rays and the central ray is $a \sin \theta/2$,
which is equal to $\lambda/2$. If θ is small, $ah/R = \lambda$, or $h = \lambda R/a$; thus, $h = 0.5\ \lambda/\sin U$.

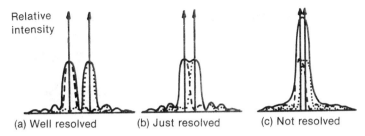

Figure 11.3 The Rayleigh criterion for resolution.

RESOLVING POWER

If two objects are to be resolved, they must have some distinguishing property —that is, they must exhibit **contrast.** In addition, the diffraction properties outlined in Worked Example 11.1, are important. The precise position at which it is possible to resolve two objects is somewhat uncertain, but the **Rayleigh criterion** is usually adopted (see Figure 11.3). This states that the patterns are distinguishable when the first minimum of one diffraction pattern falls on the central maximum of the other.

The Rayleigh criterion implies (from Worked Example 11.1) that a slit can resolve two objects if they subtend an angle greater than U. In other words, the resolving power of the slit is 0.5 $\lambda/\sin U$.

By similar arguments, the resolving power of a circular aperture is defined by 0.61 $\lambda/\sin U$. For the case where a medium of refractive index n is between the object and the lens (see Figure 11.4), the resolving power is 0.61 $\lambda/n \sin U$. The factor $n \sin U$ is called the **numerical aperture** (NA) of the lens.

Worked Example 11.2   Effect of n *and* λ *on Resolving Power*

For very high magnification, oil is often placed between the lens and the first (objective) lens of optical microscopes, and blue light is used. Can you suggest why?

Solution

The refractive index of oil is higher than that of air (by a factor of about 1.5). Thus, the numerical aperture can be increased, giving 50% improvement in resolving power. Blue light ($\lambda \approx 400$ nm) will give another 50% improvement in the resolving power, compared with red light ($\lambda \approx 600$ nm).

Figure 11.4   The numerical aperture (NA) of a lens is $n \sin U$.

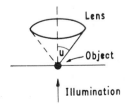

Figure 11.5   A magnetic lens for the electron microscope.

INSTRUMENTATION

Lenses

The phenomenon of refraction causes a ray of light to deviate as it passes through a prism. The angle of deviation depends on the refractive index of the prism and the angle of incidence of the light ray. A lens can be considered as a series of prisms that causes a parallel pencil of rays to focus at a point a distance f from the center of the lens. The distance f is termed the **focal length.**

It is a simple matter (see problem 1) to derive the formula

$$1/u + 1/v = 1/f$$

where u is the object length, v is the image distance, and f is the focal length of the lens (u and v are measured from the lens).

Lenses usually have a variety of **aberrations** associated with them, which means that a point object does not give rise to a point image. **Chromatic aberrations** arise when different wavelengths are refracted through the lens differently, which gives each wavelength a different focal length. This problem can be eliminated by using a lens made from different combinations of glass. **Spherical aberrations** arise because the lens geometry is such that there are different f-values for rays striking different parts of the lens. These can be corrected for by careful grinding of the lens. Another method of reducing spherical aberrations is to reduce the aperture. However, as we have seen, this reduces the resolving power.

Electrons can be deflected by both electric and magnetic fields. A typical lens for electrons is shown in Figure 11.5. (Note that such a lens gives a smooth function on any dimension scale in contrast to a solid, which would be very coarse on a nanometer scale.) Aberrations are a severe problem in the electron microscope. A result of these aberrations is that typical numerical aperture values for the electron microscope are small (around 0.008). For optical lenses, NA values of 14 are possible; thus, the improvement in resolution is less than might be expected in going from wavelengths of around 400 nm for light to around 0.002 nm for electrons. Note also that small NA values lead to a relatively high range over which the image is approximately in focus—a large "depth-of-focus" compared to the wavelength.

Figure 11.6 An optical instrument that magnifies increases the angle subtended by the object at the eye in the ratio θ_1/θ_0.

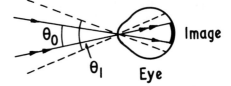

Image

Eye

Magnification

The important factor that determines magnification is the ratio of the angles θ_0 and θ_1 that subtend the object before and after processing in the optical instrument (see Figure 11.6). The angle θ_0 can be obtained by assuming that the object is at the nearest point that can be readily focused by the eye (about 250 mm). If this near-point is at distance D and the height of the object is h, then $\theta \approx h/D$ for small angles.

Worked Example 11.3   Magnification of a Microscope

Show that the magnification of the compound microscope in Figure 11.7 is the product of the magnification of the eyepiece and the magnification of the objective. (The positions of the images can be constructed by noting that rays parallel to the principal axis of the instrument go through the focal points, and rays going through the center of the lens are undeviated.)

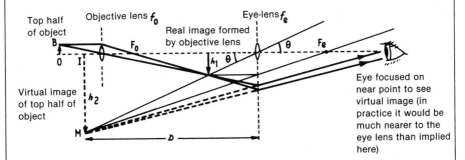

Figure 11.7   A ray diagram for a compound microscope.

Solution

The magnification of the microscope is

$$\frac{\theta_1}{\theta_0} = \frac{h_2/D}{h/D} = \frac{h_2}{h} = \left(\frac{h_2}{h_1}\right)\left(\frac{h_1}{h}\right)$$

$$= \text{magnification of eyepiece} \times \text{magnification of objective}$$

Figure 11.8   An illumination scheme for a
light microscope using a condenser lens.

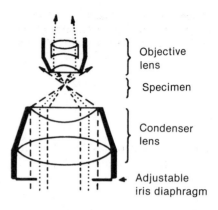

Objective
lens

Specimen

Condenser
lens

Adjustable
iris diaphragm

The Light Microscope

The diagram in Worked Example 11.3 illustrates the principles of the microscope. The eye lens enlarges the real image produced by the objective lens. However, the resolution is determined only by the objective lens.

Typical magnifications for optical microscopes are objective lens, $\times$ 40; and eyepiece, $\times$ 10; with overall magnification of $\times$ 400. The focal length of such an objective lens is about 4 mm when the distance between the surface of the lens and the sample is about 0.4 mm.

In addition to the magnification system, the illumination of the sample is very important. A typical arrangement is shown in Figure 11.8. The main purpose of the condenser lens is to achieve the maximum NA value of the objective lens. The NA of these two lenses must be similar for this to be achieved.

The Electron Microscope

A schematic diagram of the **electron microscope** (EM) is shown in Figure 11.9. The electron gun consists of a filament (wire), which is made of tungsten and which is heated by passing a current through it. The heated filament emits electrons, which are attracted to a positive anode by a very high voltage (up to 100,000 V). A narrow beam of electrons is thus produced. In addition to the objective lens, which is equivalent to the one in the light microscope, there is usually an intermediate lens and a final projector lens, which projects the image on a photographic plate or a fluorescent screen. Typical magnification factors are $\times$ 100,000.

In addition to the transmission type of electron microscope, there are also **scanning electron microscopes** (SEM). In the SEM, a fine beam of electrons is scanned back and forth across the sample surface in a controlled way. This scanning produces secondary electrons, which are collected (see Figure 11.10). An

Figure 11.9 A schematic diagram of the transmission electron microscope.

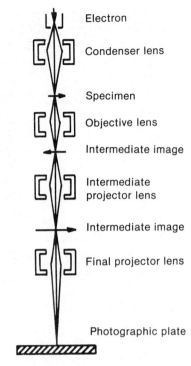

Electron

Condenser lens

Specimen

Objective lens

Intermediate image

Intermediate projector lens

Intermediate image

Final projector lens

Photographic plate

ELECTRON MICROSCOPE (transmission type)

image of the sample can be built up by plotting the intensity of the secondary electrons. This instrument gives beautiful images that give 3-D effects (see Figure 11.11), but the resolution is limited to around 3 nm.

The EM must be evacuated because electrons interact strongly with all materials. The supporting grid for the specimens also interacts with the radiation in the transmission microscope and causes problems.

The irradiation levels in the EM are very high—normal exposures are equivalent to doses of about 10^8 Gy. This causes significant damage to delicate biological structures.

Figure 11.10 The principle of the scanning electron microscope (SEM).

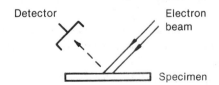

Detector

Electron beam

Specimen

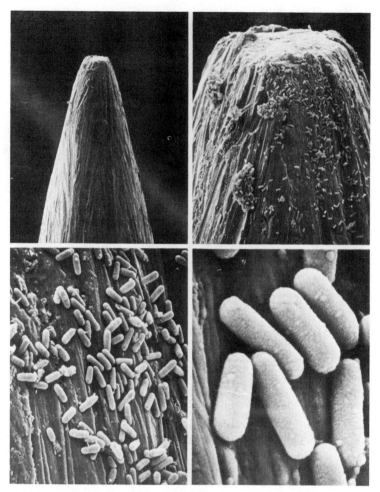

Figure 11.11   SEM micrographs of a pinhead at progressively higher magnification. Bacteria are clearly seen at high magnification (from the Science Photo Library).

THE PREPARATION OF SAMPLES
FOR MICROSCOPY

The preparation of samples for the light microscope is relatively straightforward: The sample is placed on a glass slide, possibly stained (see next section), and covered.

Sample preparation for the EM is considerably more difficult. The procedure usually involves "fixing" the sample using a cross-linking reagent such as glutaraldehyde and then cutting very thin sections of the material with a microtome.

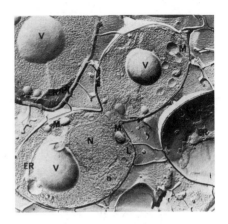

Figure 11.12 Example of freeze etching showing a budding yeast cell. The labels refer to cellular organelles as follows: PL, plasmalemma; ER, endoplasmic reticulum; V, vacuole; M, mitochondrion; L, lipid granule; N, nucleus (magnification ×10,000) (from H. Moor, *Int. Rev. Cytol.* 25(1969):391).

These sections, which may be as thin as 80 nm, are floated on water and then lifted off by a fine copper grid. Sections may also be cut from frozen tissue using refrigerated microtomes. These elaborate procedures are necessary mainly because of the harsh treatment (vacuum plus irradiation) received by the sample in the EM.

It is also common to make a **replica** of the surface for EM. This is done by coating the sample with a thin layer of platinum and a carbon layer for support. Both the metal and the carbon are deposited by vacuum evaporation. However, the metal layer is often applied with the evaporation source at an angle to the surface of the specimen. This technique, known as **shadow casting,** can give extra information about the dimensions of a particle (see problem 11.2). Replica-formation methods are harsh to the sample because water has to be removed for the evaporation technique; this often causes biological structures to collapse. Some improvement can be obtained by maintaining the sample at very low temperatures. These methods are sometimes called **freeze etching.** An example is shown in Figure 11.12.

Example: The Freeze-fracture Method Applied to Plasma Membranes.   If a suspension of cells or a small piece of tissue is frozen rapidly in liquid nitrogen (77 K) and the sample is subjected to a blow from a knife, the sample often fractures down the middle of the membrane, exposing the interior membrane surface. The reason for this is that the forces holding the outer and inner halves of the membrane together are relatively weak. For transmission EM, a replica of the fractured surface is made by evaporating a thin layer of carbon on the exposed surface. The shape of this carbon replica is further emphasized by evaporating a thin layer of platinum on the surface, but this time at an angle, to produce a shadow. The biological material is then digested away and a thin carbon-platinum

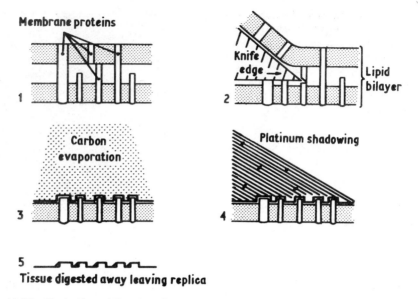

Figure 11.13   Illustration of the procedures used to obtain freeze-fracture micrographs (adapted from C. R. Hopkins, *Structure and Function of Cells*, Philadelphia: Saunders, 1978).

replica remains. The EM images of such surfaces have given a great deal of information about the nature of the distribution of proteins in membranes. The procedure is outlined in Figure 11.13.

CONTRAST IN MICROSCOPY

The Light Microscope

The difference in apparent brightness from different regions of a sample defines the contrast. In light microscopy, contrast usually arises from different absorption in different parts of the sample; this contrast can be very high, especially with staining (see further). The fact that support systems and solvents that are transparent to light can readily be found is an aid to the production of good contrast. In addition to staining, different detection methods can be used—for example, **polarization microscopes,** which use polarized light, and **phase-contrast** and **interference microscopes,** which detect differences in refractive index. The **fluorescence microscope,** which uses selective stains (fluorescent probes) and detects light at a different wavelength from the source illumination, has excellent contrast properties.

Worked Example 11.4   The Phase-Contrast Microscope

A typical arrangement for the phase-contrast microscope is shown in Figure 11.14. Because this arrangement is sensitive to refractive index differences in a sample, it is possible to obtain marked improvement in contrast by this method. Can you suggest why this arrangement works and what the image will look like? (*Hint:* Consult Chapter 8, "Refractive Index" and remember that the specimen scatters light.)

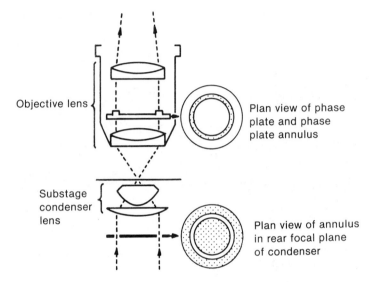

Objective lens

Plan view of phase plate and phase plate annulus

Substage condenser lens

Plan view of annulus in rear focal plane of condenser

Figure 11.14   A typical arrangment of lenses, masks, and phase plates in the phase-contrast microscope.

Solution

Consider first the situation with a uniform specimen. In addition to transmitting direct rays, shown as dotted lines in the diagram, the specimen scatters light. This scattered, or deviated, light does not pass through the phase-plate annulus. In addition, it is 90° out of phase with respect to the direct beam (see Figure 2.13). However, the direct beam, which is a hollow cone of light produced by the annular mask, is retarded by 90° by the annular phase plate. There is, therefore, now a 180° phase shift between the direct and scattered rays. If the amplitude of the direct rays is reduced by a thin film of metal on the annulus, the net result is a relatively dark background caused by cancellation of the two rays (see Figure 11.15). With a non-uniform specimen, variations in refractive index result in phase changes between the direct and deviated rays. High refractive index regions will appear bright in the image because the condition for destructive interference is removed.

(continued)

Worked Example 11.4   (continued)

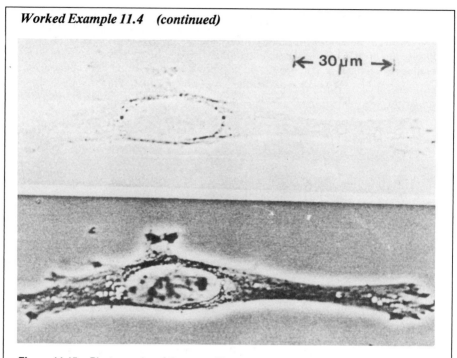

Figure 11.15   Photographs of the magnified image of a cell in a light microscope with normal illumination (top) and with phase contrast.

The Electron Microscope

The main interaction between electrons and matter is scattering. The scattering is very strong, about 10^4 times that for X-rays, but the difference between the scattering from different atoms is less than for X-rays. Heavy metals, for example, scatter much less strongly than would be expected from their atomic number (see Chapter 8). This lack of contrast is a very important feature in EM. Contrast is also reduced because of scattering from the supporting grid. The method of underfocusing helps to produce contrast to some extent and will be discussed later in this chapter.

STAINING PROCEDURES

The Light Microscope

With the light microscope, the main purpose of staining is to introduce differential contrast in the sample. The main staining method is to use dyes that bind to different cellular components. For example, a negatively charged dye, such as eosin, tends to bind to cytoplasmic components at pH 6. Nucleic acid components, on the other hand, tend to be negatively charged at this pH, so they bind

Figure 11.16 T4 virus negatively stained with uranyl acetate (courtesy Michael Moody).

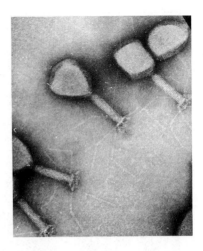

positively charged dyes, such as methylene blue. Stains can also bind via hydrophobic rather than electrostatic interactions. They may also be carried on antibodies specifically raised against a particular macromolecule. Similar selective dye procedures are used in the fluorescent microscope (see following material).

The Electron Microscope

The intrinsic contrast in the EM is very low, therefore staining is usually essential in this form of microscopy. The most common stains are metals such as tungsten, lead, uranium, and osmium tetroxide, which also acts as a fixative. Uranyl salts sometimes bind preferentially to macromolecules, while lead salts tend to bind to membranes. If the sample of interest, e.g., a virus, takes up the heavy metal it is said to be **positively stained.** If, instead, the heavy atoms bind to the background and not to the object of interest, the sample is said to be **negatively stained** (see Figure 11.16).

Worked Example 11.5   Shape and Molecular Weight of DNA

If a solution of DNA in $1M$ ammonium acetate and cytochrome *c,* (a basic protein) is floated on a weak ammonium acetate solution, a film of denatured cytochrome *c* spreads across the surface with molecules of DNA bound to it. This film can be removed on a grid and stained with uranyl acetate. Contrast can be further enhanced by shadow casting. The DNA is coated with cytochrome *c* and projects above the protein film. This method, known as the **Kleinschmidt procedure,** gives excellent contrast and the DNA can be readily seen and its length and shape (e.g., circular, supercoiled, or double-stranded) can be determined (see Figure 11.17).

(continued)

Worked Example 11.5   (continued)

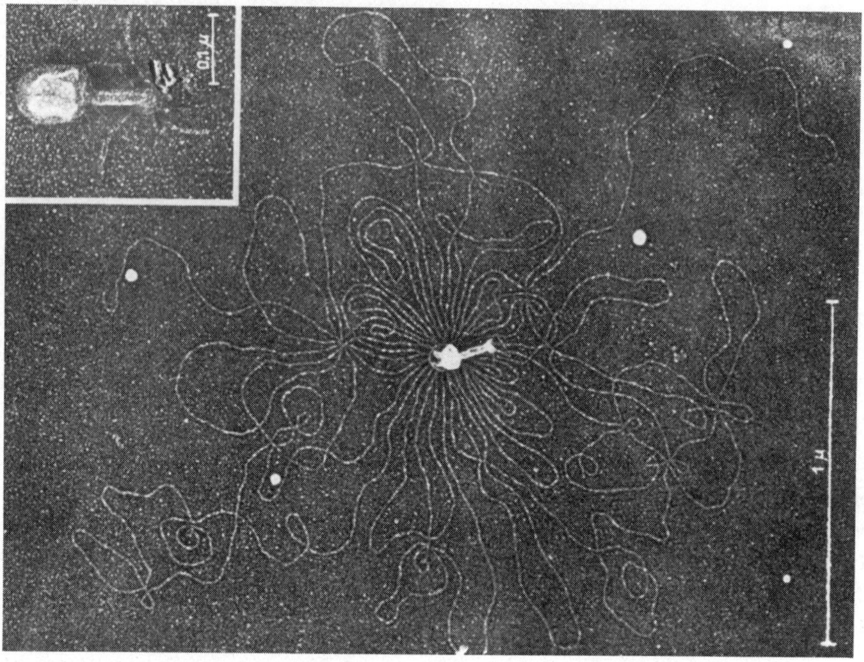

Figure 11.17   A micrograph, obtained using the Kleinschmidt procedure, of DNA spilling out of T_2 phage that has been subjected to osmotic shock (magnification × 64,000). The inset shows a normal virus at higher magnification (from A. K. Kleinschmidt et al., *Biochim. Biophys. Acta*, 61(1962):851).

In a sample of double-stranded bacteriophage DNA, the length was observed to be 15 μm. What is the molecular weight of the DNA? (The average molecular weight of a nucleotide pair is about 650 daltons and the helix contains about 10 nucleotides per turn, with a repeat every 3.4 nm.)

Solution

The length of 10 nucleotides (6500 daltons) is 3.4 nm. Thus, 15 μm is equivalent to $(15 \times 10^3/3.4) \times 6.5 \times 10^3 = 2.9 \times 10^7$ daltons.

ADVANCED TECHNIQUES IN ELECTRON MICROSCOPY

Resolution of around 0.3 nm can be achieved from good electron microscopes on nonbiological samples. However, for biological materials, typical resolution achieved is in the range 1.5–5 nm. The reasons for the low resolution are (1) preservation of the sample by fixatives, (2) severe damage caused by radiation, and (3) low contrast without staining. Some of these limitations may be overcome by various methods.

Underfocusing to Produce Phase Contrast

Phase contrast is the dominant mechanism for obtaining contrast in thin-stained and unstained specimens in EM. This technique requires constructive or destructive interference between the scattered and unscattered waves ($\pm \pi$). But because the phase shift between these two waves is $\pi/2$, there is no effective contrast normally.

There are, however, two effects that can give rise to phase shifts: One is spherical aberration of the lens; the other is an out-of-focus image. If the objective lens is deliberately weakened, it is possible to balance spherical aberration and underfocusing effects to produce a phase shift of π between the scattered and unscattered waves, thus optimizing the contrast. For negatively stained samples, underfocusing of around 500 nm is used for this purpose.

Optical Filtering Techniques

Many of the problems associated with EM images arise from a poor signal-to-noise ratio. Noise is generated by factors such as the stain, the supporting film, radiation damage, and random background on the photograph. This noise can be reduced in cases where the sample consists of a periodic or symmetrical array. Such arrays are common in biological samples. It is then possible to improve the signal-to-noise ratio by taking an average over the whole electron micrograph. One of the best ways of analyzing linear periodicities is to use the **optical diffractometer** originally used by Taylor and Lipson (see also Chapter 12).

Figure 11.18 illustrates how an optical diffraction pattern is obtained from an electron micrograph. This diffraction pattern consists of a series of spots that are

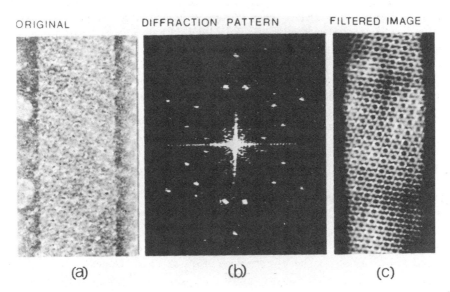

ORIGINAL   DIFFRACTION PATTERN   FILTERED IMAGE

(a)   (b)   (c)

Figure 11.18   Illustration of the use of optical filtering (from R. Crowther and A. Klug, *Ann. Rev. Biochem.*, 44(1975):161–182).

Figure 11.19 Different projections of the object can be obtained by tilting with respect to the transmission beam.

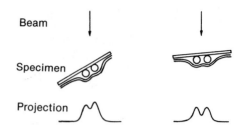

related to the original structure. The spots are superimposed on a background of noise. A suitable mask can be made, which consists of holes or slits of the correct size to match the diffraction pattern but remove the noise. This mask can be used to reconstruct a filtered image of the original micrograph. Note that the diffraction pattern is the Fourier transform of the image.

Image Reconstruction

If electron micrographs are recorded with the specimen tilted at different angles with respect to the electron beam, then a series of different projections can be obtained (see Figure 11.19). A 3-D image can be reconstructed from these projections. This method is particularly effective in EM because of the high depth of focus normally present (low numerical aperture).

There are a variety of ways in which these projected images can be recombined to give the image. The most common way is to digitize the images and perform a series of Fourier transformations. The resolution of the reconstructed image depends, among other factors, on the number of projections obtained. If the object has high symmetry, fewer tilt angles are required. The method works best in such cases, since radiation damage limits the number of projections that can be collected. In some cases, it is possible to combine projections from different specimens.

Combination of Electron Diffraction with Microscopy

In Chapter 12, the work of Unwin and Henderson on bacteriorhodopsin is discussed in which electron diffraction data are combined with electron micrographs to obtain 0.7-nm resolution on a membrane protein structure.

THE FLUORESCENCE MICROSCOPE

A schematic view of a fluorescence microscope with vertical illumination is shown in Figure 11.20. With a suitable choice of filters, only the fluorescent light emitted by the sample reaches the eyepiece of the detection camera. Transmitted-illumination types of fluorescence microscope are also in use, but the vertical-illumination scheme is superior in many ways.

Figure 11.20 A schematic view of the optical system in a vertical-illumination fluorescence microscope (adapted from S. de Petris, *Methods in Membrane Biol.,* 9 (1978):1).

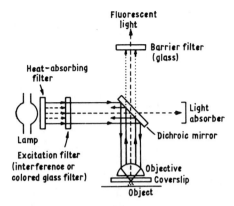

Several fluorescent dyes can be used at very low concentrations to label specific regions of a cell. These often have no apparent effect on the biological properties of the cell. For example, the orange fluorophore tetramethyl rhodamine isothiocyanate (TMRITC) can be used to label membrane proteins, *N*-phenyl-1-naphthylamine (NPN) gives blue fluorescence in lipids, and acridine orange can be used to label DNA.

An interesting application is the monitoring of molecules that perform a specific function in the cell. For example, fluorescently labeled actin has been injected into cells to follow the changes in actin distribution during a variety of cellular processes, including pinocytosis and amoeboid movement.

Worked Example 11.6   Fluorescein Labeling of Cells

Fluorescein produces apple-green fluorescence, while FDA does not (see Figure 11.21). It is FDA, however, that is added to cell suspensions when a probe for intracellular environment is required. Can you explain this?

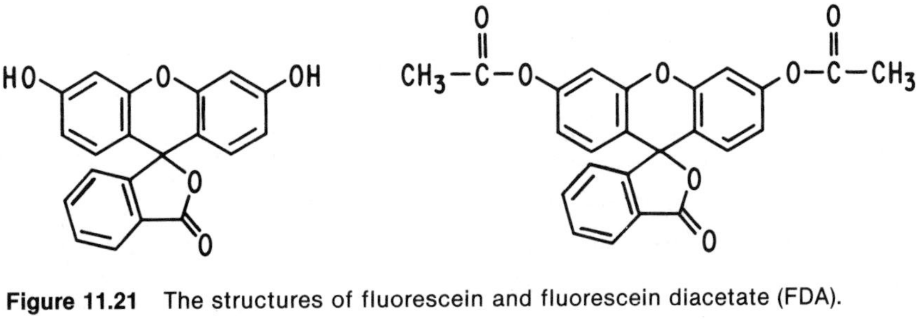

Figure 11.21 The structures of fluorescein and fluorescein diacetate (FDA).

Solution

Fluorescein is polar and does not cross intact cell membranes. FDA is nonpolar, so it crosses the membrane. Once FDA is inside the cell, endogenous esterase activity produces the probe fluorescein, which is then trapped inside the cell.

Worked Example 11.7   Measurement of Diffusion by Fluorescence Photobleaching Recovery

The surface of a single fibroblast cell was labeled with a fluorescein isothiocyanate (this attaches covalently to proteins). A small area of the surface ($\sim 10^{-7}$ cm^2) was briefly exposed to an intense laser pulse, thereby irreversibly "bleaching" the fluorophores in that region. (The light from the laser causes the fluorescein molecules to undergo a photochemical reaction by exciting their electrons to an unstable energy state. The resulting molecule does not fluoresce.) The optics of the microscope were arranged so that the fluorescence emission from the irradiated region could be monitored as a function of time (the laser was used to provide the excitation source but at very low power). The fluorescence was observed to increase approximately exponentially with time with a half-life for recovery of 1 min. Can you estimate the diffusion coefficient for the fluorophores diffusing back into the irradiated region? (A result from gas kinetic theory is that the mean free path length x of a molecule is $\sqrt{2Dt}$.)

Solution

The area of the irradiated spot is $\pi r^2 = 10^{-7}$ cm^2. A very rough estimate of the diffusion coefficient can be obtained if we assume that half the intensity will reappear when the square of the mean free path

$$\bar{x}^2 = (10^{-7})/(2\pi) = 2Dt$$

therefore,

$$D = \frac{10^{-7}}{2\pi \times 2 \times 60} \text{ cm}^2 \cdot \text{s}^{-1}$$
$$= 1.3 \times 10^{-10} \text{ cm}^2 \cdot \text{s}^{-1}$$

(For more realistic equations to solve this problem see, for example, R. Cherry, *Biochem. Biophys. Acta,* 559(1979):289–327.)

PROBLEMS

1. Prove that the relationship between focal length f, image distance from lens v, and object distance from lens u is $1/f = (1/u + 1/v)$. Assume that the angles of incidence on the lens are small.

2. Determine the wavelength of an electron of mass m (9.1×10^{-31} kg) and charge e (1.6×10^{-19} C) in an accelerating voltage V of 50,000 V. Use the relationship that equates the kinetic energy ($mv^2/2$) and the potential energy (eV) of the electron as well as the de Broglie equation, which relates the wavelength (λ) of a particle to its momentum (mv) by the equation $mv = h/\lambda$.

3. A shadow-cast electron micrograph of a mixture of spherical polystyrene particles (diameter 80 nm) and of a species of spherical virus was obtained. The apparent length of the polystyrene particles was 185 nm, while the apparent length of the virus was 105 nm. What is the diameter of the virus?

Figure 11.22

4. The images illustrated in Figure 11.22 were obtained in EM micrographs of an assembly of enzymes. What structure is the assembly likely to have?

5. A sample of microtubules consists of tubes 25 nm in diameter made up from tubulin monomers, with 13 such monomers around the perimeter of the tube. What sort of EM micrographs do you expect to observe?

6. The polarization of the fluorescence emitted by FDA-labeled blood lymphocytes is observed to increase after the cells are treated with phytohemaglutinin, a plant lectin. Can you suggest why this might occur?

CHAPTER 12

Diffraction

OVERVIEW

1. Diffraction patterns can be detected when a wave is scattered by a periodic structure whose dimensions are comparable to the wavelength of the wave. X-rays and neutrons are of suitable wavelength to detect atomic detail in molecules. The molecules must be in an ordered array such as a crystal or a fiber.

2. The diffraction patterns can be interpreted directly to give information about the size of the unit cell (hence the molecular weight), the symmetry of the molecule and in the case of fibers, information about periodicity, e.g., pitch of a helix.

3. The determination of the complete structure of a molecule cannot be carried out from the diffraction pattern alone because it is necessary to have phase information as well as the intensity and frequency information that is available. The phase can be determined using the method of multiple isomorphous replacement, where heavy metals are incorporated into the diffracting crystals.

4. Once the electron (from X-rays) or nuclear (from neutrons) scattering density of a macromolecule has been calculated, the final coordinates are determined using knowledge about the primary structure of the macromolecule and are refined by processes that include comparisons of calculated and observed diffraction patterns.

5. The biological information derived from diffraction has been enormous. The 3-D structures of about 100 different crystalline proteins have been determined and there is little evidence to suggest that the structures so determined are different in the living cell. This detailed information has greatly enhanced our understanding of enzyme mechanism and specificity, ligand-induced conformational changes, and the role and evolution of protein structures. Nucleic acids have also been studied both in the fibre and the crystalline form yielding a wealth of information on base pairing and structure. Membranes, membrane proteins, and viruses containing both RNA and protein have also been studied.

INTRODUCTION

The diffraction of radiation by crystals and fibers has provided an enormous amount of valuable information about biological structures. Light, X-rays, neutrons, and electrons have all been used to produce diffraction patterns. The most widely used have been X-rays because of their availability and penetration as well as the fact that their wavelength is small compared with the molecules studied. Diffraction patterns can be understood in terms of Fourier transforms and the scattering theory outlined in Chapter 8.

In the formation of an image by a lens, a diffraction pattern is formed in the focal plane and the image is formed from this diffraction pattern. (This idea was used in discussing the resolving power of a microscope in Chapter 11.) With X-rays, the diffraction pattern is formed by the radiation scattered from the crystal or fiber, but the image-formation stage must be carried out by a reconstruction procedure. This procedure is necessary because it is not possible to make lenses suitable for X-rays, since the refractive index of most substances is essentially unity. In addition, other focusing devices, such as mirrors, are relatively coarse compared with the dimensions of the molecules to be studied. In practice, the reconstruction is done using a computer.

PHYSICAL PRINCIPLES OF DIFFRACTION METHODS

Diffraction

Diffraction is the name given to phenomena occurring when a wave passes an object and "spreads" beyond the object. The case of diffraction at a slit has been discussed elsewhere. Different diffracted waves interfere with each other and produce an **interference** pattern. Interference patterns are most informative when the wavelength of the applied radiation is less than the dimensions of the diffracting object and when the object consists of a *regular array* or *lattice*.

Lattices

A **crystal lattice** is a regularly repeated arrangement of atoms. The **unit cell** of a lattice is the smallest and simplest unit from which the three-dimensional periodic pattern can be produced. The unit cell may be made up from several identical **asymmetric units.**

These ideas are illustrated for a two-dimensional lattice in Figure 12.1. In (a) the lattice is shown as an array of points. In (b), (c), and (d), three different unit cells are illustrated. The crystal lattice is made up by placing (b), (c), or (d) at every lattice point. In (c) and (d), the unit cell is made up from two and four asymmetric units, respectively.

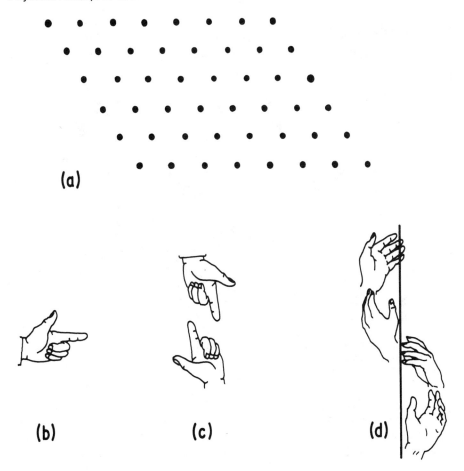

Figure 12.1   An illustration of (a) a two-dimensional array of lattice points, (b) a unit cell with no symmetry, (c) a unit cell with two asymmetric units related by a rotational axis, and (d) a unit cell with four asymmetric units related by a fourfold screw axis. A crystal lattice can be formed by placing (b), (c), or (d) at every lattice point in (a).

It should be noted that the asymmetric units in (c) and (d) are related by symmetry. For (c), this is a twofold rotation axis perpendicular to the page. For (d), it is a fourfold screw rotation in the plane of the page. (An object has an n-fold rotation axis if the structure appears identical after being rotated through an angle of $360°/n$ about the axis.)

The asymmetric unit is the "brick" from which the crystal is made. It is not necessarily the same as the molecular brick. In some cases, the asymmetric unit comprises more than one molecule, while in others it is made up from part of a molecule.

Note that unit-cell dimensions are usually labeled by the letters a, b, and c, while the dimensions of the molecule are usually labeled x, y, and z.

Worked Example 12.1   Unit Cells, Asymmetric Units, and Symmetry

Describe the unit cell, asymmetric unit, and symmetry of the structure shown in
Figure 12.2.

Figure 12.2   Unit cells, asymmetric units, and symmetry (from F. M. C. Crick and J.
C. Kendrew, *Adv. Prot. Chem.*, 12(1957):134).

Solution

The most obvious unit cell to take contains two mermaids. The asymmetric unit
is one mermaid, and there is twofold rotational symmetry between the two asymmetric units.

The Reciprocal Lattice

The diffraction pattern has a reciprocal relationship with the object. With two
slits, for example, the closer together the slits are, the farther apart are the mini-
ma in the diffraction pattern. A crystal is a 3-D array of atoms in *real space,* and
its diffraction pattern is a 3-D array in *reciprocal space.* The labels used to define
the array in reciprocal space are *h, k,* and *l.*

Fourier and Optical Transforms

An elegant way of describing diffraction patterns is to use Fourier transforms,
which describe the mathematical relationships between the diffraction patterns
and the objects (see Appendix VII). This procedure allows interconversions to be
made between real space with coordinates (*xyz*) and reciprocal space with coordinates (*hkl*).

Fourier transformation can be performed by a digital computer as a series of
calculations, or by an optical device, which can carry out the process by means of

Figure 12.3 Examples of optical transforms showing the masks and the resulting
diffraction patterns in (a) one molecule and
(b) a lattice of molecules (from C. A. Taylor
and H. Lipson *Optical Transforms* (London:
Bell and Sons, 1964)).

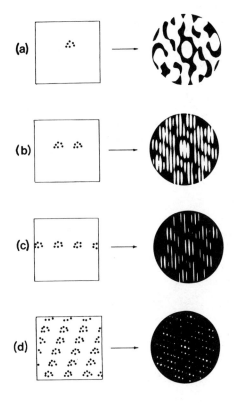

a lens. A mask of an object can be used to generate a diffraction pattern, or a mask of a diffraction pattern can be used to generate an image of the object by a device called an optical diffractometer. It is useful also in electron microscopy (see Chapter 11). The transformations carried out are called **optical transforms.** Some examples of optical transforms are given in Figure 12.3.

Convolution

It is often useful to consider a crystal lattice to be made up from two separable parts: the lattice and the molecule or unit cell at each point on the lattice. A mathematical method suitable for bringing about this separation is **convolution.** If the lattice function in Figure 12.1(a) is called F_1, and if the hand in Figure 12.1(b) is called F_2, then the function representing the crystal lattice, where a hand is placed at every lattice point, is the convolution of F_1 and F_2—that is, $F_1 * F_2$. The convolution theorem (see Appendix VII) states that the convolution of two functions in one domain (e.g., real space) is the multiplication of these two functions in the Fourier transform domain (e.g., reciprocal space). A practical example of this is seen in Figure 12.3, where the various arrays of molecules in real space give rise to the product of the diffraction pattern of the lattice times the diffraction pattern of the molecule.

Worked Example 12.2   The Unit Cell in Reciprocal Space

The Fourier transform of a row of points separated by a distance x is a row of lines, separated by $1/x$ and running perpendicular to the row of points. Using the convolution theorem, calculate the diffraction pattern from a 2-D lattice of points with a unit cell as shown in Figure 12.4(a).

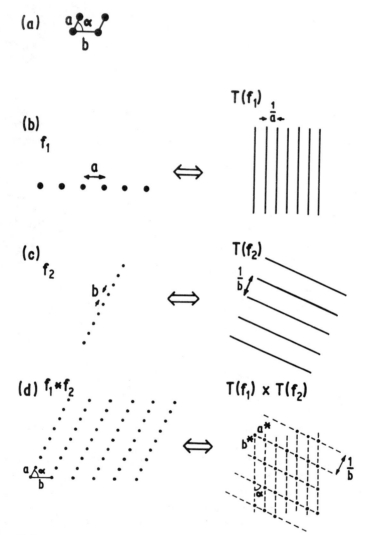

Figure 12.4

(continued)

Worked Example 12.2 (continued)

Solution

A row of points separated by a (f_1) gives a diffraction pattern $T(f_1)$, as shown in Figure 12.4(b). The row separated by $b(f_2)$ gives a similar result, as shown in Figure 12.4(c).

The 2-D lattice is the convolution of f_1 and f_2 ($f_1 * f_2$), since f_2 is placed at every point in f_1. The diffraction pattern of the 2-D lattice is $T(f_1)$ multiplied by $T(f_2)$. Two sets of lines multiplied together give zero everywhere except at the intersections. The diffraction pattern, or reciprocal lattice, is thus as shown in Figure 12.4(d).

The unit cell in reciprocal space thus has sides of lengths $a^* = 1/(a \sin \theta)$ and $b^* = 1/(b \sin \theta)$. (The unit-cell dimensions in reciprocal space are usually labeled $a^* b^* c^*$.)

The Bragg Equation

So far, we have discussed diffraction patterns in terms of Fourier transforms and the convolution theorem. There are alternative ways of visualizing the problem, one of which is to consider the diffracted beams as if they were *reflected* from planes passing through points in a crystal lattice. This view was first proposed by Bragg and led to the famous equation bearing his name,

$$n\lambda = 2d \sin \theta$$

where n is an integer, d is the spacing between the lattice planes in the crystal, and θ is the angle of incidence of the beam with wavelength λ. (see Figure 2.12(b)). The Bragg equation gives an alternative way of deriving the diffraction pattern of the lattice; it does not give a description of the diffraction from the molecules in the unit cell.

EXPERIMENTAL MEASUREMENT OF DIFFRACTION

Sources of Radiation

The most common radiation used for diffraction studies are X-rays, neutrons, and electrons. Electron sources were discussed in Chapter 11.

It is important to have intense sources of X-rays for work on large biological structures. A common method of producing X-rays is to strike a target of a metal with electrons accelerated in a high voltage. A commonly used wavelength is 0.154 nm, which is produced from a copper target. Recently, excellent diffraction patterns have been obtained by using the intense X-rays available from a synchrotron radiation source. In this system, accelerated electrons are injected into an electron-storage ring and these orbiting electrons emit synchrotron radiation.

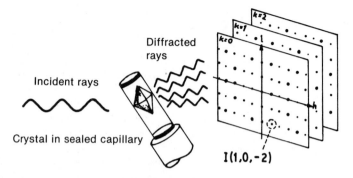

Figure 12.5 Photographic collection of data from an enzyme crystal shown schematically. In any given experiment, the film records the blackness of reflections on one section of the 3-D X-ray diffraction pattern. The 1, 0, −2 reflection is circled (after D. Eisenberg, *The Enzymes*, 2d ed., 1(1970):1).

Neutrons are produced in a nuclear reactor. Like all particles, they obey the de Broglie relationship $\lambda = h/p$. The momentum p depends on the temperature of the reactor (see problem 3). Typical wavelengths are 0.1–1 nm.

The Sample in the Diffraction Experiment

The usual samples in a diffraction experiment are fibers or crystals. Fibers often occur naturally, but orientation can be induced by applying shear, magnetic fields, and the like. The fiber is usually mounted perpendicular to the axis of the beam.

Crystals must have dimensions of about 0.2 mm in two directions for normal sources, but they can be considerably smaller if synchrotron radiation is used. The growing of suitable crystals is one of the most difficult and tedious stages in diffraction studies. Very high ionic strengths are often required to salt out the crystals. The usual method of mounting the crystal is shown in Figure 12.5. The crystal is in a sealed capillary, but sometimes it is kept in the **mother liquor**—the solution from which the crystal was formed.

Detection of the Diffraction Pattern

We have indicated that the diffraction pattern of a crystal is a 3-D array of points in reciprocal space. The spots can be detected by a photographic film, as shown in Figure 12.5.

In practice, photographs from stationary crystals are not used. Instead, the crystal is rotated in a defined way to generate the diffraction pattern. A common method of doing this is to use a system where both crystal and photographic film precess about the incident beam. A precession photograph of a lysozyme crystal showing the (h0l) plane is reproduced in Figure 12.6.

Figure 12.6 A precession photograph of a tetragonal crystal of lysozyme (courtesy C. C. F. Blake).

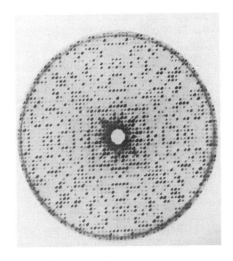

In addition to photographic film, which is suitable for X-rays and electrons, various other radiation detectors and counters can be used for X-rays and must be used for neutrons. It is important to note that all these detectors record only the intensity of the scattered radiation, not the amplitude. The intensity is an average rate of arrival of energy and is proportional to the square of the amplitude. This leads to difficulties when attempts are made to reconstruct the image from the diffraction pattern (see later discussion). Note also that X-rays and electrons are scattered by the electrons in the crystal, while neutrons are scattered by the nuclei.

INTERPRETATION

We have discussed the formation of the patterns and the way in which the information is collected. Now we come to the interpretation of the patterns in terms of molecular structure. This is equivalent to a discussion of reconstruction procedures that are carried out by a lens in a microscope. As we shall see, the complete reconstruction is quite complex. However, there is information that can be obtained directly from the diffraction pattern with a minimum of computation. We discuss this information first.

Direct Interpretation of Diffraction Data

Determination of Molecular Weight. It is clear from Worked Example 12.1, that the dimensions of the unit cell can be obtained from the reciprocal lattice. This information can be used to obtain limits on the size and molecular weight of the molecule, as shown in the following worked example.

Worked Example 12.3   Calculation of Molecular Weight from Unit-Cell Volume

A crystal of a protein has unit-cell dimensions 7.02 nm × 4.23 nm × 8.54 nm. The density of the native crystal is measured as 1.29 g·ml^{-1} and the water content is found to be 36% (w/w). Calculate the apparent molecular weight, assuming that the axes are orthogonal (at right angles).

Solution

The volume V of the unit cell is $a \times b \times c$. Thus, $V = 7.02 \times 4.23 \times 8.54 \times 10^{-27}$ m^3 = 2.54×10^{-25} m^3.

The mass of the unit cell is $1.29 \times 2.5 \times 10^{-25} \times 10^6 = 3.28 \times 10^{-19}$ g. Since 36% of the crystal is water, the mass of the protein in the unit cell is $0.64 \times 3.28 \times 10^{-19} = 2.1 \times 10^{-19}$ g.

If there is one protein per unit cell, then mol. wt. = $2.1 \times 10^{-19} \times \mathfrak{N}$, where $\mathfrak{N}$ is Avogadro's number (6.02×10^{23}). Thus, the apparent molecular weight is 126,000 daltons. Note that the real molecular weight could be less or more than this depending on the number of molecules in the unit cell.

The Symmetry of the Molecule.   Macromolecules often display some symmetry. The best known example of this is probably hemoglobin, where two pairs of $\alpha\beta$ subunits are related by a twofold axis of rotation. Such symmetric molecules often crystallize in such a way that the rotation axis of the molecule is also a rotation axis of the crystal. In such a case, information about the symmetry of the molecule can be obtained from the diffraction pattern.

The symmetry of crystals is usually defined by seven different systems. These systems define the relationship between the three axes of the unit cell (a, b, and c). For example, the cubic system has a threefold axis along the diagonals of the cube, while the hexagonal system has a sixfold axis.

In addition to the relationship between the axes, the symmetry relating the contents of the unit cell can be defined by its **space group.** If a unit cell is made up from a number of identical units (the asymmetric unit), then these units may be related to each other by certain symmetry operations, for example, by a rotation axis or a screw axis. (Other types of symmetry operation are not possible for biological molecules, which are chiral.) A unit cell with a fourfold screw axis is shown in Figure 12.1(d).

There is a direct relationship between the symmetry of the crystal lattice and the reciprocal lattice. For example, the crystal lattice must have either an n-fold rotation or an n-fold screw axis if the reciprocal lattice has an n-fold axis of symmetry. A screw axis can be distinguished from a rotation axis in the diffraction pattern, since the former causes certain spots to have zero intensity systematically.

Worked Example 12.4   Symmetry from Diffraction Patterns

In one crystal form, an enzyme crystallizes with a unit cell containing eight asymmetric units, but there are only four molecules per cell. In another crystal form of the same molecule, it was deduced from the diffraction pattern that the crystal had a threefold rotation axis. Can you deduce anything about the symmetry of the molecule from these data?

Solution

In the first crystal form, there must be a twofold rotation axis, since there are two asymmetric units per molecule. A twofold plus a threefold rotation axis is characteristic of a hexamer. (In fact, the enzyme aspartate *trans*-carbamylase from *E. Coli* has been found to behave in this manner.)

Diffraction Patterns from Fibers.   Many biological molecules are long and occur naturally as fibers or can be induced to line up in one direction. With such an oriented array, it is possible to obtain an informative diffraction pattern. The usual method of obtaining the diffraction pattern is to direct the X-ray beam at right angles to the direction of the fiber.

The fiber is made up of molecules that are oriented *randomly* about the common direction, which causes an averaging of much of the diffraction pattern. The regular repeats in the structure, however, are not averaged out. This means that no information is obtained about the structure at the atomic level (e.g., the sequence of bases in a DNA strand), but the regular features (e.g., a helix) can often be discerned.

To analyze fiber patterns properly, one must use complicated mathematical functions called Bessel functions. We shall not discuss them here, but shall simply indicate the expected diffraction pattern for a helix.

For a continuous helix, it can be shown that the diffraction pattern is as shown in Figure 12.7. The pitch *P*, the radius *r*, and the pitch angle β are indicated on the figure. Note the layer lines in the diffraction pattern, separated by

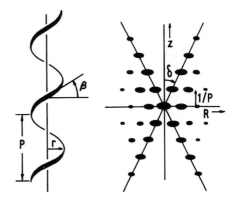

Figure 12.7   A continuous helix and its corresponding diffraction pattern (its Fourier transform). The coordinates of the helix, *z* and *r*, are in nanometers; the coordinates of the diffraction pattern, *Z* and *R* are in reciprocal nanometers (from R. E. Dickerson, *The Proteins*, 2d ed. (New York: Academic Press, 1964)).

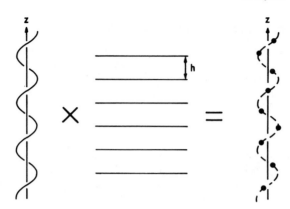

Figure 12.8   The product of a continuous helix and a set of planes separated by *h* is a discontinuous helix, as shown.

$1/p$, and the "cross" formed by the strong innermost reflections on each layer line. The half-angle, δ, of this cross is approximately equal to β.

Real biological fibers do not have an even distribution of scattering centers. A better approximation to a biological helix is one with a certain number of points per turn. To construct the expected diffraction pattern for such a helix we can use the convolution theorem. A discontinuous helix can be made by multiplying a continuous one by a set of planes, as illustrated in Figure 12.8. The Fourier transform of the discontinuous helix is thus the convolution of the Fourier transforms of the continuous helix and a set of planes. The effect of making the helix discontinuous is to introduce new origins for the cross on each of the points corresponding to the Fourier transform of the set of planes, which are $1/h$ apart. Note that there are lines on the meridian ($R = 0$) when $Z = 1/h$. This is illustrated by the optical transforms shown in Figure 12.9.

Figure 12.9   (a) Optical transform of a continuous helix. (b) Optical transform of a helix with 10 points per turn. (c) Optical transform of a helix with 5 points per turn (from K. C. Holmes and D. M. Blow, *Meth. Biochem. Anal.*, 13(1965):113).

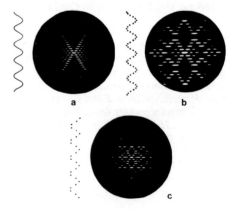

Worked Example 12.5
Diffraction Pattern of Moist DNA

What information can you deduce from the diffraction pattern shown in Figure 12.10, which shows the X-ray diffraction pattern of moist DNA. The numbers refer to spacings, in nanometers, corresponding to the regular features of the DNA fiber.

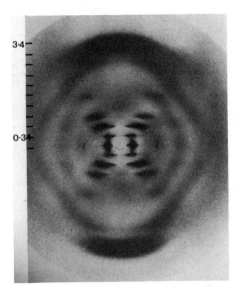

Solution

The cross pattern is characteristic for a helix. There are strong reflections on the meridian ($R = 0$) line at 0.34 nm, while the separation of the layer lines is 3.4 nm. This suggests that there are 10 bases per turn of the helix (see Figures 12.8 and 12.9), with a helix pitch of 3.4 nm.

Figure 12.10   Diffraction pattern of moist DNA (from R. E. Dickerson, *The Proteins*, 2d ed. (New York: Academic Press, 1964)).

Determination of Molecular Structure

We shall now discuss in more detail the way in which molecular structure is reconstructed from the diffraction pattern. This process is usually carried out in several stages and these will be described in the following sections. First, we shall outline some mathematical procedures, then discuss the "phase problem" and its solution. Finally, we shall discuss the way in which molecular structure is determined from computed scattering density and the methods used to refine structures.

Fourier Synthesis, Fourier Transformation, and Patterson Functions.   In this section we will define some mathematical procedures useful in the reconstruction process.

The crystal and the diffraction patterns are periodic systems and any such system can be simulated by the superposition of a set of sinusoidal waves. This operation is known as *Fourier synthesis*. Figure 12.11 shows two examples of such a synthesis. Five different frequency components are summed to yield a representation of two one-dimensional crystals—one containing a molecule with two atoms, the other a molecule with three atoms. Note (1) that the amplitude and phase of the higher frequency components are different in the two cases and (2) that the resolution of the image depends on the number of higher-frequency components included in the summation. The molecules are poorly resolved when only components 1, 2, and 3 are summed.

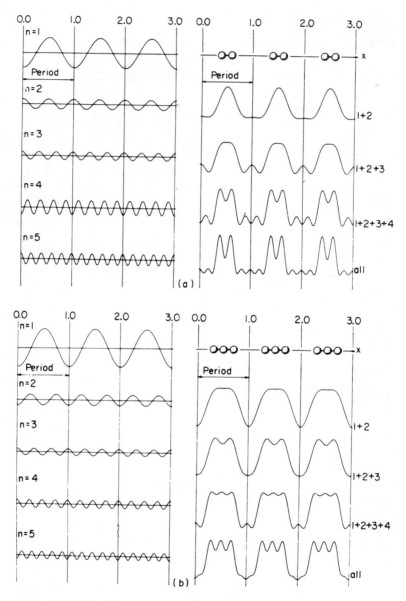

Figure 12.11   Two different representations of one-dimensional crystals with (a) a diatomic molecule in the unit cell and (b) a triatomic molecule in the unit cell. On the right are shown waves of different amplitude and phase. Summations of these different components are shown below the molecules. Note that the same frequencies are used in both (a) and (b), but the amplitudes and phases are different (from J. Waser, *J. Chem. Educ.*, 45(1968):446.)

The structure of a molecule can thus be synthesized by summing waves, but how do we determine the frequency, amplitude, and phase of the waves to be summed? The information must be derived from the diffraction pattern and the best method of doing this is by Fourier transformation. The mathematical form of a Fourier transformation is given in Appendix VII. Although the form given there can be extended to three dimensions, it is more convenient to use summations rather than integration. The equation that relates the scattering density $\rho(xyz)$, the real space domain, to the diffraction pattern $F(hkl)$, the reciprocal space domain, is then

$$\rho(xyz) = K \sum_h \sum_k \sum_l F(hkl)\exp[-2\pi i(hx + ky + lz)] \qquad (12.1)$$

where K is a scaling factor that depends on experimental variables, such as the volume of the unit cell. We shall write this equation in the convenient shorthand form

$$\rho(xyz) \iff F(hkl)$$

Note that it is also possible to calculate a diffraction pattern from $\rho(xyz)$, since Fourier transforms have an inverse property (see Appendix VII).

There is a problem in the evaluation of equation 12.1, which will be discussed in the next section. The measurable parameter in a diffraction pattern is $I(hkl)$, the intensity of the reflection at position (hkl). The magnitude of the structure factor $F(hkl)$ can be determined from $I(hkl)$, but the phase cannot be determined. Thus, equation 12.2 cannot be calculated directly from the diffraction pattern.

Another equation that is very useful in interpreting diffraction patterns is the **Patterson function,**

$$P(uvw) \iff |F(hkl)|^2 \qquad (12.2)$$

where the bars indicate absolute values. Since $|F(hkl)|^2$ is a measurable quantity, this calculation can be performed directly from the diffraction pattern, unlike equation 12.1.

The problem with the Patterson function $P(uvw)$ is that it is much more complicated than $\rho(xyz)$. For a structure with n "atoms", equation 12.2 generates n^2 peaks in the image, with n peaks superimposed at the origin, and $(n^2 - n)$ peaks elsewhere. The positions of the scattering atoms can be deduced by placing, in turn, each atom of the structure at the origin. For large molecules, this function produces very complex patterns, which cannot be interpreted.

In many cases, greatly simplified images can be obtained by taking differences. **Fourier difference maps** arise from calculations of

$$\rho'(xyz) \iff [F_1(hkl) - F_2(hkl)]$$

where F_1 and F_2 are structure factors obtained under slightly different conditions. These calculations are useful, for example, where structure factors are observed

---

Worked Example 12.6   The Patterson Function

Draw the Patterson image expected for a triangular molecule.

Solution

The Patterson image of a triangle is as shown in Figure 12.12.

Figure 12.12   Structure   Patterson function

---

in proteins with (F_1) and without (F_2) a bound ligand. In Fourier difference maps, knowledge about phase is necessary. Patterson difference maps arise from the Fourier transform of $|(F_1 - F_2)|^2$, for example, with (F_1) and without (F_2) a heavy-metal atom.

The Phase Problem and Its Solution.   We have mentioned several times the fundamental problem that arises because phase information is lost as a result of the fact that the experimental parameter is $I(hkl)$ rather than $F(hkl)$. It is shown in Appendix III that a convenient notation for a wave with a phase angle ϕ and amplitude a_0 is $a_0\exp(i\phi)$. Thus, we can rewrite $F(hkl)$ as $|F(hkl)|\exp(i\phi_{hkl})$. $|F(hkl)|$ is proportional to $[I(hkl)]^{1/2}$, but ϕ_{hkl} is unknown. It is also shown in Appendix III that a wave can be represented as a vector in a complex plane with an amplitude $|a_0|$ and phase angle ϕ. Applying this picture to structure factors, we see that any measured structure factor is known to lie somewhere on a circle, as shown in Figure 12.13. Note that $F(hkl)$ is really a vector.

Two methods that can be used to determine phase angles are outlined below. The first is by far the most widely used.

Figure 12.13   Representation of a structure factor of amplitude $|F|$ in a "complex" plane. The angle ϕ is unknown.

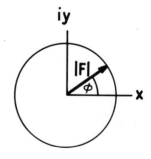

Figure 12.14 Isomorphous replacement. Two precession photographs of triclinic lysozyme crystals are superimposed, slightly out of horizontal register. The left spot of each pair is from a native lysozyme crystal; the right spot is from a crystal that has been diffused with $HgBr_2$. This is a photograph of the (0kl) plane of the reciprocal lattice. Note the differences in intensities (from R. Dickerson, in *The Proteins*, 2d ed., vol. 2, ed. H. Neurath (New York: Academic Press, 1964)).

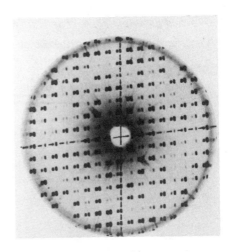

Multiple Isomorphous Replacement. The **multiple isomorphous replacement** method depends on finding a way of attaching an atom or group of atoms that scatter strongly to one or more specific sites on a crystal. It is important that the crystal should otherwise retain the same (*iso*) form (*morphe*).

For X-rays, derivatives of mercury, uranium, platinum, gold, and lead have been among the many reagents used. (Remember that the X-ray scattering is proportional to the atomic number; see Chapter 8.) Isomorphorus crystals have been made both by diffusing the reagent into a performed crystal and by crystallizing the protein in the presence of the reagent.

The effect on the diffraction pattern of suitable isomorphous replacement is to change the intensities (see Figure 12.14). The changes in intensity can be measured and a Patterson difference map can be obtained from the Fourier transform of $(\Delta F)^2$. This difference map can often be used to identify the positions of the heavy atoms, since it is relatively simple compared to a normal Patterson map (see problem 4).

Let us denote the structure factor of the original crystal as the vector F_P, and the structure factor of the crystal with the heavy atom as F_{PH}. The structure factor of the contribution from the heavy atom can be written as F_H. The scattering from the crystal with the heavy atoms is the sum of that from the original crystal plus that from the heavy atom. Thus,

$$F_{PH} = F_P + F_H$$

The vector F_H is calculated from a Patterson difference map or anomalous scattering data.

The worked example below shows how a diagram (known as a **Harker diagram**) can be used to determine the phases angle of F_P.

Worked Example 12.7   The Harker Diagram

First, draw F_H as a vector in the complex plane (take, for example, $\phi = 45°$, $F_H = 1$ unit). Next, draw a circle of radius F_P (e.g., 5 units), taking the end of F_H as the origin. Draw another circle of radius F_{PH} (e.g., 5.5 units) with its center at the beginning of F_H. What do you conclude about possible phase angles for F_P?

Solution

As shown in Figure 12.15, there are two possible phase angles for F_P. It is necessary to make at least one more isomorphous replacement in order to obtain F_H', and F_{PH}'. This will lead to a "unique" (within experimental error) solution for the F_P phase angle.

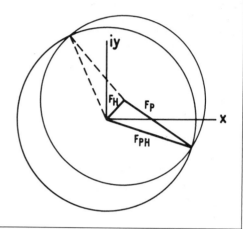

Figure 12.15

Anomalous Scattering. The anomalous scattering method has been much less widely used than isomorphous replacement, but it has been used recently in a complete structure determination of crambin and also to locate heavy atoms in a structure. In addition, the technique has an interesting physical basis that relates to much of our discussions in earlier chapters.

So far in this chapter, we have assumed in our discussions that all the scattering atoms scatter with the same phase. We showed in Chapter 8, however, that the phase of the scattering undergoes a phase shift of 180° as the applied wavelength λ goes from the condition $\lambda << \lambda_0$ to $\lambda >> \lambda_0$, where λ_0 is the resonant wavelength of the scatterer.

If a crystal contains some atoms that scatter in a different phase from others because they are near an absorption (or resonance) frequency, then this can be detected in the diffraction pattern.

Consider the scattering from a pair of atoms, as shown in Figure 12.16. If the two atoms scatter in the same phase, then the path difference will be the same for the reflection at (hkl) as it is for the reflection with negative coordinates denoted $(\bar{h}\bar{k}\bar{l})$. (This is known as **Friedel's law.**) Now consider the case where one of the atoms (e.g., the filled circle) is an anomalous scatterer and scatters with a path difference of q compared with the open circle. The path difference for the (hkl) reflection will now be $2p + q$, while the path difference for the $(\bar{h}\bar{k}\bar{l})$ reflection will be $2p - q$. Since the intensity of the diffraction depends on the path differences between different scatterers, the result of anomalous scattering is that

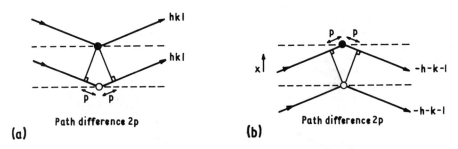

Figure 12.16 Scattering from a pair of scatterers. If the scattering from the filled circle and the open circle has the same phase then the intensity of the reflection at position h,k,l will be the same as at position $-h, -k, -l$ since the path difference ($2p$) is the same for these two reflections.

$F(hkl) \neq F(\bar{h}\bar{k}\bar{l})$. It can be shown that difference maps calculated from $\Delta F = |F(hkl)| - |F(\bar{h}\bar{k}\bar{l})|$ can be used to calculate phase angles. Anomalous scattering has also been used to define the positions of heavy atoms.

Worked Example 12.8   Anomalous Scattering in Crambin

The structure of crambin, a small hydrophobic protein, was determined by W. A. Hendrickson and M. Teeter (*Nature,* 290 1981:107) using the anomalous scattering of the sulfur in the three disulfide bonds. The wavelength of the X-rays used was 0.154 nm and the nearest absorption band of sulfur corresponds to 0.52 nm. How much anomalous scattering do you expect when $\lambda = 0.229$ nm and 0.071 nm?

Solution

One would expect the anomalous scattering to increase as the applied radiation approaches the absorption band.

Determination of Molecular Structure from $\rho(xyz)$.   Calculations of $\rho(xyz)$ are presented as contour maps of scattering density. An example is shown in Figure 12.17 for a benzene molecule.

It is clear that a benzene ring can be fitted very well to this map if the ring is tilted with respect to the direction of the projection shown. For large molecules, however, it is not always so simple. The fitting procedure also depends on the quality of the map, which depends on the quality of the crystal, the phase determinations, and the resolution.

The effect of resolution on the map is illustrated in Figure 12.18. The electron-density maps calculated from the diffraction pattern within the three circles in (a) are shown in (b), (c), and (d) with resolutions of 0.4 nm, 0.2 nm, and

Figure 12.17 Nuclear-density contours for benzene determined by neutron methods (courtesy of G. E. Bacon *Endeavour*, 25 (1969):129). The greater density for the top and bottom hydrogens results from the fact that in this projection, protons from two molecules are almost superposed at these two positions. Dotted contours indicate negative density, continuous contours positive density.

0.1 nm, respectively. The diffraction pattern was calculated from the projection of two strands of poly-L-alanine in an antiparallel β-sheet onto a plane. It can be seen that while it is difficult to fit the molecular structure with any precision at 0.4-nm resolution, it is very clear at 0.1-nm resolution. In practice, 0.6-nm resolution is often used as a starting point in data analysis because α-helixes can just be observed. To identify the images of groups of atoms, 0.2 nm is barely sufficient.

In practice, it is nearly always necessary to know the sequence of amino acids or bases in the macromolecule before a model can be constructed, since a model can then be made from a good map with a resolution of about 0.28 nm. The procedure for fitting the sequence to the electron density has usually been done with a box—first used by F. Richards—in which a half-silvered mirror is used to superimpose a wire model and contour maps on transparent sheets. Computer graphics and computer searches are increasingly being used to simplify this stage of the analysis.

It should be noted that these procedures assume the structures of the side chains and backbone of the molecule and allow only for rotations about bonds to give the best fit to the contour maps.

Refinement. Once a structure $\rho(xyz)$ has been calculated, then the expected diffraction pattern, $F(hkl)$, can be calculated from this structure using equation 12.1. An estimate of how good the agreement is between the observed data (F_0) and the calculated data (F_{calc}) is then given by

$$R = \frac{|F_0| - |F_{calc}|}{|F_0|}$$

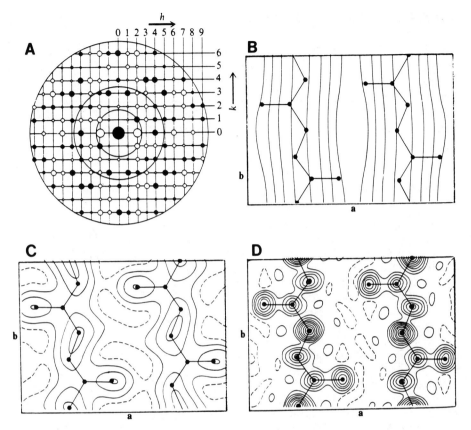

Figure 12.18 Electron-density maps at different resolution: (a) Shows the calculated diffraction pattern for a two-dimensional crystal formed from the projection of strands of polyalanine in an antiparallel sheet. The circles contain the data required to perform an analysis to 0.4-nm, 0.2-nm, and 0.1-nm resolution. Open circles represent $F(hk) < 0$. The size of each point represents $|F(hk)|$. (b) An electron-density map calculated at 0.4-nm resolution, (c) at 0.2-nm resolution, and (d) at 0.1-nm resolution (from R. Fraser and T. McRae in *Physical Principles and Techniques of Protein Chemistry*, Part A, ed. S. J. Leach (New York: Academic Press, 1969)).

An R-value of around 0.25 means that most atoms are precisely placed, whereas an R-value greater than 0.45 suggests a very poor fit. Errors in the structure can arise in a variety of ways—both in measurement of $|F_0|$ and in the phase angles. The error in F_0 is usually around 5–10% while the errors in phase are higher around 30%, because of uncertainties such as nonisomorphism of crystals.

It is possible to improve a calculated structure by minimizing differences between the calculated and observed diffraction patterns. It is often assumed that all the errors arise from the phase angles. Some refinement procedures also use energy minimization. These refinement procedures, which can be very time consuming, obviously work best on high-resolution data: for example, 0.15 nm.

PROTEIN CRYSTALLOGRAPHY

Perhaps the most spectacular success of diffraction techniques has been the determination of the 3-D structures of about 100 different proteins. This wealth of information has had a considerable impact on our understanding of protein structure and function. The contributions to our knowledge include information about enzyme specificity and catalysis, conformational changes, the folding of proteins, their evolution, and the prediction of their structures from amino acid sequences.

Worked Example 12.9   Convergent Evolution and the Charge-Relay System in the Serine Proteases

Figure 12.19 shows three residues at the active site of the proteolytic enzyme chymotrypsin. The —OH group of Ser-195 is found to be very reactive, probably due in part to the so-called charge-relay system, which involves interactions between Asp-102, Ser-195 and His-57. A very similar situation occurs in the bacterial enzyme subtilisin, where there is also a reactive serine in a similar orientation to a histidine and an aspartate. Except for this similarity, however, the 3-D structures and the amino acid sequences of the two enzymes are very different. Can you suggest why this might be?

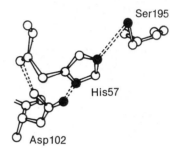

Figure 12.19   Conformation of three residues at the active site of chymotrypsin (from D. M. Blow and T. A. Steitz, *Ann. Rev. Biochem.,* 39(1970):86).

Solution

The vertebrate enzyme chymotrypsin and the bacterial enzyme subtilisin have reached the same solution to the problem of cleaving peptide bonds. This appears to be an example of convergent evolution.

Worked Example 12.10   Specificity in the Serine Proteases

The serine proteases (a) chymotrypsin, (b) trypsin, and (c) elastase prefer to cleave peptide bonds when they are next to side chains that are (a) aromatic or bulky and nonpolar, (b) lysine or arginine, and (c) small and uncharged. X-ray crystallography studies show that the "pocket" next to the point of cleavage is related to the schematic diagrams shown in Figure 12.20. Can you suggest which enzyme has which pocket?

(continued)

Worked Example 12.10   (continued)

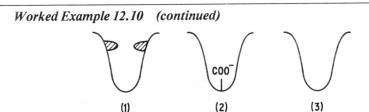

(1)   (2)   (3)

Figure 12.20

Solution

   The enzymes (a), (b), and (c) have pockets (3), (2), and (1), respectively. (This is, in fact, an example of divergent evolution, where mutations of an ancient gene gave rise to these three enzymes.)

Worked Example 12.11   Cooperative Binding of O_2 to Hemoglobin

Figure 12.21 shows a schematic drawing of the oxy (Fig 12.21(a)) and deoxy (Fig 12.21(b)) conformations of the hemoglobin tetramer which was deduced from X-ray diffraction studies. In the deoxy form, the molecule 2,3-diphosphoglycerate (DPG) is bound between the β-subunits. It is possible to give an explanation for the observed cooperative binding of O_2 on the basis of conformational changes and changes in electrostatic interactions (salt bridges) between the two forms of the hemoglobin molecule. Also, DPG binds selectively to the deoxy form. One particular structural change that occurs is that His-β-146 is relatively free in the oxy form while it is close to Asp-β-94 in the deoxy form. Can you suggest how this might be linked to the fact that a drop in pH from 7.2 to 7.0 favors the formation of the deoxy form?

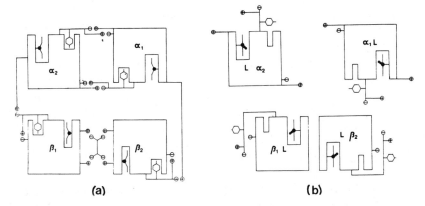

(a)   (b)

Figure 12.21   Schematic views of the (a) deoxy and (b) the oxy forms of the tetramer of hemoglobin (from M. F. Perutz, *Nature*, 228(1970):726). Recent results by Shaanan have suggested that the tyrosines do not move as much as indicated here (Perutz, private communication).

(continued)

Worked Example 12.11   (continued)

Solution

The pK$_a$ of a histidine in free solution is about 6.8. The acidic form of the histidine side chain is positively charged, while the basic form is neutral. Thus, the proximity of the negative aspartate stabilizes the acidic form. In fact, the pK$_a$ of His-β-146 increases from about 6.8 to the oxy form to 7.8 in the deoxy form (see Figure 6.34). An alternative and equivalent statement is that the acidic form favors the formation of the salt bridge between His-β-146 and Asp-β-94.

Methods Used to Depict 3-D Protein Structures

There is a problem in handling the large amount of detailed information that is given by coordinates determined using X-ray crystallography. One method is to have a 3-D model of the protein structure constructed from some suitable material, such as wire, wood, or plastic. However, the use of computer graphics with sophisticated programs that allow the rotation and expansion of regions, together with calculations of protein surfaces, is becoming a very powerful technique.

It is also desirable to represent the structure on paper and this has been done by various kinds of drawings (see illustrations in this chapter). An alternative method is to use **stereo drawings,** which allow one to visualize the 3-D structure from a pair of pictures. This can be done with a stereoscope, but with practice it can also be done with the unaided eye. It is a skill well worth learning, since it is very common to present structures in this form. The technique requires looking at both pictures at once, one with each eye, and superimposing a focused image. A good description of the technique is given in W. B. Wood et al., *Biochemistry: A Problems Approach* (Menlo Park, Calif.: Benjamin/Cummings, 1981).

Try to obtain stereo views of the enzyme hexokinase from the stereo pairs given in Figure 12.22. T. A. Steitz and colleagues have shown that this enzyme closes around the substrates ATP and glucose when they bind. In Figure 12.22(a), there is no substrate, but in Figure 12.22(b), ATP is bound. In Figure 12.22(c), a more detailed view is shown with a few of the amino acids around ATP and glucose. The glucose can be seen to interact with an aspartate, while the phosphate groups of ATP interact with a lysine residue [see T. A. Steitz et al., in *Mobility and Function of Proteins and Nucleic Acids,* (London: Pitman (CIBA Foundation 93, 1982), p. 25. It may help initially to place a white card between your eyes and each picture so that each eye sees only one picture.

Example: Neutron Diffraction to Study Hydrogen–Deuterium Exchange in Proteins.   The main advantage of using neutrons instead of X-rays in diffraction studies is the ease with which hydrogen atoms can be detected and distinguished

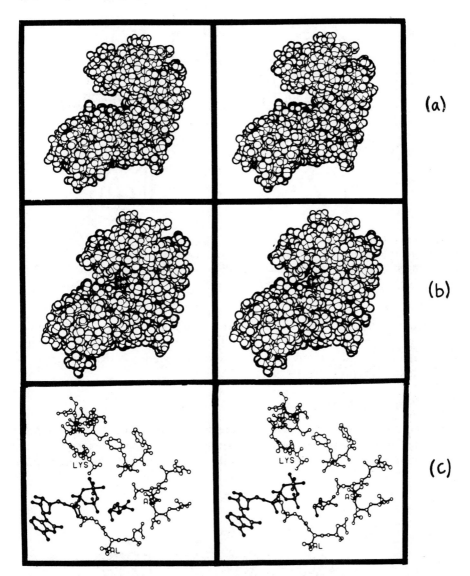

Figure 12.22 Stereo views of the enzyme hexokinase, which phosphorylates glucose using ATP; (a) without ATP, (b) with ATP, and (c) an expanded view of the active-site region (from C. M. Anderson et al., *Science*, 204(1979):375).

from deuterium. Figure 12.23 shows parts of the scattering-density maps of trypsin taken from a neutron diffraction study. In (a), the crystal had been soaked in D_2O for 1 y before data collection was begun. This treatment caused many of the H atoms to exchange with D atoms in the peptide bonds. The unexchanged control is shown in (b). Note that the dotted line signifies negative density, which

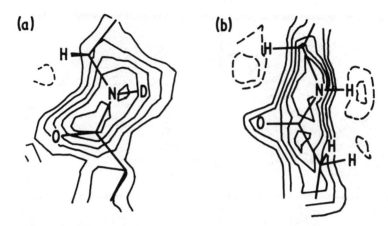

Figure 12.23   Sections of a neutron-density map taken in the plane of a peptide group. In (a) the peptide group is fully exchanged while in (b) it is unexchanged. Note that the dotted lines indicate negative density (adapted from A. A. Kossiakoff, *Nature,* 296(1982):713).

arises from the negative scattering cross section of H (see Chapter 8). A complete analysis of the H–D exchange throughout the protein is thus possible; the results have been interpreted in terms of conformational flexibility in the protein (see A. A. Kossiakoff, *Nature,* 296(1982):713.

Classification of Proteins

It is sometimes convenient to classify proteins into different types, based on the percentage of a given type of secondary structure. Drawings of examples of the classes antiparallel α, parallel α/β and antiparallel β are shown in Figure 12.24.

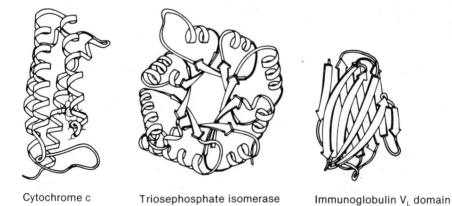

Cytochrome c   Triosephosphate isomerase   Immunoglobulin V_L domain

Figure 12.24   Drawings of three different classes of protein (J. S. Richardson, *Adv. Prot. Chem.,* 34(1981):167).

Figure 12.25 The effect of isotropic vibration on the scattering by a carbon atom. The atomic scattering is shown as a function of Q. Vibration causes the scattering at large angles to decrease. This reduces the intensity of reflections in the outer regions of the diffraction pattern (adapted from J. P. Glusker and K. N. Trublood, *Crystal Structure Analysis: A Primer,* (Oxford: University Press, 1972)).

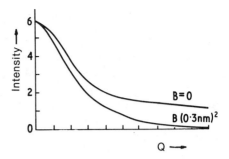

Temperature Factors and Protein Mobility

We have seen (see Chapter 8) that the X-ray scattering factors (f) for individual atoms are proportional to the atomic number and fall off with Q or $\sin(\theta/2)/\lambda$, where θ is the deviation of the diffracted or scattered beam from the incident beam. Such a relationship is shown for a stationary carbon atom in Figure 12.25. If the scattering atom is undergoing thermal vibration then it is bigger and the curve will fall off faster with increasing Q. Curves of this kind, shown in Figure 12.25, can be approximated by the relationship $f \exp(-2B \sin^2\theta/\lambda^2)$, where B is called the **temperature factor.** It is equal to $8 \pi^2 \overline{U}^2$, where $\overline{U}^2$, is the mean square amplitude of the atomic vibration.

Temperature factors can be determined from good crystal diffraction data and from refinement procedures; thus, $\overline{U}^2$ can be estimated. Figure 12.26 shows a plot of $\overline{U}^2$ for each of the main chain atoms in hen (HEW) and human (HL) lysozymes. The correspondence of the high values obtained around residues 47, 72, and 101 suggests that the plots represent characteristic motion in the lysozyme molecules. These regions of high displacement generally lie on the "lips" of the active site cleft of lysozyme. It is important to note that it is not possible to distinguish between thermal motion and static disorder from single measurements of this kind.

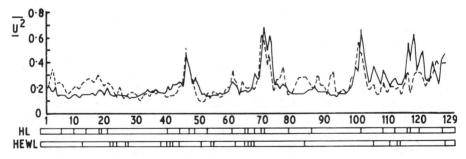

Figure 12.26 A plot of the main chain $\overline{U}^2$ values against residue number for HL (full line) and HEW (broken line) (from P. J. Artymiuk et al., *Nature,* 280(1979):563).

The Validity of Protein Crystallography

The preceding discussion and examples should have given you an idea of the enormous amount of detailed information that has been derived from studies of protein crystals. It seems appropriate to ask, however, if the structures of proteins in the living biological system are the same as in a crystal, which contains high concentrations of salt. The answer to this question appears to be yes, they are the same, although some small changes probably occur. In general, it appears that the crystal lattice forces are weaker than the forces that hold the protein together.

The evidence for the validity of the structures determined comes from a variety of sources, including the fact that some enzymes retain their activity in the crystal form. In addition, solution studies using methods such as NMR, scattering, and chemical modifications are nearly always consistent with crystal studies. It must be remembered, however, that the detailed structural information obtained from these other methods is almost never as precise as from diffraction data.

Another problem with crystallography is the relatively long time scale required both to collect and to analyze the data, although this particular problem is being alleviated by modern radiation sources and computing methods. This still means, however, that molecular motion cannot be detected directly, and changes in structure induced by changes in solution conditions such as pH, temperature, and ligand concentration may be hard to analyze.

CRYSTALLOGRAPHY OF SYSTEMS OTHER THAN PROTEINS

In addition to the success of protein crystallography, studies of crystals of other systems have yielded detailed information about many molecules and molecular assemblies. Examples include the complete 3-D structure of t-RNA, the structure of a histone–DNA complex from chromatin, and the membrane-bound protein bacteriorhodopsin.

Crystallography of Nucleotides and Polynucleotides

The Watson-Crick base-pairing scheme in DNA is consistent with the hydrogen-bonding scheme observed in some crystals of mixtures of bases. For example, an electron-density map obtained from the crystalline complex of 9-ethylguanine with 1-methyl cytosine is shown in Figure 12.27(a). This can be compared with the structure of the G–C Watson-Crick pair proposed from studies of DNA fibres (see Figure 12.27(b)).

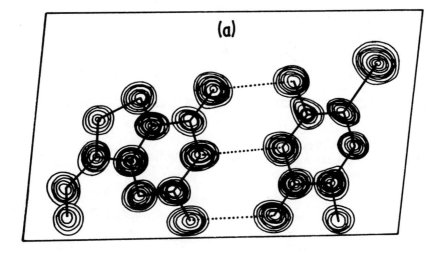

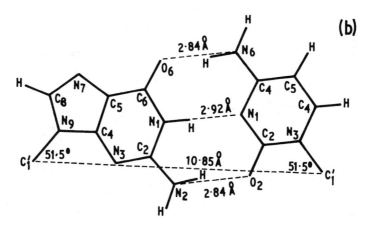

Figure 12.27   (a) The electron density obtained from mixed crystals of 9-ethylguanine with 1-methyl cytosine (from H. M. Sobell et al., *Proc. Nat. Acad. Sci.* U.S.A., 49(1963):885). ; (b) the Watson-Crick base pairing for G–C proposed from fiber diffraction studies of DNA by M. H. Wilkins and S. Arnott, *J. Mol. Biol.*, 11(1965):391). (Diagram Courtesy of the Biophysics Department,Kings College, London)

It is now widely accepted that DNA in low ionic strength aqueous solution is in the *B* conformation with 10 bases per turn in the helix and a rise per residue of 0.34 nm (see Worked Example 12.5). Various other conformations are, however, possible and several models have been proposed. Some of the evidence for these proposed structures comes from diffraction studies of crystals of oligonucleotides such as (dA–dT)$_2$ and (dC–dG)$_3$ (see, for example, R. D. Wells et al., *Prog. Nucl. Acid Res.,* 24(1980):167).

Worked Example 12.12   Neutron Diffraction Studies of the Nucleosome

Chromosomes from higher organisms are composed of a substance called chromatin, which consists of protein and DNA. The protein part is made up largely of histones, whose function appears to be to fold DNA in an efficient manner. It is possible to degrade chromatin using an enzyme that attacks DNA—a nuclease. This procedure, if carefully carried out, produces a particle that consists of about 146 base pairs of DNA and an octamer of histones.

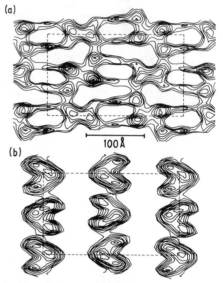

These core particles can be crystallized and both X-ray and neutron diffraction studies have been carried out on these crystals (see R. D. Kornberg and A. Klug, *Scientific American*, 244, 1981:48–60). A low-resolution (20 Å) neutron diffraction map is shown in Figure 12.28. (The phases were calculated from electron microscopy data; see Chapter 11.) Map (a) was calculated from data collected with the crystal in 39% D_2O buffer. Map (b) was obtained when the crystal was in 65% D_2O. What can you deduce about the structure of chromatin from these maps?

Solution

With 65% D_2O, the scattering from DNA is "matched" to the solvent (see Chapter 8); thus, scattering is observed from the protein alone. The histone octamer appears to be wedge shaped. With 39% D_2O, the observed scattering is predominantly from the DNA. It is clear that the DNA is wound around the histone core.

Figure 12.28   Neutron diffraction of the nucleosome (from G. A. Bentley et al., *J. Mol. Biol.* 145(1981):771).

Worked Example 12.13   Tobacco Mosaic Virus

Tobacco mosaic virus (TMV) particles are rodlike, 300 nm long and 18nm in diameter. TMV contains 2140 protein subunits, each with a molecular weight of 17.2 k daltons. It also contains a single strand of RNA 6400 nucleotides long. Figure 12.29 shows the radial density distribution obtained by interpretation of X-ray fiber diffraction data from an oriented gel of TMV and from helixes of polymerized TMV protein without the RNA. What can you deduce about the structure of the virus from these data? The protein polymer exhibits polymorphism in that it can form discs as well as helixes. It is also known that assembly of the TMV occurs by threading the RNA into preformed protein discs. The discs have been crystallized and the protein structure determined. Can you suggest why no density

(continued)

Worked Example 12.13   (continued)

is observed in the disc form at the position corresponding to 2.5 nm in
Figure 12.29?

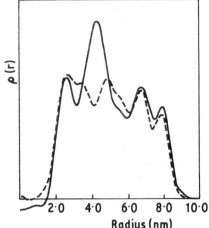

Figure 12.29 Radial density distribution of
TMV (continuous line) and the helical form
of the repolymerized protein without RNA
(broken line) (from K. C. Holmes, *Mobility
and Function in Proteins and Nucleic Acids*,
(London: Pitman, CIBA Foundation
Symposium, 1982) p. 116).

Solution

Figure 12.29 is consistent with that of a hollow virus, with the RNA located
primarily at 4.0 nm. The helix formed by the protein is otherwise very similar to the
form in the virus. In the discs, there is apparently a very flexible part of the polypeptide chain around 2.5 nm that gives very little density because of movement or
disorder in the crystal. This suggests a mechanism whereby the RNA can "wriggle"
into its correct position in the protein polymer and then order this part of the
polypeptide chain in the intact virus.

**Example: Electron Diffraction Studies of Two-Dimensional Crystals of Bacte-
riorhodopsin.** Membrane-bound proteins are obviously of great importance in
biology, yet no X-ray crystallographic analysis of these proteins has yet been
carried out because of the difficulty of obtaining suitable crystals. In some cases,
however, ordered arrays can be obtained in the plane of the membrane. A
structure of a membrane protein was obtained in such a case, using electron
diffraction and electron microscopy techniques. A resolution of 0.7 nm was
achieved.

The purple membrane of *Halobium halobium,* a salt-requiring bacterium,
contains bacteriorhodopsin, a 26,000 molecular weight protein that converts light
into a *trans*-membrane proton gradient. Crystalline sheets of this membrane with
diameters as large as 1 μm can be obtained with 75% of the mass in bacteriorhodopsin. The crystal lattice is one unit cell thick (4.5 nm).

Henderson and Unwin studied these sheets using glucose to protect the
specimen and very low electron doses to reduce radiation damage. Electron

(a)

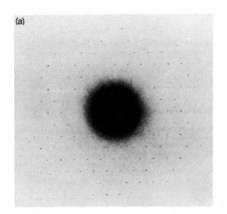

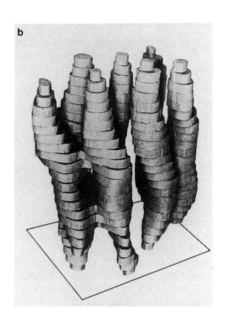

Figure 12.30   The electron diffraction pattern observed from the purple membrane is shown in (a). This pattern was analyzed to give a 0.7-nm resolution map of the structure of bacteriorhodopsin. A model built from this map is shown in (b) (from R. Henderson and P. N. T. Unwin, *Nature*, 257 (1975):28).

micrographs and electron diffraction patterns of unstained membranes were analyzed. The almost featureless micrographs were recorded with weak phase contrast provided by defocusing (see Chapter 11). They were improved by averaging the signals over thousands of unit cells using Fourier transformations.

The images produced were used to provide the phases, while the amplitudes of the structure factors were obtained from the diffraction patterns. The micrographs and diffraction patterns in Figure 12.30(a) were collected for various tilt angles of the membrane to the electron beam to obtain the 3-D structure illustrated in Figure 12.30(b). Seven α-helixes 3.5–4.0 nm long can be observed to run approximately normal to the plane of the membrane.

The advantage of using the diffraction pattern is that details of the structure can be detected that are unresolved in the image. The data can also be collected at relatively low electron dosage, thus minimizing radiation damage. (Staining protects the sample from radiation damage, but this procedure obliterates fine molecular detail and limits the resolution obtainable to about 2 nm.)

PROBLEMS

1. In the determination of the structure of myoglobin by X-ray diffraction, 400 reflections were measured to obtain a resolution of 0.6 nm. Approximately how many reflections would be needed to obtain a resolution of 0.2 nm?

2. Crystals of a protein of 55,000 daltons have unit-cell dimensions $a = 6.65$ nm, $b = 8.75$ nm, and $c = 4.82$ nm. If the crystals have a density of 1.26 g•ml^{-1} and are composed of approximately 50% water by weight, calculate the number of protein molecules in the unit cell (assume the axes are orthogonal). If there are eight asymmetric units in the unit cell, how many subunits does the protein have?

3. Calculate the wavelength of neutrons that emerge from a reactor at a temperature of about 100°C with a velocity of 2.8×10^3m•s^{-1}(mass of neutron = 1.67×10^{-27} kg).

4. The rules for calculating the positions of the peaks in a Patterson function were given in the section on Fourier synthesis. The amplitude of the peaks can also be calculated by multiplying the amplitude of the peak at the origin by the peak amplitude at the other positions. For example, the intensities in the triangle in Worked Example 12.6 would appear as $A \times C$, $B \times C$, and so forth, as shown in Figure 12.31. Using this information, what do you expect the appearance of a Patterson map of a molecule consisting of five carbon atoms and two mercury atoms?

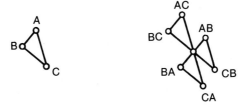

Figure 12.31   The intensities observed in a Patterson function.

5. Sometimes regions of a polypeptide chain give very weak electron-scattering density in X-ray studies of protein crystals, while other regions are clearly observed. Can you suggest a reason for this?

6. From X-ray crystallography studies, D. C. Phillips and colleagues postulated that when a polysaccharide molecule binds to lysozyme, the cleavage takes place between binding sites labeled D and E. They used model building in this work and proposed that the sugar ring in site D was distorted from its normal position. Can you suggest why they could not observe this distortion directly?

7. Glycogen phosphorylase is involved in the breakdown of glycogen and is subject to a series of sophisticated control mechanisms, including phosphorylation and conformational changes. Crystals of phosphorylase b shatter in the presence of glucose-6-phosphate (G-6-P), a substrate of the enzyme. Can you suggest why this might be?

8. The electron-density profile of an oriented phospholipid bilayer, derived from
    X-ray diffraction data, is shown in Figure 12.32. Give an explanation for the
    shape of the profile.

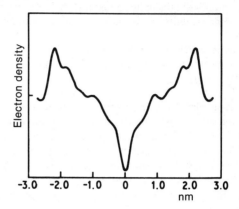

Figure 12.32   The electron-density profile of a lipid bilayer.

CHAPTER 13

Other Spectroscopic Methods

This chapter briefly presents the principles of several interesting spectroscopic methods not dealt with in the previous chapters. Some of these are very new; others are continually being developed and improved. Technical advances in physics, engineering, and chemistry are often applied to biology soon after their inception. Examples of rapid advances have been developments in laser technology and X-ray sources such as the synchrotron; many others are described both here and elsewhere in the book.

MÖSSBAUER SPECTROSCOPY

Mössbauer spectroscopy involves the absorption of a gamma ray by a suitable nucleus. The absorption energy corresponds to separations in the energy levels of the observed nucleus. Nuclei, which are made up from various particles, such as the neutron and the proton, have energy levels similar to those in atoms, which are made up from electrons and nuclei. The separations between these nuclear levels are very large and very high energies are involved.

The separations between the nuclear energy levels are sensitive to the environment. For example, if the charge distribution of the nucleus is nonspherical (this occurs when the spin quantum number is greater than $\frac{1}{2}$; see Chapter 6) and if there is an electric field gradient at the nucleus (from the electrons in a chemical bond, for example), then the energy levels will be split. The magnitude of this **quadrupole splitting** will reflect the symmetry of the electronic distribution around the Mössbauer nucleus.

In addition to quadrupolar effects, there are two other important interactions that affect a Mössbauer spectrum. One is the so-called **isomer shift**, which is a small effect that arises when electrons have some charge density at the nucleus.

This shift influences the position of the spectrum, rather than the splitting, and can give information about changes in the number and distribution of electrons around the nucleus—for example, the oxidation state. The third kind of environmental interaction is the **magnetic interaction,** in which applied or local magnetic fields can remove the degeneracy of the nuclear energy levels, just as in NMR.

In general, it is difficult to find a gamma-ray source of the correct energy to cause transitions between nuclear energy levels. The usual source nucleus used is ^{57}Co, which loses a gamma ray by radioactive decay. The energy of the gamma ray emitted depends on the energy difference between the excited and ground state of the source nucleus, which is not a variable. However, the energy can be varied by mechanically moving the source relative to the absorber, which causes a Doppler shift in the frequency of the emitted radiation (higher frequency if the source moves toward the absorber, lower if it moves away from it). It is thus possible to obtain a spectrum by recording the transmission of gamma rays through an absorbing sample as a function of the Doppler velocity.

The most commonly studied nucleus is ^{57}Fe. The natural abundance of ^{57}Fe is only 2%, but this is sometimes increased by enrichment. The relative motion of source and sample that causes the Doppler shift is usually a few millimeters per second and is often achieved by attaching the source to a microphone coil. The counts of gamma rays transmitted at each Doppler velocity are averaged many times to improve signal-to-noise ratio, and the spectra are collected over wide temperature ranges, from 4.2 K to over 300 K. Often, the sample is placed in a strong magnetic field to observe magnetic interactions.

Example: Mössbauer Spectroscopy of Deoxymyoglobin Crystals.   Figure 13.1 shows the Mössbauer spectrum of deoxymyoglobin crystals at 245 K. Note that the quadrupolar splitting, which arises from the low symmetry of the iron environment, gives rise to two Lorentzian lines, which can be well fitted to the experimental data. Note also the scale of millimeters per second and the isomer shift, which causes the spectrum to be symmetrical about $0.9 \text{ mm} \cdot \text{s}^{-1}$ rather than $0.0 \text{ mm} \cdot \text{s}^{-1}$. These results have been interpreted to mean that the iron lies above the heme plane. The iron was found to be farther out of the plane in hemoglobin than in myoglobin. Recent analysis of this spectrum has also revealed that there is a significant contribution from the mobility of the protein, which has been interpreted as being linked to the mobility of solvent water.

Reference.   S. A. Fairhurst and L. H. Sutcliffe, *Prog. Biophys. Molec. Biol.,* 34(1978):1.

PICOSECOND SPECTROSCOPY

Very short, high-intensity light pulses can now be produced from lasers by diverting one of a train of pulses to strike the sample of interest. These pulses allow events that occur on a picosecond (10^{-12} s) time scale to be investigated. This

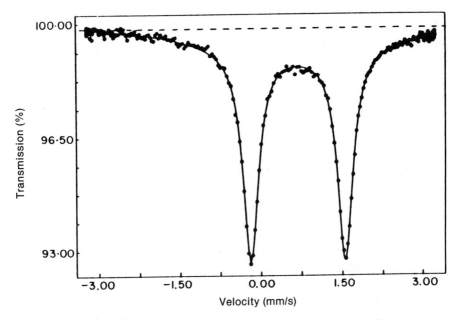

Figure 13.1 A Mössbauer spectrum of deoxymyoglobin with a theoretical fit to two Lorentzian lines (from F. Parak, E. W. Knapp, and D. Kucheida, J. Mol. Biol., 161(1982):177).

technique has been used to follow some chemical events in photosynthesis and in visual pigments.

Example: Photodissociation of Myoglobin. Myoglobin and hemoglobin bind CO, but the bound molecule can be made to dissociate by irradiation with suitable light. The spectra of myoglobin with and without ligand are different, as shown in Figure 13.2. The dissociation can be monitored by observing the absorbance at 420 nm as a function of time after the excitation pulse. The time constant for this process was observed to be about 3 ps. The maximum at 440 nm

Figure 13.2 The spectra of CO—myoglobin (MbL) and myoglobin + CO (Mb + L).

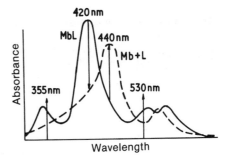

corresponds to the unbound form of the protein, and the time constant for the appearance of this band was observed to be about 11 ps when the photodissociation was caused by excitation pulses at both 355 nm and 530 nm. These experiments suggest that there is some sort of intermediate state with a lifetime of 7–10 ps on going from CO-bound to unbound myoglobin.

References.   A. H. Reynolds, S. D. Rand, and P. Rentzepis, *Proc. Nat. Acad. Sci. U.S.A.*, 78(1981):2292.

PHOTOACOUSTIC SPECTROSCOPY

Most optical spectroscopy is carried out by observing the absorbance of a clear solution. It is also possible to measure the scattered or reflected light from a clean surface. A third possibility is now available—a technique known as photoacoustic spectroscopy (PAS), which allows spectra to be obtained from powders, tissues, gels, and suspensions. In this technique, the sample, which is usually solid, is placed in a closed cell together with a nonabsorbing gas, such as air, and a sensitive microphone. The sample is irradiated with light of a known wavelength, as usual, but the intensity of the light is modulated by some means, such as a chopper. If the sample absorbs light, electrons reach an excited state and then lose this energy by nonradiative processes, which cause heating of the sample. The heating, which is periodic because the incident light is modulated, produces pressure and volume changes in the gas, which can be detected by the microphone.

Photoacoustic spectroscopy has several advantages, among which are its insensitivity to scattering, since this is an energy-conserving process; its ability to give optical absorption data on opaque materials; and its simple detection system—no photoelectric device is required to detect the signal.

Example: Photosynthesis in Chloroplasts.   The PAS spectra of two samples of chloroplast membranes prepared from lettuce leaves are shown in Figure 13.3. One of these preparations is known to be photosynthetically active. In one sample

Figure 13.3   The photoacoustic spectra of photosynthetic and poisoned chloroplast membranes (from D. Cahen, S. Malkin, and E. I. Lerner, *FEBS Lett.*, 91(1978):339).

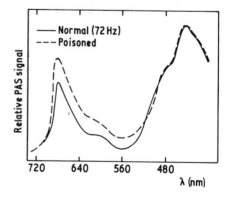

the chloroplasts have been poisoned by 3-(3,4-dichlorophenyl)-1,1-dimethylurea, the compound known as DCMU. At 680 nm, the poisoned chloroplasts give a signal that is significantly stronger than that of the unpoisoned sample. (The photosynthetic activity is known to be low at 440 nm; thus, comparisons can be made between samples of different photosynthetic activity by normalizing the spectra at 440 nm.) Photosynthesis, a chemical process, might be expected to compete with the nonradiative transitions that cause the heating detected by PAS. The experiment illustrated in Figure 13.3 demonstrates that this is indeed the case. Note that the frequency of modulation of the light beam is a variable in PAS. In this particular experiment, modulation frequencies of both 72 Hz and 770 Hz were used and differences between the observed PAS spectra were explained by the kinetics of inhibition.

Reference. A. Rosencwaig, *Ann. Rev. Biophys. Bioeng.*, 9(1980):31).

OPTICAL DETECTION OF MAGNETIC RESONANCE

In organic molecules, phosphorescence is a result of emission from the lowest triplet state (see Chapter 5), which is paramagnetic, with $S = 1$. This magnetic state is formed by two unpaired electrons, one in the ground state and the other in the excited state. There is a strong dipole-dipole interaction between these two electrons, which results in the splitting of the triplet-state levels even in zero field. This is called zero-field splitting (see Chapter 7). The magnitude of this zero-field splitting (ZFS) gives information on the molecule and its environment.

It is possible to detect the EPR signal from molecules in their excited triplet state and thus to obtain information about the ZFS, but greater sensitivity is achieved if the signals are detected optically. The phosphorescence emitted in the triplet state is, in general, different from each level. Thus, if transitions are induced between the levels by an applied microwave frequency, changes in the emission can be detected by a technique known as optical detection of magnetic resonance (ODMR), which is illustrated in Figure 13.4. The increased sensitivity of this technique is due to the greater ease of detection of emitted photons than of

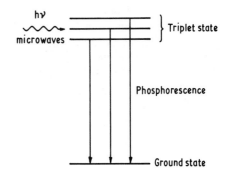

Figure 13.4 Illustration of the phosphorescence emission from a triplet state in zero field. Often this emission comes predominantly from one of the three energy levels. Microwave radiation of a suitable frequency can induce transitions between the triplet energy levels, which changes their populations and hence the intensity of the phosphorescence.

microwave absorption. All ODMR experiments must be carried out on frozen samples because the phosphorescence is quenched otherwise. They are also usually carried out in zero field.

Example: ODMR of a Carcinogen Bound to DNA. In experiments carried out on calf thymus DNA covalently bound to a derivative of 1,2 benzopyrene, a potent carcinogen, it was found that the ODMR spectra were very much like those of a chromophore in a simple solvent system and not like those of a chromophore intercalated between the base pairs of DNA.

References. A. L. Kwiram and J. B. A. Ross, *Ann. Rev. Biophys. Bioeng.,* 11(1982):223; S. M. Lefkowitz et al., *FEBS Lett.,* 105(1979):77.

X-RAY ABSORPTION SPECTROSCOPY AND EXTENDED X-RAY ABSORPTION FINE STRUCTURE (EXAFS)

Many biological systems contain metal ions. The technique of X-ray absorption spectroscopy, which is *selective* for a particular element, allows individual metal ions to be probed. Structural information can be derived from the measurement of the variation in the X-ray absorption, as a function of the incident photon energy, at and above the *absorption edge* of the metal. The experiments require a tunable source of X-rays—which is available from **synchrotron** radiation. The samples may be either liquids or solids.

The X-ray absorption spectrum of a metal in a complex can be divided into two distinct regions. The first is the **absorption edge,** or lower energy-transition region, in which $1s$ core electrons are promoted to "bound" unoccupied orbitals of the metal (e.g., $3d$, $4s$, $4p$, . . .). The shape and position of the absorption edge contain information about the types of coordinated ligands, their geometry (site symmetry), and possibly the oxidation state of the metal. The transitions are observed as relatively sharp absorptions, superimposed on which is a steeply rising absorption trend (see Figure 13.5), which arises from the ionization of the $1s$-electrons. This ionization produces photoelectrons.

The emitted photoelectrons can be considered as spherical waves radiating from the absorbing metal atom. These waves may be scattered by neighboring atoms in such a way that some of these waves will arrive back at the metal atom. The resulting interference between the outgoing and backscattered waves modulates the X-ray absorption of the metal atom. This modulation is manifested as a second, distinct oscillatory region in the spectrum—the extended X-ray absorption fine structure or **EXAFS region,** which can be measured up to 1 keV *above* the absorption edge.

Figure 13.5 Absorption of X-rays by Fe in a lyophilized sample of *Peptococcus aerogenes* rubredoxin. The sharp peak at low energy is the $1s \rightarrow 3d$ transition of the Fe. The broader peaks above the edge are the EXAFS arising mainly from the back scattering of the four ligands around the Fe (from R. G. Shulman, P. Eisenberger, and B. M. Kincaid, *Ann. Rev. Biophys. Bioeng.*, 7(1978):559–78). (*Note:* Sometimes only the EXAFS part of the spectrum is presented as a function of photoelectron wavenumber (10^8 cm^{-1}), k, rather than electronvolts).

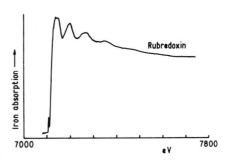

Rubredoxin

Iron absorption →

7000   eV   7800

Analysis of EXAFS can give information about the number and type of atoms around the metal ion and their corresponding interatomic distances. Since each element has a unique value for its energy of ionization, it is possible to study one kind of atom in the presence of many others by tuning the X-ray energy to the appropriate edge. Typical values of the photon energies for edges vary from about 3 to 9 keV for K to Cu, to about 20 keV for Mo (for further details and references see C. D. Garner, and S. S. Hasnain, in *Progress in Biophysics and Molecular Biology* (1984), in press. The *Journal of the American Chemical Society* also contains many of the biological applications of EXAFS).

Worked Example 13.1   X-ray Absorption Studies on Cytochrome c Oxidase

Cytochrome *c* oxidase is a membrane-bound protein that mediates electron transfer from reduced cytochrome *c* to molecular oxygen. The protein can be isolated in a relatively pure form and contains two heme Fe atoms and two Cu atoms. Suggest how X-ray absorption or EXAFS studies could be used to study the role of the metals in the electron-transfer process. (The redox state of cytochrome oxidase can be experimentally varied by the addition of nonmetal-containing compounds.)

Solution

The selectivity of X-ray absorption allows the Cu and Fe atoms to be independently studied. The absorption-edge studies should give information on the redox state of the metals and also on the nature and geometry of ligands around the metal sites. The redox state and ligand geometry of the metal atoms could therefore be compared in the oxidized and reduced states of cytochrome oxidase. EXAFS could, additionally, provide information on the distance between individual metal sites (i.e., Fe and Cu). It might even be possible to decide whether the oxygen is bound by one or more metal atoms simultaneously (see for example, L. Powers et al., *Biophys. J.*, 34(1981):465–498).

CIRCULAR POLARIZATION OF
THE LUMINESCENCE (CPL)

The phenomenon of circular dichroism (CD) results because optically active molecules absorb right- and left-circularly polarized light to different extents. CD reflects the asymmetry of the molecule in its *ground electronic state*. It measures the difference in the probability of transitions, caused by right- and left-circularly polarized light, from the ground state to the excited state.

The fluorescent (or phosphorescent) light emitted by optically active molecules (from excited states) will also be circularly polarized to some extent. The circular polarization of the luminescence—fluorescence—(CPL) is a measure of the optical activity of chromophores in their *electronically excited state*. It is the emission analogue of CD.

CPL is expressed by the emission anistropy factor ($g_{em} = \Delta f / f / 2$), where Δf is the intensity of the circularly polarized component in the fluorescence and f is the total fluorescence of any given wavelength. By convention g_{em} is *positive* for left-handed circular polarization.

Since only fluorescent (or phosphorescent) chromophores can contribute to the CPL spectrum, the information obtained has more specificity than that obtained from CD. CD probes those chromophores in an asymmetric environment in the ground state. CPL probes only those chromophores that are both fluorescent and in an asymmetric environment in their electronically excited state. For example, in proteins, several chromophores can contribute to the CD spectrum. These include the peptide bond, S—S cross-links, and aromatic side chains. From Chapter 5 (on fluorescence), we can deduce that in proteins, CPL will be a very selective probe for the asymmetric environment of tryptophan residues and to some extent tyrosine residues. Note, however, that there may be slight differences in the electronic environment of a fluorescent chromophore between the ground and excited states. The difference between the CD and CPL may reflect this (for an authoritative review, see I. Z. Steinberg *Ann. Rev. Biophys. Bioeng.*, 7(1978):113–37; see, also, the article by N. Steinberg, A. Gafri, and I. Z. Steinberg *J. Am. Chem. Soc.*, 103(1981):1636–1640 and references therein).

Worked Example 13.2   Probing an Antibody Combining Site Using CPL, CD, and Fluorescence

An antigenic group R was attached via a spacer group to the dansyl chromophore (Figure 13.6). CD, CPL, and fluorescence measurements were made on complexes formed between the dansyl derivatives of R and anti-R antibodies. Explain the following observations:

(a) None of the dansyl derivatives showed any detectable CD or CPL when free

(continued)

Worked Example 13.2   (continued)

in solution. (b) No measurable CD spectra were obtained in the dansyl absorption region, 300–400 nm, for any complexes in which $n = 1$ to $n = 6$. (c) With $n > 6$, the fluorescence spectrum was the same as that of the free fluorophore in water but lacked circular polarization.

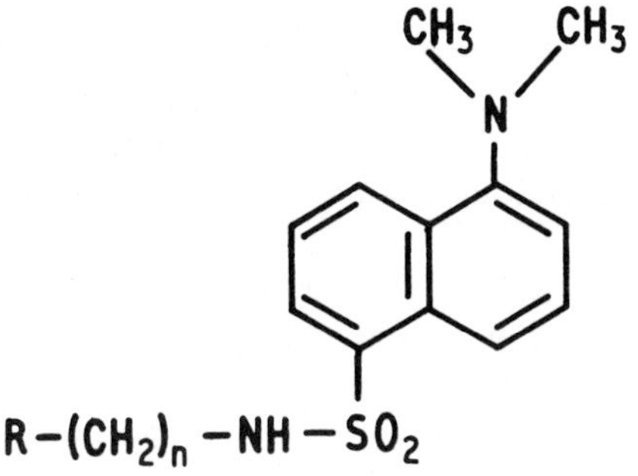

Figure 13.6

$$CH_3 \quad CH_3$$

$$R-(CH_2)_n-NH-SO_2$$

Solution

(a) The lack of CD or CPL of the free dansyl derivatives shows that the chromophore is not optically active either in its ground or excited state. (b) The CPL spectra for the complexes ($n = 1$ to $n = 6$) show that the dansyl derivatives acquire optical activity when bound to antibody. However, the corresponding absence of CD spectra of the complexes shows that a change in the mode of interactions between the dansyl chromophore and its binding site occurs upon electronic excitation. (Therefore, care is needed in making general conclusions resulting from CPL measurements in such cases). (c) With $n > 6$, the results imply that the dansyl group protrudes out of the site into the aqueous solvent. This type of observation helps to give information on the depth of the combining site.

ATOMIC SPECTROSCOPY

Atomic spectroscopy is used in biological systems for the analytical determination of metals, which are often in trace amounts (nanomolar). The experimental procedures can be divided into three main categories: (1) atomic emission, (2) atomic absorption, and (3) atomic resonance fluorescence.

Atoms have only quantized electronic energy levels (there are no vibrational or rotational levels in atoms). The wavelengths absorbed or emitted by atoms in the gas phase are highly discrete and monochromatic—a typical spectral bandwidth is $\sim 10^{-3}$ nm. The narrow bandwidth combined with the fact that each element has a unique spectrum make techniques based on atomic spectroscopy highly specific.

In atomic emission spectroscopy (AES), the sample is atomized by a source (such as a flame or an electrical device). The gaseous atoms of the element are thermally excited (by collisions in the source) to excited electronic states. As the atoms return to their ground state, they emit light of characteristic wavelength or wavelengths. The intensity is compared with that of a standard of known concentration. This method is often used for detecting sodium and potassium.

In atomic absorption spectroscopy (AAS) there are two distinct sources. One atomizes the sample and the other provides the spectrum of the element to be analyzed. The gaseous atoms of the sample absorb their characteristic radiation from the second source and the resulting decrease in intensity is measured. This method can now be used for almost all elements.

In atomic resonance fluorescence spectroscopy, radiation is absorbed as in atomic absorption and reemitted as fluorescence. However it differs from conventional fluorescence (see Chapter 5) in that the absorption and reemission occur at the same wavelength. As yet, this method has not been widely used (for further details, see C. Veilon and B. L. Vallee *Methods in Enzymology,* 54 (1978):446–484. E. J. Wood, *Spectroscopy for Biochemists,* ed. S. R. Brown (London: Academic Press, 1980)).

APPENDIX I

Mathematical Functions

Complex Variables

A complex number c is of the form $a + ib$, where a and b are real numbers, and $i = \sqrt{-1}(i^2 = -1)$ is the imaginary unit. The magnitude of $c(|c|)$ is defined to be cc^* where c^*, the complex conjugate of c, is $a - ib$:

$$|c| = (a + ib)(a - ib) = a^2 + b^2.$$

Exponential and Trigonometric Functions

To define the exponential, sine, and cosine functions, we use the infinite power series

$$e^x = 1 + x + \frac{x^2}{2!} + \frac{x^3}{3!} + \cdots$$

$$\sin x = x - \frac{x^3}{3!} + \frac{x^5}{5!} - \frac{x^7}{7!} + \cdots$$

$$\cos x = 1 - \frac{x^2}{2!} + \frac{x^4}{4!} - \frac{x^6}{6!} + \cdots$$

where $n! = n(n - 1)(n - 2) \cdots 2 \cdot 1$.

It follows from these definitions that

$$e^{ix} = \cos x + i \sin x$$
$$e^{-ix} = \cos x - i \sin x$$

Other useful relationships are

$$\cos^2 x + \sin^2 x = 1$$
$$\cos(x_1 + x_2) = \cos x_1 \cos x_2 - \sin x_1 \sin x_2$$

$$\cos x_1 + \cos x_2 = 2 \cos\left(\frac{x_1 + x_2}{2}\right)\cos\left(\frac{x_1 - x_2}{2}\right)$$

Vectors

A quantity that is completely determined by its magnitude is called a **scalar** quantity. A quantity that is determined by its magnitude and direction is called a **vector**. It is usually represented by a boldfaced letter, such as **V**.

A vector **V** is often written as $a\mathbf{i} + b\mathbf{j} + c\mathbf{k}$, where **i**, **j**, and **k** are unit vectors along the x-axis, y-axis, and z-axis, respectively. Its magnitude is $(a^2 + b^2 + c^2)^{1/2} = |V|$.

A **scalar product** of two vectors $\mathbf{V}_1$ and $\mathbf{V}_2$ is $|V_1|\,|V_2|\cos\theta$, where θ is the angle between $\mathbf{V}_1$ and $\mathbf{V}_2$.

A vector (or cross) product of two vectors $\mathbf{V}_1$ and $\mathbf{V}_2$ is written as $\mathbf{V}_1 \times \mathbf{V}_2$ and is defined to be $|V_1|\,|V_2|(\sin\theta)\mathbf{1}$, where θ is the angle between $\mathbf{V}_1$ and $\mathbf{V}_2$, and **1** is a unit vector perpendicular to the plane of $\mathbf{V}_1$ and $\mathbf{V}_2$ and so directed that a right-handed screw axis in the direction of **1** would carry $\mathbf{V}_1$ into $\mathbf{V}_2$.

APPENDIX II

Oscillators

Simple Harmonic Oscillation

Oscillating systems are ones that exhibit periodic or cyclic motion. Familiar examples are the pendulum, a mass on a helical spring, and a tuned electrical circuit. A simple harmonic oscillator is one in which the frequency of oscillation (v) is independent of the amplitude of the oscillation (a).

Consider a mass (m) on a spiral spring, as shown in Figure A.1. The strength of the spring is defined by a force constant (k) measured in units of newtons per meter. The force exerted on the mass after it is displaced a distance (x) from its equilibrium position is $\mathbf{F} = -kx$. The negative sign indicates that the force is to the left in the diagram. A mass subjected to a force undergoes acceleration according to Newton's first law, and acceleration is defined as d^2x/dt^2, which is more conveniently written as $\ddot{x}$. Thus, we have the equation

$$m\ddot{x} = -kx \tag{A.1}$$

This is a differential equation that can be solved by a variety of standard methods. In this case, however, it is instructive to solve it by considering the circular motion of a particle (P) rotating at a constant angular speed (ω). For such uniform motion in a circle (radius r), the particle is always changing direction and

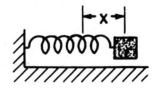

Figure A.1 A mass on a spiral spring displaced a distance x with respect to its equilibrium position.

Figure A.2   A model for solving the differential equation. (1) A point *P* moves with constant angular velocity. The projection on *AB* is *x*.

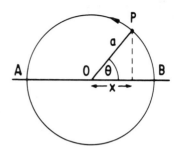

experiences an acceleration of $\omega^2 a$ toward the center of the circle. The component of this accelaration along the line *AB* in Figure A.2 is $\omega^2 a \cos \theta$. Thus

$$\ddot{x} = -\omega^2(a\cos\theta) = -\omega^2 x \qquad (A.2)$$

It can be seen that the solution to this equation, which describes the motion of *P* projected onto *AB*, is $x = a\cos(\omega t + \phi)$, where ϕ defines the position of *P* when $t = 0$. For example, if *P* is at *B* when $t = 0$, then $\phi = 0$. If *P* is directly above 0 at $t = 0$, then $\phi = 90°$. The angle ϕ is called a **phase angle**.

Note that the time period for one revolution of *P* is $2\pi/\omega$. Thus, the frequency is $\nu = \omega/2\pi$.

Equation A.2 has the same form as equation A.1 with $\omega = (k/m)^{1/2}$. Thus, the frequency of oscillation of the spring is $\nu = (k/m)^{1/2}/2\pi$.

Damped and Forced Oscillations

The energy of most simple harmonic oscillators decreases with time, that is, it is **damped**. The oscillation is then characterized by an additional parameter, a decay constant (τ). These decays are often exponential (as shown in Figure A.3) and the wave is multiplied by $\exp(-t/\tau)$.

If some external driving force is imposed on the oscillating system, two further phenomena are of interest: (1) The response of the system is a maximum when the driving oscillator has the same period as the oscillator of interest, a phenomenon known as **resonance**; (2) The forced oscillation has the same frequency as the driving oscillator, whatever its natural frequency, but the *phase* of the response depends on whether the driving frequency (ν) is greater than or less than the natural frequency (ν_0).

Figure A.3   A damped exponential oscillation characterized by amplitude a_0, frequency ν, phase ϕ, and a decay constant τ.

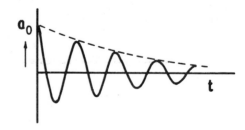

Figure A.4   Barton's pendulums illustrating resonance and phase shifts. The pendulums A–E are light relative to H, the driving pendulum. If H oscillates perpendicular to the plane of the paper, A to E respond. Pendulum C, which has the same length as H, shows the largest amplitude oscillations (it resonates) and lags behind H by a quarter-period. Pendulums D and E tag behind H by about half a period, while A and B are approximately in phase with H.

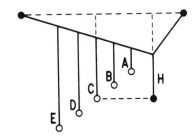

The reader can demonstrate these two phenomena by setting up Barton's pendulums, as shown in Figure A.4. A pendulum of the same length as the driving pendulum will respond most strongly, and there will be a phase difference between those pendulums shorter than and those longer than the driving pendulum.

The differential equation for a damped, forced oscillator can be obtained by adding terms to equation A.1. Thus,

$$m\ddot{x} = -kx - f\dot{x} + F_0\cos(\omega t)$$

where $-kx$ is the restoring force from the spring, $-f\dot{x} = -f\,dx/dt$ is the damping force opposing the motion, and $F_0\cos(\omega t)$ is the applied driving force. The general solution to this differential equation is

$$x = \dot{x}\cos(\omega t) + \ddot{x}\sin(\omega t)$$

where

$$\dot{x} = F_0 \frac{m(\omega_0^2 - \omega^2)}{m^2(\omega_0^2 - \omega^2)^2 + f^2\omega^2}$$

and

$$\ddot{x} = F_0 \frac{f\omega}{m^2(\omega_0^2 - \omega^2)^2 + f^2\omega^2}$$

A plot of the functions $\dot{x}$ and $\ddot{x}$ versus ω is shown in Figure A.5. These have the same form as dispersion and absorption curves observed when electromagnetic waves interact with matter (see Chapter 2). It can be shown that the width of the absorption line is related to both $1/f$ and $1/\tau$ (see Appendix VII).

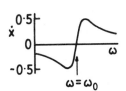

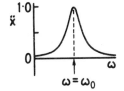

Figure A.5   Plots of the functions $\dot{x}$ and $\ddot{x}$.

Note that the absorption function ($\ddot{x}$) is 90° out of phase with respect to the driving oscillation ($\sin(\omega t)$ rather than $\cos(\omega t)$), while the dispersion function ($\dot{x}$) is in phase when $\omega << \omega_0[\dot{x} \rightarrow (F_0/m\omega_0^2)]$ and 180° out of phase when $\omega >> \omega_0[\dot{x} \rightarrow -F_0\, m/(m\omega^2 + f^2)]$ (remember that $-\cos\theta = \cos(180 + \theta)$).

APPENDIX III

Waves and Their Superposition

Oscillations of the type described in Appendix II represent wave motion. Undamped standing waves are characterized by three parameters; the amplitude a, the frequency ν ($= \omega/2\pi$), and the phase ϕ (waves that propagate at velocity c have a similar form; see Chapter 2). The equation describing such a wave is

$$x = a \cos(\omega t + \phi)$$

A widely observed phenomenon is that waves superpose—the net disturbance is the vector sum of the disturbances that would have been produced by the individual waves. This superposition gives rise to the observation of **interference**.

To study the more interesting effects of the superposition it is sufficient to consider waves that differ only in phase. Let two such waves be

$$x_1 = a_0\cos(\omega t) \quad \text{and} \quad x_2 = a_0\cos(\omega t + \phi)$$

The sum of these is (see Appendix I)

$$x_r = 2a_0\cos\left(\omega t + \frac{\phi}{2}\right)\cos\left(\frac{\phi}{2}\right)$$

This resultant wave has angular frequency ω, amplitude $2a_0\cos(\phi/2)$, and phase $\phi/2$. Note that x_r has the same frequency as x_1 and x_2, and that the amplitude is $2a_0$ when $\phi = 0°$, and zero when $\phi = 180°$. This result is, of course, obvious from a consideration of the graphical sum of two waves.

In general, the summation of waves, even those differing only in phase and amplitude, is tedious using the preceding notation. A much simpler notation is to use vectors. The wave is represented by a vector with amplitude $|a_0|$ and a direction defined by phase angle ϕ. Such a vector has components $a_0\sin \phi$ and $a_0\cos \phi$ along the 0° (x) and 90° (y) phase-angle axes, respectively. An even more convenient notation arises if the two axes are considered in terms of complex numbers,

Figure A.6   A representation of the amplitude and phase of a wave in the complex
plane.

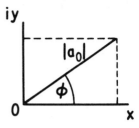

with the y-axis considered as complex. Such a scheme is shown in Figure A.6.
Note that $x = a \cos \phi$ and $y = a \sin \phi$; thus, we can write the vector as

$$a_o e^{i\phi} = a_o\cos \phi + ia_o\sin \phi$$

using the definitions in Appendix I.

It is thus convenient to call the 90° phase-angle axis the "imaginary" axis,
and the 0° axis the "real" axis. It must be stressed that this is merely a convenient
way of representing these axes; each component is, in fact, real.

Worked Example

Using vector notation, prove the relationship derived for the sum of two waves with
the same amplitude but differing in phase by an angle.

Solution

Draw a vector length a_o at some arbitrary angle ϕ to the x-axis. Then draw
a vector length a_o parallel to the x-
axis, i.e., with zero phase angle. The
amplitude of the resultant vector is
$2a_o\cos (\phi/2)$ and the phase angle is
$\phi/2$, as shown in Figure A.7.

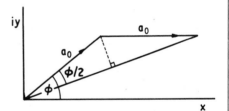

Figure A.7

APPENDIX IV

Wave Mechanics

When dealing with an object such as a tennis ball, we know that we can describe its behavior precisely by measuring quantities such as position and momentum (mass and velocity). In dealing with elementary particles, such as the electron, we must modify our way of thinking. According to **Heisenberg's uncertainty principle**, it is impossible to measure precisely both the position and momentum of an electron simultaneously. (In fact, the uncertainty in position (Δx) and momentum (Δp) are related by the equation $\Delta x \Delta p > \hbar/2\pi$. In 1900, Planck hypothesized that light was emitted or absorbed in packets called **photons**, or **quanta**, with energy $E = h\nu$, where ν is the frequency of light (i.e., a wave property). In the 1920s, it was realized that not only light but also electrons could act as both particles and waves, and these various ideas began to be rationalized.

One of the most powerful ways of reconciling these theories is to use **wave mechanics**, which is a mathematical way of describing the wave behavior of particles. Consider a particle of mass m with potential V and kinetic energy $mv^2/2 = p^2/2m$, where v is its velocity and p its momentum. The law of conservation of energy means that

$$p^2/2m + V = E \qquad \text{(A.1)}$$

where E is the total energy.

Let us now regard p not as the momentum but as a **momentum operator**. (An operator is a symbol that tells us to do something, e.g., $\sqrt{}$, log, or d/dx.) In wave mechanics, p^2 in equation A.1 is replaced by

$$-\hbar^2 \left(\frac{d^2}{dx^2} + \frac{d^2}{dy^2} + \frac{d^2}{dz^2} \right)$$

The term in parentheses is usually called ∇^2, the Laplacian operator. Its function is to define the kinetic energy of the particle at the point x,y,z in space. We now redefine the left-hand side of equation A.1 as

$$\mathcal{H} = \frac{\hbar^2}{m}\nabla^2 + V$$

where $\mathcal{H}$ is called the Hamiltonian of the system. This is merely the wave-mechanical transcription of the kinetic energy plus the potential energy of all parts of the system. To complete our transcription of the conservation of energy equation, we regard E as another operator

$$\mathbf{E} = i\hbar\frac{d}{dt}$$

where $i = \sqrt{-1}$. The term $\mathbf{E}$ is still related to the total energy of the system.

The operators $\mathbf{E}$ and $\mathcal{H}$ must have something to operate on, so we introduce a wave function (ψ) on each side of the transcribed equation. This leads directly to the time-dependent, one-particle Schrödinger wave equation

$$\mathcal{H}\psi = E\psi$$

The term ψ describes, mathematically, how the particles of the system behave as waves, while ψ^2 is a measure of the probability that a particle is in a particular region of space. Solutions of the wave equation for electrons in atoms have given insight into the possible distributions of the electron and this had led to the concept of orbitals (see Appendix V).

For a particular wave function, E can be found by imposing certain **boundary conditions** on the system—that is, the solution is made to satisfy some condition at some chosen point in space. For example, E values can be calculated for an electron in a chemical bond. In this situation, solution of the wave equation is possible only for discrete (quantized) energy values.

APPENDIX V

Atomic and Molecular Orbitals

The spatial distribution and energy of the electrons in atoms can be characterized in terms of a three-dimensional wave function, or **orbital**. (In fact, the square of the function at any point gives the electron density.) These orbitals can be described by the three quantum numbers n, l, and m, which come from the solution of the Schrödinger equation. For historical reasons, orbitals described by $l = 0$, 1, 2, 3 are labeled s, p, d, f. The shapes of s, p-, and d-orbitals are shown in Figure A.8.*

In molecules, the orbitals describing the energy levels of the electrons can be assumed to arise from a combination of atomic orbitals. The shapes of some of these are shown in Figure A.9. The σ molecular orbital results from the addition of two s-type atomic orbitals. The increase of electron density between the nuclei (given by the square of the function) represents the formation of a bond it is called a **bonding orbital**. In contrast, the σ^* orbital results from the subtraction of two s atomic orbitals and has the electronic charge mainly outside the nuclei. The orbital is described as **antibonding**. The σ^* orbital is a state of higher energy than two separate atoms, while the converse is true for the σ orbital. Similarly, molecular orbitals made up from a linear combination of p-orbitals can give rise to σ_p and σ_p^* bonds from "head-on" overlap of the p_z orbitals, or π bonds from parallel overlap of the p_y and p_x orbitals. Note that the molecular orbitals that are cylindrically symmetrical about the internuclear axis are called σ-orbitals (analogous to the s-orbital, the atomic orbital of highest symmetry).

---

*The conventional way of drawing d-orbitals suggests that the d_{z^2}-orbital has a very different shape from the other four. However this difference is only apparent. It is not possible to choose five d-orbitals that are independent and have the same shapes. The d_{z^2}-orbital can in fact be written as a linear combination of orbitals that have just the same shape as the other d-orbitals.

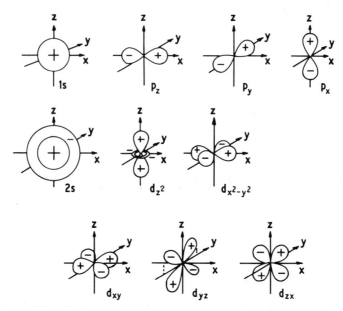

Figure A.8   The shapes of *s*-, *p*-, and *d*-orbitals.

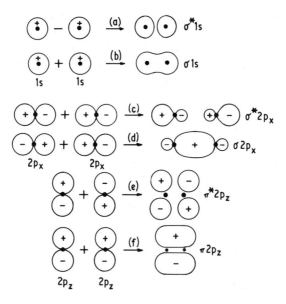

Figure A.9   Symmetry of molecular orbitals formed from atomic orbitals illustrating sigma (a)–(d) and pi (e)–(f) orbitals and bonding (b), (d), (f) and antibonding (a), (c), (e) orbitals. (Note that when a positive function is added to a negative function, there is canceling, which reduces the value of the function possibly to zero, when it is called a **node**. The electron density will be zero at the nodal points.)

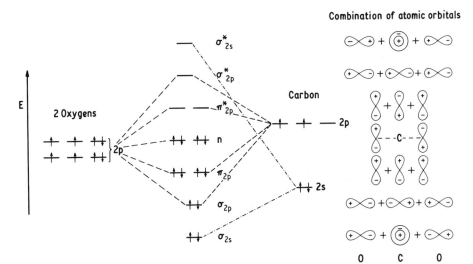

Figure A.10   Relative energies and electron distribution of molecular orbitals for CO_2. The interaction of oxygen $2p$ atomic orbitals with carbon $2s$ and $2p$ atomic orbitals results in formation of σ and π bonding molecular orbitals (but note that the highest occupied molecular orbital has no contribution from the carbon atomic orbitals and is therefore nonbonding). (The energy level of the $2p$ atomic orbitals on the oxygens is lower than that for carbon because of the increased nuclear charge on the oxygen, which results in the electrons' being more tightly held.

Many molecules also contain electrons that are not directly involved in bonding; these are called **nonbonding** or n-electrons and are mainly located in the p-type atomic orbitals of oxygen, nitrogen, sulfur, and the halogens. As n-electrons do not form bonds, there are no antibonding orbitals associated with them.

Each molecular orbital can contain up to two electrons with opposed spins, as illustrated in Figure A.10, which shows the relative values of electronic energy levels for a simple molecule that has σ-, π-, and n-orbitals, formed from the appropriate combinations of atomic orbitals.

APPENDIX VI

Some Concepts of Crystal Field Theory

Electronic spectra in transition metal complex generally arise from transitions involving the d-electrons. EPR spectra of transition metal complexes are determined by their electronic energy states.

A simple treatment of d-orbitals that allows the absorption spectrum of a transition metal complex to be related to the properties of metal ions and ligands is called **crystal field theory**. The theory supposes that the ligands impose an electric field on the central metal ion. The symmetry of the field reflects the electrical nature of the ligand, particularly any repulsion between the electrons of the ligands and those in the d-orbitals of the metal.

The electronic energy state of a transition metal ion complex follows from a consideration of the effect of the ligand field on the d-orbitals of the metal ion. In the free ion, the five d-orbitals have the same energies (they are degenerate); the ligand field removes this degeneracy by changing the energies of some of the d-orbitals relative to the others.

For instance, consider an octahedral field in which six negative charges are placed symmetrically on each of the x-, y-, and z-axes. The d-electrons in the d_{xy}, d_{yz}, and d_{xz} orbitals will experience less repulsion from the negative charge than will the $d_{x^2 - y^2}$ and the d_{z^2} orbitals, which lie along the axes. The consequence is that these two orbitals are more destabilized than the other three. This is illustrated on the energy level diagram in Figure A.11(a). The splitting between the levels is given the symbol Δ in an octahedral field. Similar arguments lead to the energy level diagram for a tetrahedral field, Figure A.11(b) (in which the splitting between the levels is $4/9\Delta$; see Figure A.11(b).)

The value of Δ varies with ligand, which leads to the concept of the spectrochemical series. The order of field strength is generally $I^- < Br^- < Cl^-$

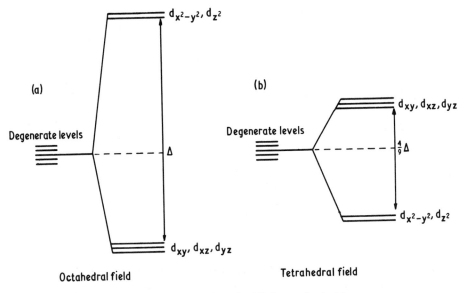

Figure A.11   The crystal field splitting of the d-orbital energies in (a) an octahedral and (b) a tetrahedral field.

$< SCN^- < F^- < OH^- < H_2O < NCS^- < NO_2^- < CN^- \sim CO$. Thus, for example, complexes with CN^- ligands have a larger Δ than those with halogen ligands.

Each d-orbital can contain up to two electrons with opposed spins. The d-orbitals initially fill up on the principle that electrons tend to remain unpaired as long as possible (thus minimizing electron-electron repulsion). Figure A.12 illustrates this principle for an octahedral field in which the metal ion supplies an increasing number of d-electrons. The situation becomes complex when the metal ion contributes the fourth d-electron. Whether this electron pairs up in one of the lower d-orbitals or goes into a higher energy level depends on the value of Δ. If Δ is large, the spin-paired arrangement is favored. Such complexes are termed low spin. Conversely, high-spin complexes are favored when the disadvantages of occupying a high-energy orbital are overcome by the reduction in electron-electron repulsion that would occur in the doubly filled orbital.

The crystal field description of the electronic states has the d-electrons localized on a central metal ion, and the electron pairs of the ligand localized in orbitals confined to the ligand. More sophisticated descriptions of the electronic states of transition metal complexes involve molecular orbitals spread over both. The occurrence of low-spin and high-spin complexes is then accounted for in terms of the energy splittings that occur on the formation of bonding and antibonding orbitals.

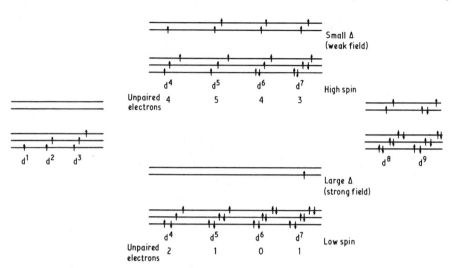

Figure A.12   Weak- and strong-field arrangements of d-electrons for octahedral complexes. Note there is only one possible arrangement of electrons for d^1, d^2, d^3, d^8, and d^9 configurations. Both high- and low-spin possibilities exist for d^4 to d^7. Change of ligand can produce complexes with different magnetic properties by altering Δ.

APPENDIX VII

Fourier Transforms and Convolution Functions

The **Fourier transform** of a function $f(x)$* is defined as

$$F(S) = \int_{-\infty}^{\infty} f(x)e^{-i2\pi xs}\, dx$$

which has the reversible property

$$f(x) = \int_{-\infty}^{\infty} F(s)_e^{i2\pi sx}\, ds$$

Examples of such Fourier transform pairs are shown in Figure A.13(a) and A.13(b). Physical examples of the pair in Figure A.13(a) are the diffraction pattern $[F^2(h)]$ of a slit $[f(x)]$ and the frequency response $[F(\omega)]$ of a pulse $[f(t)]$. A physical example of the Fourier transform pair in Figure A.13(b) is the response of an NMR spin system to a pulse $[f(t)]$ and the absorption spectrum $[F(\omega)]$.

The **convolution** of two functions $F(s)$ and $G(s)$ is a function* with the property

$$F(s)*G(s) = f(x) \times g(x)$$

where $\times$ is simple multiplication, and $f(x)$ and $g(x)$ are the Fourier transforms of $F(s)$ and $G(s)$, respectively.

Examples of convolution include:

1. A crystal can be considered as the convolution of a lattice and a unit cell. Thus, the observed diffraction pattern is the product of the diffraction pattern of the lattice and the diffraction pattern of the unit cell.

---

*It is sometimes convenient to drop the 2π term in the exponents of the exponentials. A factor $1/2\pi$ then appears outside the integral on the right-hand side of the equation for $f(x)$.

2. If $f(x)$ is the function $e^{-x/a} \cos s_0 x$, and $g(x)$ is the function $e^{-x/b}$, then the Fourier transform of the product of $f(x)$ and $g(x)$ is the convolution of $F(s)$ and $G(s)$. This is the function $C/[1 - (s - s_0)^2 C^2]$, where $1/C = 1/a + 1/b$.

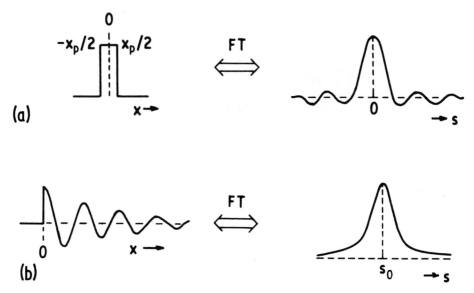

Figure A.13   Two Fourier transform pairs: (a) a step function of width x_p transforms to a $(\sin y/y)$ function passing through zero when $x_p s/2 = n\pi$, (b) a transient oscillation beginning at $x = 0$, with angular frequency S_0 and decay constant a transforms to a line shape with width at half-height $\Delta S = 1/a$.

A practical example of this last is the manipulation of line shapes in NMR spectra by multiplying the time-domain signals by exponential functions; this process leads to an increase in absorption linewidth for decaying exponentials and a decrease for increasing exponentials.

APPENDIX VIII

Magnetic Properties of Matter

A moving charge generates a magnetic field. For example, at the center of a loop of wire of N turns, of radius r and current I, a field $B = $ constant $\times IN/r$ is generated.

The magnitude of the field generated depends on the material within the loop. If the material is air, then the field generated is defined as B_0. For other materials,

$$B = B_0 + B_M$$

where B_M is the additional field (flux density) set up by the material. It is negative for a diamagnetic material and positive for a paramagnetic one.

The **magnetic susceptibility** of the material is defined as $\chi = B_M/B_0$. This gives $B/B_0 = 1 + \chi$.

The flux density, B_M, can also be written as $\mu_0 M$, where M is the **magnetization**, which is defined as the magnetic moment per unit volume, and μ_0 is a constant—the permeability of free space (air).

APPENDIX IX

Dipoles and the Interaction Between Them

Consider first an isolated point charge (Q). The electric field at some point a distance r from Q is given by

$$E = \left(\frac{1}{4\pi\epsilon_0}\right)\frac{Q}{r^2}$$

where ϵ_0 is the permittivity of free space. This means that the field falls off radially as the square of the distance from Q. This is a well-known result of elementary physics.

Consider an **electric dipole** consisting of two charges $+Q$ and $-Q$ separated by a distance $2a$ (see Figure A.14). The **electric dipole moment** of this dipole is defined as $2aQ = \mathbf{p}$ ($\mathbf{p}$ has a direction defined as going from $-Q$ to $+Q$).

The field E at any point P comprises the fields from $+Q$ and $-Q$. It can be shown that E falls off radially as $1/r^3$. In addition, the field strength depends on the *position* with respect to the direction of the dipole. In general, the field at any point P at a distance r and some angle θ to the $\mathbf{p}$ direction (see Figure A.15) can be calculated to be

$$E_p = \left(\frac{\mathbf{p}}{4\pi\epsilon_0}\right)\frac{(3\cos^2\theta - 1)}{r^3}$$

Similarly, the **magnetic dipole moment** (μ) is defined as $2a$ times the pole strength, and its direction is from south to north. The field produced by a dipole μ is

Figure A.14 Two charges separated by $2a$ generate a dipole $\mathbf{p} = 2aQ$.

Figure A.15   The field generated by the dipole **p** at position P is related to r and θ.

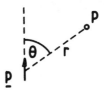

$$\mathbf{B} = \left(\frac{\mu\mu_0}{4\pi}\right) \frac{(3 \cos^2\theta - 1)}{r^3}$$

where μ_0 is the permeability of free space.

THE INTERACTION BETWEEN DIPOLES

Two Parallel Dipoles

The energy of interaction in the case of two parallel dipoles follows directly from the preceding and is

$$\left(\frac{\mu_1\mu_2\mu_0}{4\pi}\right) \frac{(3 \cos^2\theta - 1)}{r^3}$$

Two Dipoles in a General Conformation

If α is the angle between μ_1 and μ_2, and if β and γ are the angles between μ_1 and $\mathbf{r}$ and μ_2 and $\mathbf{r}$, respectively, then the energy of interaction between two dipoles in a general conformation (see Figure A.16) is

$$\left(\frac{\mu_1\mu_2\mu_0}{4\pi}\right) \left(\frac{(3 \cos \beta \cos \gamma - \cos \alpha)}{r^3}\right)$$

which reduces to the special case of two parallel dipoles when $\alpha = 0$, $\beta = \gamma$.

Figure A.16   The interaction between two dipoles μ_1 and μ_2 can be described by the angles α, β, and γ, as shown.

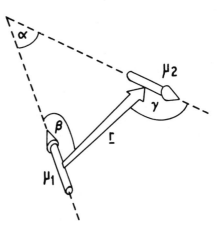

APPENDIX X

Spectra of Interacting Dimers

Let two identical interacting transition dipoles μ_1 and μ_2 have directions and separation as defined in Figure A.16. Let the frequency and transition probability of the monomer spectrum be ν_0 and D_0 respectively.

The absorption bands of the monomers are split into two by the dipole-dipole interaction (see Figure A.17). The two new bands are formed from transitions to two excited states, denoted by plus and minus. The plus state is formed when the two chromophores of the dimer oscillate in phase, the minus state when they are out of phase (see Figure A.18).

The transition probabilities (sometimes called transition dipole moments) for the two states are given by

$$D_\pm = D_0(1 \pm \cos \alpha)$$

Note that the total intensity $D_+ + D_- = 2D_0$ for any angle α. The relative intensity, however, can vary strongly with α, that is, with molecular geometry.

The frequencies of the new absorption bands are

$$\nu_\pm = \nu_0 \pm \nu_{12}$$

where ν_{12} is a measure of the dipole-dipole interaction between μ_1 and μ_2

Figure A.17 The absorption spectrum of a monomer is split into two bands corresponding to transitions to the ν_+ and ν_- state when two monomers are near enough for their transition dipoles to interact.

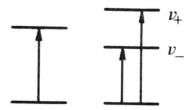

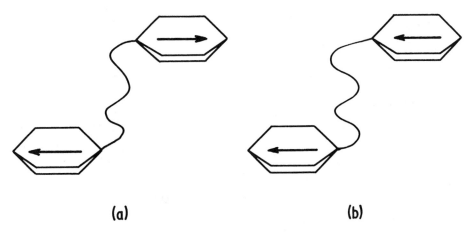

(a)   **(b)**

Figure A.18   Illustration of the orientation of transition dipoles in two chromophores that correspond to (a)the minus state and (b)the plus state. Note the
inversion of the direction of the dipole in the upper chromophore in the minus
state. The relative molecular geometry of the chromophores is irrelevant for this
definition of phase. The direction of the dipole in the lower chromophore could
also have been reversed to change the phase.

$$\nu_{12} = D_0 \; \frac{(\cos \alpha - 3 \cos \gamma \cos \beta)}{r^3}$$

(see Appendix IX).

Optical activity of a molecule involves both magnetic and electric transition
dipole moments. Such interactions can arise from certain geometries of a dipole
pair. The dominant contribution to the rotational strengths R_+ and R_- of the two
states arising from the dipolar interaction is usually given by

$$R_{\pm} = \pm \left(\frac{\pi}{2\lambda}\right) r \; \cos \delta (\mu_1 \times \mu_2)$$

This equation has been written using vector notation for convenience, where
$\mu_1 \times \mu_2$ is a cross product (see Appendix I). The angle δ is the angle between r and
the perpendicular to the plane formed by μ_1 and μ_2. In this case, the product is a
measure of the helical twist in going from μ_1 and μ_2.

Worked Example

Using all the preceding information, calculate $\nu_{\pm}$, $D_{\pm}$, and $R_{\pm}$ for the interacting
dimer formed by μ_1 and μ_2 with the following relative geometries defined by α, β,
and γ.

(a) $\alpha = 0$, $\beta = \gamma = 90°$, i.e., two parallel dipoles (see Figure A.19(a)).
(b) $\alpha = 0$, $\beta = \gamma = 0$, i.e., two parallel dipoles along the same axis (see Figure
   A.19(b)).
(c) $\alpha = 30°$, $\beta = \gamma = 90°$ (see Figure A.19(c)).   *(continued)*

Worked Example   (continued)

Solutions

(a) $\cos \alpha = 1,$ $\sin \alpha = 0,$
         $\cos \beta = \cos \gamma = 0$

therefore,

$$D_\pm = D_0 \pm D_0$$
$$(\text{i.e., } D_+ = 2D_0, D_- = 0)$$

$$\nu_\pm = \nu_0 \pm \frac{D_0}{r^3}$$

$$R_\pm = 0$$

(b) $\cos \alpha = 1,$ $\sin \alpha = 0,$
         $\cos \beta = \cos \gamma = 1$

therefore

$$D_\pm = D_0 \pm D_0$$

$$\nu_\pm = \nu_0 \mp \frac{2D_0}{r^3}$$

$$R_\pm = 0$$

(c) $\cos \alpha = 0.866,$ $\sin \alpha = 0.5$
    $\cos \gamma = \cos \beta = 0$

but we now have to consider δ,
since $\sin \alpha \neq 0$

$$\delta = 90°,   \cos \delta = 1$$

therefore,

$$D_\pm = D_0(1 \pm 0.866)$$
$$(\text{i.e., } D_+ = 1.866, D_- = 0.134)$$

$$\nu_\pm = \nu_0 \pm \frac{0.866 D_0}{r^3}$$

$$R_\pm = \pm \frac{\pi D_0 r}{\lambda}$$

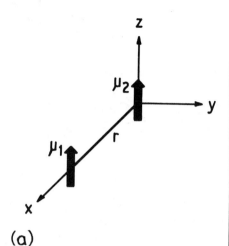

(a)

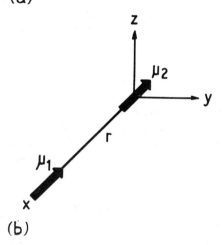

(b)

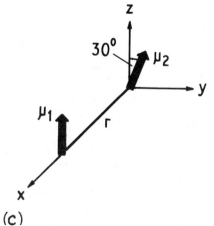

Figure A.19   (c)

Worked Example   *(continued)*

The solutions are illustrated graphically in Figure A.20. Note how the positions, the intensities, and rotational strengths vary.

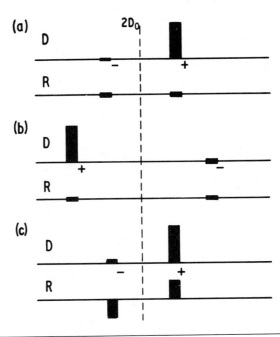

Figure A.20

General References

Some general references on the topics discussed in the different chapters are listed here. More specific references have been given in the text and in figure captions. Reviews on the topics discussed appear regularly in series such as *Annual Reviews of Biochemistry, Progress in Biophysics and Molecular Biology, Annual Review in Biophysics and Bioengineering, Methods in Enzymology,* and *Quarterly Reviews in Biophysics.* Examples of microscopy appear in histology texts and reviews.

Chapter 2

Atkins, P. 1982. *Physical Chemistry.* Oxford University Press.
Brown, S. R., ed. 1980. *Spectroscopy for Biochemists.* London: Academic Press.
Cantor, C. R. and Schimmel, P. R. 1980. *Biophysical Chemistry.* pt. 2. San Francisco: W. H. Freeman & Company Publishers.
Freifelder, D. M. 1982. *Physical Biochemistry.* San Francisco: W. H. Freeman & Company Publishers.
Marshall, A. G. 1978. *Biophysical Chemistry.* New York: John Wiley.
Whellan, P. M. and Hodgson, M. J. 1978. *Essential Principles of Physics.* London: Murray.

Chapter 3

This topic is treated well in several of the references given for Chapter 2 and also in:

Parker, F. S., 1983. Application of Infra-red, Raman and Resonance Raman Spectroscopy in Biochemistry. New York and London: Plenum Press.

Chapter 4

This topic is treated well in several of the references given for Chapter 2.

d'Albis, A. and Gratzer, W. B. 1974. Electronic spectra and optical activity of proteins. In *Companion to Biochemistry,* ed. A. T. Bull, J. R. Lagnado, J. O. Thomas, and K. F. Tipton. New York: Longman. p. 163.

Kronman, M. J. and Robbins, F. M. 1970. Buried and exposed groups in proteins. In *Fine Structure of Proteins and Nucleic Acids,* ed. G. D. Fasman and S. N. Timasheff. New York: Marcel Dekker. p. 271.

Chapter 5

Chen, R. F. and Edelhoch, H. 1975. *Biochemical Fluorescence.* vols. 1 and 2. New York: Marcel Dekker.

Steiner, R. F., ed. 1983. Excited States of Biopolymers. New York: Plenum Press.

Stryer, L. 1978. Fluorescence energy transfer as a spectroscopic ruler. *Ann. Rev. Biochem.* 47:819.

Weber, G. 1972. Uses of fluorescence in biophysics: Some recent developments. *Ann. Rev. Biophys. Bioeng.* 1:553–570.

Chapter 6

Dwek, R. A.; Campbell, I. D.; Richards, R. E.; and Williams, R. J. P., eds. 1977. *NMR in Biology.* London: Academic Press.

Dwek, R. A. 1973. *NMR in Biochemistry.* Oxford: Clarendon Press.

Gadian, D. G. 1982. *NMR of Living Systems.* Oxford: At the University Press.

Jardetzky, O. and Roberts, G. C. K. 1981. *NMR in Molecular Biology.* New York: Academic Press.

Knowles, P. E.; Marsh, D.; and Rattle, H. W. T. 1976. *Magnetic Resonance of Biomolecules.* New York: John Wiley.

Wüthrich, K. 1976. *NMR in Biological Research.* New York: Elsevier North-Holland.

Chapter 7

Cammack, R. C. and Watts, A. 1983. *ESR in Biochemistry.* New York: Academic Press.

Knowles, P. E.; Marsh, D.; and Rattle, H. W. T. 1976. *Magnetic Resonance of Biomolecules.* New York: John Wiley.

Wertz, J. and Bolton, J. 1974. *Electron Spin Resonance: Elementary Theory and Applications.* New York: McGraw-Hill.

Chapter 8

Berne, B. J. and Pecora, R. 1976. *Dynamic Light Scattering with Applications to Biology, Chemistry, and Physics.* New York: John Wiley.

Kratky, O. and Pilz, I. 1978. Solution X-ray scattering. *Quart. Rev. Biophys.* 11:39.

Latimer, P. 1982. Light scattering and absorption as methods of studying cell population parameters. *Ann. Rev. Biophys. Bioeng.* 11:129.

Stuhrmann, H. B. and Millar, A. 1978. Small angle scattering of biological structures. *J. Appl. Crystallogr.* 11:325–345.

Chapter 9

Carey, P. R. 1982. *Biochemical Applications of Raman and Resonance Raman Spectroscopies.* New York: Academic Press.

Carey, P. R. 1978. Resonance Raman spectroscopy in biochemistry and biology. *Quart. Rev. Biophys.* 11:309.

Parker, F. S., 1983. Application of Infra-red, Raman and Resonance Raman Spectroscopy in Biochemistry. New York and London: Plenum Press.

Chapter 10

Adler, A. J.; Greenfield, N. J.; and Fasman, G. D. 1973. Circular dichroism and optical rotatory dispersion of proteins and polypeptides. *Meth. Enz.* 27D:675.

Holmquist, B. and Vallee, B. L. 1978. Magnetic circular dichroism. *Meth. Enz.* 49:149.

Tinoco, I.; Bustamante, C., and Maestre, M. F. 1980. The optical activity of nucleic acids and their aggregates. *Ann. Rev. Biophys. Bioeng.* 9:107.

Chapter 11

Amos, L. A.; Henderson, R.; and Unwin, P. N. T. 1982. Three-dimensional structure determination by electron microscopy of two-dimensional crystals. *Prog. Biophys. Mol. Biol.* 39:183.

Crowther, R. A. and Klug, A. 1975. Structural analysis of macromolecular assemblies by image reconstruction from electron micrographs. *Ann. Rev. Biochem.* 44:161.

Fisher, H. W. and Williams, R. C. 1979. Electron microscopic visualization of nucleic acids and of their complexes with proteins. *Ann. Rev. Biochem.* 48:649.

Spencer, M. 1982. *Fundamentals of Light Microscopy.* Cambridge University Press.

Wischnitzer, S. 1981. *Introduction to Electron Microscopy.* 3d. ed. Elmsford, N.Y.: Pergamon Press.

Chapter 12

Amos, L. A.; Henderson, R.; and Unwin, P. N. T. 1982. Three-dimensional structure determination by electron microscopy of two-dimensional crystals. *Prog. Biophys. Mol. Biol.* 39:183.

Blake, C. C. F. 1979. X-ray Crystallography of Biological Macromolecules. In *Companion to Biochemistry,* ed. A. T. Bull, J. R. Lagnado, J. O. Thomas, and K. F. Tipton. London: Longman.

Blundell, T. L. and Johnson, L. N. 1976. *Protein Crystallography.* New York: Academic Press.

Glusker, J. P. and Trueblood, K. N. 1972. *Crystal Structure Analysis.* Oxford: At the University Press.

Greenhough, T. J. and Helliwell, J. R. 1983. The uses of synchrotron X-radiation in the crystallography of molecular biology. *Prog. Biophys. Mol. Biol.* 41:67–123.

Holmes, K. C. and Blow, D. M. 1966. The Use of X-ray Diffraction in the Study of Protein and Nucleic Acid Structure. New York: Interscience.

Solutions to Problems

Chapter 2

1. In Figure 2.3 we noted that a circularly polarized wave was made up from two superimposed E oscillations shifted in phase by $\pi/2$ with respect to each other. As shown in the figure, an E-wave in the yz-plane, advanced $\pi/2$ with respect to the xz-wave results in a circularly polarized wave rotating counterclockwise as we look to the source. It is clear that a yz-wave, retarded $\pi/2$ with respect to the xz-wave, results in a clockwise rotation. The sum of the two rotating waves will cancel the yz-wave since there are two waves out of phase. The net result is a plane polarized wave in the xz-plane.

2. (a) The wavenumber requires λ to be expressed in centimeters.

$$600\,\text{nm} = 600 \times 10^{-9}\,\text{m} = 600 \times 10^{-7}\,\text{cm}$$
$$\text{therefore wavenumber} = 1.67 \times 10^4\,\text{cm}^{-1}$$

The frequency is related to the wavelength by $\nu = c/\lambda$.

Thus, $\nu = 3 \times 10^8/600 \times 10^{-9} = 5 \times 10^{14}\,\text{s}^{-1}$.

(b) From the formula $E = h\nu$, we obtain $E = 6.63 \times 10^{-34} \times 5 \times 10^{14}(\text{J}\cdot\text{s} \times \text{s}^{-1}) = 3.3 \times 10^{-19}\,\text{J}$.

This is the energy associated with *one* atom. If we wish to express it as an energy per mole, we must multiply this number by Avogadro's number (6.03×10^{23}). Thus, $E = 1.99 \times 10^5\,\text{J}\cdot\text{mol}^{-1} = 199\,\text{kJ}\cdot\text{mol}^{-1}$.

This represents a substantial amount of energy, which can be utilized, for instance, in the initiation of photosynthesis.

3. We noted in the section on "What is Matter?" that the energies of a particle restricted to a box of width a are proportional to $1/a^2$. In a conjugated ring system the electrons are extensively delocalized, i.e., a is large; hence, the energy levels are closer together than for smaller rings.

4. It is clear that the refractive index for X-rays is such that this radiation would, effectively, pass straight through the lens; thus, it is not possible to focus X-rays by a lens. An additional complication is that most materials also absorb X-rays. This inability to focus X-rays has important consequences in the determination of structures by X-rays.

5. As illustrated, the two transition dipole moments have the same phase. The resultant moment will thus bisect the $H-C-H$ angle. Such a transition moment is said to be polarized in this direction.

6. When a molecule changes its rotational energy state, it rotates faster or slower. If the molecule has no permanent electric dipole moment, transitions will not involve a displacement of charge. Molecules with no permanent electric dipole moment cannot therefore generate rotational absorption spectra. In this example, only HCl and H_2O have a permanent electric dipole and can give rise to rotational spectra. (CO_2 and N_2 are linear and symmetrical and have no charge separation.)

7. Most proteins contain tryptophan, which has by far the largest intensity of the three aromatic amino acids and has a maximum around 280 nm. We shall see in Chapter 4 that the observed absorption is directly proportional to concentration.

8. The γ constants are proportional to μ; thus, the ratio of the electron-spin moment to the nuclear-spin moment is 658. In a field of 1 T, the absorption frequency for the electron is 28,017 MHz, and for the nucleus of the hydrogen atom (the proton) it is 42.58 MHz.

9. The probability of spontaneous emission is proportional to ν^3. The ratio of the probabilities is

$$\frac{\text{Electronic}}{\text{Rotation}} = \frac{(10^{15})^3}{(10^9)^3} = 10^{18}$$

Spontaneous emission is therefore very important at high frequencies (short wavelengths). It can be calculated that for an electronic transition the spontaneous radiative emission will take about 10^{-7} s. From the previous example, this would mean that it would take 10^{11} s for the radiative rotational transition.

10. We can use the relationships $\delta E = \hbar/\tau$ and $h\nu = E$ to calculate the uncertainty in frequency: $\delta\nu \sim 1/2\pi\tau$. (a) $\delta\nu = 1/(2\pi \times 10^{-13}) = 0.16 \times 10^{13}$ Hz. To convert to wavenumber, divide by c, the velocity of light (3×10^{10} cm$\cdot$s^{-1}); $\delta\nu = 50$ cm^{-1}. (b) $\delta\nu = 1/(2\pi \times 10^{-1}) = 1.6$ Hz $= 0.5 \times 10^{-8}$ cm^{-1}.

11. The molecules in solution move and tumble. The small magnetic moments exert small fluctuating fields on each other. These magnetic fluctuations cause transitions between the energy levels and allow the Boltzmann distribution to be set up.

12. In solution, many vibrations and collisions are possible, and the thermal decay processes that take the molecule to the *lowest* vibrational level of the excited state (F) are very rapid. The transfer of energy to the other electronic state (P) involved in the phosphorescence occurs via a resonant process (often called **intersystem crossing**), which usually occurs from higher vibrational levels. Thus, if the thermal decay is rapid (e.g., in solution) there may be no time to populate P significantly. In the solid, the thermal decay will be much slower and P can become significantly populated. Radiationless decay from P will also be less likely in the solid.

13. The difference between NMR and rotational energy levels is their lifetime. In NMR, the relaxation processes are fluctuating magnetic fields, and these are very inefficient compared with the collisional relaxation processes for rotational levels. Typical NMR level lifetimes are about 1 s, whereas rotational levels have a lifetime of $\sim 10^{-13}$ s.

Chapter 3

1. The selection rule requires the normal mode of vibration to result in a change in dipole moment. This is true for all the fundamental modes except the symmetric stretching, which leaves the dipole moment unchanged, so this mode is infrared *inactive*. (The two flapping modes (perpendicular bending motions) have the same frequency because of the symmetry of the molecule. They are said to be *degenerate*) (see Chapter 2). Note that the stretching frequencies are higher than the bending frequencies—it is generally easier to distort a molecule by bending than by stretching.

2. This is an example of the isotope effect on the reduced mass (μ). Treating the C—H vibration as independent of the rest of the molecule, we have initially that $1/\mu = 1/M_C + 1/M_H = 1/12 + 1/1 = 13/12$. On substitution with deuterium, $1/\mu = \frac{1}{12} + \frac{1}{2} = \frac{7}{12}$. The ratio of reduced masses is $\frac{7}{13}$. This alters the fundamental vibrational frequency by $\frac{7}{13} \times 1308$ cm^{-1} (see Worked Example 3.2), or 960 cm^{-1}.

3. There is exchange with D_2O during the reaction and the CH_3 becomes CD_3.

4. About 60% of the total RNA is base-paired under these conditions (see R. I. Cotter, and W. B. Gratzer, *Nature,* 221(1969):154–156).

5. The difference spectrum is probably that of the ribosome proteins *in situ*. The contributions of protein and RNA are additive in the ribosome spectrum. This would mean that the conformation of RNA in the ribosome is base-paired (see R. I. Cotter and W. B. Gratzer, *Europ. J. Biochem* 8(1969):352–356).

6. From the data, the pK_a is 3.7. Therefore, heparin is completely ionized under physiological conditions.

7. The displacement indicates that the exchange rates of the antibody with the reduced disulfides are much faster, reflecting an increased conformational flexibility.

8. Inspection of Table 3.1 shows that the protein probably exists in the α-helical form (for further details see S. Krimm and A. M. Dwiveldi, *Science,* 216 (1982):407–408 and references therein.)

9. The band that is perpendicular polarized must be involved in *inter-molecular* hydrogen bonds. (This probably contributes to the strength of cellulose). The obvious candidate is the proton on O6. The other two bands are probably along the chain direction and must form *intramolecular* hydrogen bonds involving O5 · · · HO3 and O2H · · · O6.

10. The random motion of the polymer molecules in solution will average out the polarization effects of the transition dipoles. However, the dipolar coupling is *intramolecular* and the splitting is not averaged out by this motion.

11. (a) The reduced masses of $^{12}C^{16}O$ and $^{13}C^{16}O$ are 48/7 and 208/29, respectively. The isotope effect on the stretching frequency of $^{12}C^{16}O$ is given by the square root of the ratio of the masses, 0.98. This gives the stretching frequency for $^{13}C^{16}O$ as 2092 cm^{-1} when free and 1902 cm^{-1} when bound to Mb. (The assumption is that the CO stretching is unaffected by being attached to Mb. This is generally the case when CO is the ligand—unlike the case when O_2 is the ligand).

(b) By inspection of the figure, we see that the rates of the two isotopes binding are different, that for $^{13}C^{16}O$ being just over half that for $^{12}C^{16}O$. The difference in zero-point energies, ΔE, is given by $\frac{1}{2}h\Delta\nu$, where $\Delta\nu$ is the difference in the Fe—CO stretching frequencies of the two isotopes. To convert this difference, 5 cm^{-1}, to s^{-1}, we have to multiply by the velocity of light. Thus, $\Delta E = \frac{1}{2} \times 6.63 \times 10^{-34} \times 5 \times 3 \times 10^{10}$ J $= 5 \times 10^{-23}$ J. Using this result in exp $(\Delta E/kT)$, we calculate 1.2 as the expected kinetic isotope effect. This is much smaller than the observed isotope effect and suggests that some other mechanism might be operative (for further details, see J. O. Alben et al., *Phys. Rev. Lett.*, 44(1980):1147.)

Chapter 4

1. The most accurate measurements will be when $I_t \approx 50\%(I_0)$.

2. There must be some specific interaction in the antibody combining site, since a blue shift would have been expected otherwise. (On the basis of model studies, the red shift is assumed to arise from the interaction of Dnp with tryptophan residues in the site, since solutions of Dnp and tryptophan also give a red shift).

3. (a) This implies that the three tryptophan residues are fully exposed to the solvent.

(b) This implies that all six residues are partially exposed. This problem illustrates the ambiguities that can arise with this method, which measures only the *average* exposure. In lysozyme, (b) was found to be the case, so all six residues are partially exposed.

4. The positive denaturation difference spectra result from the destacking of bases and a decrease in hypochromicity, since the chromophores no longer interact. To estimate the number of base pairs in a given RNA, we note that the denaturation difference spectra have quite different shifts. We can then assume that each base pair contributes independently to the absorbance changes. As a result, the denaturation difference spectrum of a given RNA is characteristic of the proportion of A–U and G–C base pairs disrupted (for further details, see, for example, J. R. Fresco, L. C. Klotz, and E. G. Richards, *Cold Spring Harbor Symp. Quant. Biol.*, 28 (1963):83–93.)

5. The band in the near-ultraviolet is very intense and probably corresponds to a charge-transfer band between Co(II) and a ligand. In the visible region, the bands come from $d \rightarrow d$ transitions of Co(II). The intensities suggest that the environment is distorted from tetrahedral symmetry (see G. S. Baldwin et al., *J. Inorg. Biochem.*, 13(1980):189–204).

6. We can determine the concentration of nitrated tyrosine on the protein from the equation $A = \epsilon c l$. Given that $\epsilon = 4100 \ cm^2 \cdot mol^{-1}$, $l = 1 \ cm$, and $A = 0.154$, by substitution, $c = 3.7 \times 10^{-5} \ mol \cdot dm^{-3}$. Since the protein concentration is $4.0 \times 10^{-5} \ mol \cdot dm^{-3}$, this means that 3.7/4.0, or 0.94, tyrosine residues are nitrated. From the plot of absorbance versus pH, the pK_a of this tyrosine is about 7.3.

7. The change in OD is caused by the oxidation of NADH to NAD^+ by the pyruvate. The curve follows a single, first-order process with a constant half-life of about 36 ms. This suggests that the four bound sites for NADH are identical and independent (see R. A. Stinson and H. Gutfreund, *Biochem. J.* 12(1971):233 and H. Gutfreund, *Enzymes: Physical Principles* (New York: John Wiley, 1975) p. 209).

8. The band can be resolved into *two* overlapping ones. The separation between the two bands probably arises from exciton splitting as a result of the formation of the dimer (see K. Bauer, J. R. L. Smith, and A. J. Schultz, *J. Am. Chem. Soc.*, 88(1966):2681-2688).

9. If the structure of the sample is known, linear dichroism measurements will give the direction of the transition dipole (see Figure 4.16). Conversely if the direction of the transition dipole is known, dichroism measurements can be used to assess the orientation of a group within a particular structure. Dichroism measurements can also help to determine whether a spectral band corresponds to a single electronic transition. If there are two overlapping transitions, it is likely that they will have different dichroic ratios (see, e.g., Figure 4.18).

10. Since different bands will be expected to have different dichroic ratios, this suggests that there is only *one* transition contributing to this absorption band.

11. Viral DNA is single stranded—e.g., $\Phi174$—while the replicative must be double stranded and has an increased hypochromism in the native structure arising from a higher degree of base stacking. (In fact, all replicative DNAs are double stranded.)

12. The charge displacement corresponding to the transfer (or partial transfer) of an electron can occur over relatively large distances. The electric transition dipole moment μ_e is therefore large. The intensity of a transition depends on μ_e^2.

Chapter 5

1. The transition time is given by 1/(frequency of the transition). Its value is typically about 10^{-15} s. The relaxation time (or lifetime) characterizes how long, on average, the molecules remain in the excited state. The rate at which the populations relax back to their original distribution with all the molecules in the lowest vibrational level in the ground electronic state is $1/\tau$. The value of τ is typically nanoseconds.

2. The rate of depopulation is given by adding all the reaction rate constants

$$\frac{1}{\tau} = k_{IC} + k_T + k_C + k_Q [Q] + k_F$$

where $k_Q[Q]$ is the pseudo-first-order rate constant. All the other rate constants are for first-order processes.

3. In the photosynthetic unit, energy transfer occurs to other chlorophyll units, thus depopulating the excited state and shortening the lifetime. The fluorescence yield in solution is given by $\Phi_F = \tau/\tau_F = 7 \times 10^{-9}/25 \times 10^{-9} = 0.28$. In the photosynthetic unit, $\Phi_F = 10^{-10}/25 \times 10^{-9} = 0.004$.

4. The pH variation of the fluorescence in the native enzyme implicates a group on the protein with $pK_a = 4.5$. Presumably, the ionization of this group alters the environment of the tryptophans and the quantum yield. (Since an increase in quantum yield is observed on ionization, this *could* suggest a more hydrophobic environment—e.g., ionization of histidine to the neutral species). Modification of Trp-69 hardly reduces the fluorescence, which shows that it does not contribute very much to the total fluorescence. By contrast, the modification of Trp-177 dramatically reduces the fluorescence intensity at high pH. This suggests that the main contribution to the fluorescence comes from Trp-177. Note, too, that there is no pH dependence after modification. The ionizable group may therefore be close to Trp-177 (for further details, see G. Lowe and S. Whitworth, *Biochem. J.*, 141 (1974):503–575).

5. There is a large blue shift (more than 30 nm) and increase in quantum yield of the ANS on binding to GDH. This suggests that ANS binds in a hydrophobic site. The addition of GTP or NADH does not alter the fluorescence, so they must bind at sites distinct from ANS. The sigmoidal effect of NADH in the presence of GTP is characteristic of an allosteric transition. ANS fluorescence is sensitive to this transition and is therefore a useful probe for detecting conformational changes and binding in this system. Note that there is no NADH effect on the ANS fluorescence in the absence of GTP, which suggests that the GTP promotes NADH binding (see G. H. Dodd and G. K. Radda, *Biochem. J.*, 114(1969): 407–416).

6. The intensity of fluorescence is proportional to the intensity of the exciting light (see Worked Example 5.4).

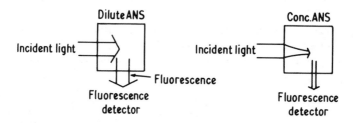

At high concentrations of ANS, the solution in the region of the cell monitored by the detector will be exposed to a lower intensity of the exciting light because of significant absorption by the intervening ANS solution (see Fig. 5.27). This is regarded as an **inner filter effect** and explains the reduction in the ANS fluorescence intensity with increasing concentration of ANS. (In fact, the behavior could be predicted by expanding the exponential in the equation in Worked Example 5.4).

7. In the fluorescence cuvette, Dnp is present in two environments—free in solution and bound to the protein. The free Dnp is farther from the protein tryptophan residues, and so it quenches less efficiently. However at high enough Dnp concentration (approximately 15 μmol·dm^{-3} here), the quenching is significant.

(This effect is often referred to as **trivial quenching**). The blank allows correction of this. When the binding is weak (see Figure 5.20(a)), the quenching by the large amount of free Dnp effectively swamps the contribution by the antibody. In practice, this limits concentrations to $<< 100 \ \mu mol \cdot dm^{-3}$.

8. From the spectra in Figure 5.21, we can measure the Tb(III) fluorescence quenched in the presence and absence of Co(II). The ratio of intensities is 0.25. Thus,

$$\frac{\phi_T}{\phi_D} = 1 - E_T, \ 0.25 = 1 - E, \ E = 0.75$$

Substituting for R_0 and E_T in $R = R_0[(1 - E_T)/E_T]^{\frac{1}{6}}$, we obtain that $R = 1.35$ nm. (This is in close agreement with the value of 1.39 nm reported from the crystal structure work (for further details, see V. G. Berner, D. W. Darnall, and E. R. Birnbaum, *Biochem. Biophys. Res. Commun.*, 66(1975):763–768).

9. (a) The tryptophan emission spectrum overlaps with the absorption spectrum of the dansyl, so there is energy transfer. The tryptophan fluorescence is therefore quenched when the enzyme substrate complex (ES) is formed, and as this complex breaks down, the fluorescence returns to its normal value. The dansyl fluorescence is enhanced initially because of the energy transfer in the ES complex and then decays as this breaks down.

   (b) The fluorescence emission of the dansyl (DNS) group overlaps the Co^{2+} absorption spectra. Energy transfer from DNS to Co^{2+} occurs and results in a 100% quenching of the DNS fluorescence. This quenching can be used to calculate the distance between DNS and Co^{2+} (if R_0 is known) (for more details on this relay energy transfer system, see S. A. Latt, D. S. Auld, and B. L. Vallee, *Biochemistry,* 11(1972):3015–3022).

10. We may assume that each Fab arm behaves as a rigid unit. The simplest interpretation of the $(Fab)_2$ data is that the value of 100 ns results from the flexibility of the Fab arms with respect to each other. Further, the presence of flexibility in the $(Fab)_2$ fragment but not the Fab would establish the site of flexibility as being in the hinge region joining them and the Fc.

11. The fluorescence lifetime (nanoseconds) provides the time scale over which we can observe dynamic events. In model I (Fig. 5.24b) the two distinct slopes correspond to those of the fully native and unfolded forms (Fig. 5.24a) which suggests that the interconversion does not occur on the nanosecond time scale. The single slope in model II (Fig. 5.24c) is intermediate between the two forms and is indicative of interconversion on the nanosecond time scale.

12. By dividing the area into eight strips, we can write

$$J = \sum_{i=1}^{i=8} f_D(\bar{\lambda}_i)\epsilon_A(\bar{\lambda}_i)\bar{\lambda}_i^4 \, \Delta\lambda \quad \text{and} \quad f_D(\bar{\lambda}_i) = \frac{F_D(\bar{\lambda}_i)}{\sum\limits_{i=1}^{i=8} F_D(\bar{\lambda}_i) \, \Delta\lambda}$$

thus,

$$J = \frac{\sum\limits_{i=1}^{i=8} F_D(\bar{\lambda}_i)\epsilon_A(\bar{\lambda}_i)\bar{\lambda}_i^4 \, \Delta\lambda}{\sum\limits_{i=1}^{i=8} F_D(\bar{\lambda}_i) \, \Delta\lambda}$$

We shall evaluate the denominator sum first.

i	$\lambda(cm)$	$F_D(\bar{\lambda}_i)(\%)$	$\epsilon_A(\bar{\lambda}_i)$
1	2.5×10^{-6}	8	2×10^{-5}
2	2.5×10^{-6}	36	9×10^{-5}
3	2.5×10^{-6}	74	1.85×10^{-4}
4	2.5×10^{-6}	97	2.43×10^{-4}
5	2.5×10^{-6}	91	2.28×10^{-4}
6	2.5×10^{-6}	77	1.92×10^{-4}
7	2.5×10^{-6}	51	1.27×10^{-4}
8	2.5×10^{-6}	37	9.25×10^{-5}

Hence the denominator term is 1.18×10^{-3} cm. Now, we shall evaluate the numerator terms (N_i).

i	$\bar{\lambda}_i(cm)$ $(\times 10^5)$	$\bar{\lambda}_i^4(cm^4)$ $(\times 10^{18})$	$\epsilon_A(\bar{\lambda}_i)$ $(mol \cdot dm^{-3})^{-1}/$ $cm^{-2} \times 10^{-4})$	$F_D(\bar{\lambda}_i)$ $(\%)$	N_i
1	4.125	2.90	0.76	8	4.41×10^{-19}
2	4.375	3.66	1.9	36	6.26×10^{-18}
3	4.625	4.58	2.9	74	2.46×10^{-17}
4	4.875	5.65	3.7	97	5.07×10^{-17}
5	5.125	6.90	3,7	91	5.81×10^{-17}
6	5.375	8.35	2.3	77	3.70×10^{-17}
7	5.625	10.01	0.95	51	1.21×10^{-17}
8	5.875	11.91	0.38	37	4.19×10^{-18}

Hence

$$J = \frac{\sum\limits_{i=1}^{i=8} N_i}{1.18 \times 10^{-3}} = \frac{1.93 \times 10^{-16}}{1.18 \times 10^{-3}}$$
$$= 1.64 \times 10^{-13} \text{ cm}^3(\text{mol dm}^{-3})^{-1}$$

Compare this with the real value of 1.84×10^{-13} cm^3 (mol $\cdot$ dm$^{-3})^{-1}$; this approximate method is seen to be fairly good.

Chapter 6

1. The aromatic resonances of tyrosine can appear as four doublets or two doublets, depending on the rotation rate of the ring about the γ-C—β-C bond (see Figure 6.41). Most proteins have a simple two-doublet spectrum because the rotation is relatively fast. This spectrum is different from those of the other three aromatic residues. The pair can be identified by decoupling experiments, where irradiation at one doublet causes the other doublet to collapse. This can be detected using difference spectroscopy. A more powerful method is to use the two-dimensional technique shown in Figure 6.21.

2. (a) From the spectra in the text, we note that the H_2O resonance is at about 5 ppm. NH resonances occur over a wide range (see Table 6.2), but typically have a value around 9 ppm. The shift between solvent and NH is thus 4 ppm, which would be a frequency of 1200 Hz on a 300-MHz instrument. This

corresponds to $\Delta\omega = 2\pi\nu \simeq 7.5 \times 10^3$ rad $\cdot$ s^{-1}. For the NH resonance to be observed, the exchange rate must therefore be less than about 10^4 s^{-1}.

(b) The relaxation rate of the ^{14}N nucleus is fast because this nucleus is quadrupolar ($I = 1$), which produces an efficient relaxation mechanism. This fast relaxation averages out the coupling. (In ESR, the ^{14}N hyperfine splitting is very large (MHz); thus, the ^{14}N relaxation rate is not fast enough to average this out.)

3. From equation 6.12,

$$1/T_2 = \frac{f_A}{T_{2A}} + \frac{f_B}{T_{2B}} + \frac{f_A f_B \Delta^2}{k_a + k_b}$$

$$= \frac{1}{3} \times 3 + \frac{2}{3} \times 2 + \frac{1}{3} \times \frac{2}{3} \times \frac{(502.6)^2}{15 \times 10^3}$$

Therefore,

$$\Delta\nu_{1/2} = 1/\pi T_2 = 6.07/\pi = 1.93 \text{ Hz}$$

4. The titrations in Figure 6.53 are characteristic for a histidine C-2 hydrogen (e.g., Figure 6.23). From the midpoint ($\delta = 8.22$) the pK_a is estimated to be 5.87 in the absence of ligand, while the pK_a in the presence of ligand is 6.2. A possible explanation for the increased pK_a in the presence of ligand is stabilization of the histidine acidic form (positively charged) by the phosphate group (negatively charged). The distance 2nm is too far for charge stabilization to be significant, therefore the effect on resonance B is presumably because of a conformational change.

5. The spectrum represents resonances from nuclei that are near the spin label in the combining site. The paramagnetic center causes a broadening of the NMR line-widths that is inversely proportional to the sixth power of the distance between the paramagnetic center and the nucleus. Thus, the most profound effect is on nuclei close to the paramagnetic center whose resonances may have become so broad that they are now undetectable in the second spectrum. However, those nuclei will show up in a difference spectrum, which in favorable cases will represent the NMR spectrum of the binding site. Note here, for instance, the high proportion of aromatic resonances in the difference spectrum, in contrast to that in the original Fv spectrum. This reflects the aromatic nature of the binding site, now firmly established by other NMR and X-ray studies.

6. We solve for r (the metal-H$_1$ distance):

$$\frac{\delta^1/\delta_{\text{ref}}}{\delta^2/\delta_{\text{ref}}} = \frac{r_2^3}{r_1^3} \frac{3 \cos^2\theta_1 - 1}{3 \cos^2\theta_2 - 1}$$

For (a) $8/1 = (r + 0.4)^3/r^3$, or $2 = (r + .4)/r$; therefore, $r = 0.4$ nm.
In (b) $1.728/1 = (r + 0.4)^3/r^3$, or $1.2r = r + 0.4$; therefore, $r = 2$nm.
Part (c) is as for (a), and (d) is as for (b) because geometric terms cancel out. This represents the range of distances over which Ln(III) shift probes are usable.

7. Using the ZnConA derivative as the diagmagnetic control, we obtain

$$1/T_{\text{1obs}} = f/T_{\text{1ZnConA}} + (1 - f)/T_{\text{1 free}}$$

from which $T_{\text{1ZnConA}} = 0.23$ s.

For the MnConA derivative, $1/0.43 = 1/T_{lobs} = 0.049/T_{1MnConA} + 1/1.08$;
therefore, $T_{1MnConA} = 0.074$ s.
Therefore, $1/T_1 = 1/T_{1MnConA} - 1/T_{1ZnConA} = 9.15$ s^{-1}.
Using this in equation 6.8 for $1/T_{1M}$, we obtain

$$9.15 = \frac{2\gamma_I^2 \mu_{eff}^2}{15r^6}\left(\frac{3\tau_c}{1 + \omega^2\tau_c^2}\right)$$

Since $\tau_c = 3 \times 10^8$ s, $\gamma_I = 0.67 \times 10^8$ rad $\cdot$ s^{-1}T^{-1}, $\omega = 2\pi \times 24.9$MHz $= 1.56 \times 10^8$ rad $\cdot$ s^{-1}, and $\mu_{eff} = 3.56 \times 10^{-30}$, we have $r = 1.04$ nm.

8. From Table 6.1, the ratio of γ for ^{1}H : ^{13}C : ^{31}P is 425.8 : 107 : 172.4 Taking ^{1}H as 1 gives us the relative sensitivity of ^{13}C and ^{31}P as 0.016 and 0.066, respectively. With the natural abundance of ^{13}C taken into account, the ^{13}C sensitivity becomes 1.76×10^{-4}. If the ^{1}H spectrum takes 1 min, the ^{31}P spectrum will take $(1/0.066)^2 = 229$ min, while the ^{13}C spectrum will take $(10^4/1.76)^2 = 3.2 \times 10^7$ min. In practice, the differences are not so great as implied, especially for ^{13}C, where the spectra are usually sharp singlets and an nOe effect is produced by ^{1}H irradiation.

9. From equation 6.13, we have for the P-creatine $\rightarrow$ ATP direction $I'_{ATP}/I_{ATP} = (1/1.5)/(1/1.5 + k_{-1}) = 0.7$; therefore, $k_{-1} = 0.67/0.7 - 0.67 = 0.29$ s^{-1}. For the ATP $\rightarrow$ P-creatine direction, $I'_{PC}/I_{PC} = 1/3.2/(1/3.2 + k_{+1}) = 0.82$; therefore, $k = .313/0.82 - 0.313 = 0.07$ s^{-1}. The relative flux is $0.07 \times 6 : 0.29 = 1.45:1$ for the forward and backward reactions.

Chapter 7

1. In (I), there are two sets of splittings: interaction with the nitrogen nucleus producing 3 lines and each being further split by the β-hydrogen into 2. In (II), the degeneracy of the two sets of splittings produces 4 lines. In (III), there are 12 lines made up as follows: 3 from the nitrogen, each further split into 2 by the β-hydrogen to give 6 lines, which are each split into 2 by the γ-hydrogen.

2. The electron is delocalized over all six carbon atoms. It will therefore interact with the six protons giving a 7-line spectrum of relative intensities 1:6:15:20:15:6:1.

3. The spectrum of phosphorylase b is that of a weakly immobilized label, i.e., fast rate and large amplitude of motion. The spectrum of creatine kinase may correspond to either (a) slow isotropic motion of the label or (b) a small amplitude of fast motion (restricted in a site on the enzyme). Since the lines are relatively broad (see phosphorylase b), it is probable that there is slow isotropic motion because the enzyme is tumbling freely in solution. The spectrum of phospho-fructokinase has two components—one moderately immobilized and one rigidly immobilized (note the outer splittings in this case). Since only one label is present, it must exist in two environments (which, in fact, correspond to the T and R states of the enzyme). On the EPR time scale, there is slow exchange between these states, so the different A values are not averaged. If the extreme values of A in the two states are taken, then the exchange rate $<< 10^8$ s^{-1}; see R. A. Dwek, *N.M.R. in Biochemistry* (Oxford: Clarendon Press, 1973).

4. (a) Cu(II) has axial symmetry, $g_{\|}$ is split by the nuclear spin of ^{63}Cu (3/2). (b) High-spin Fe(III) has axial symmetry and large ZFS (behaves as a spin system with a "fictitious" spin of $\frac{1}{2}$). (c) Low spin Fe(III) has low symmetry. (d) This looks very similar to the low-spin Fe(III), $S = \frac{1}{2}$ case. Again, the different g-values indicate that the environment of the metal ion has low symmetry. However, the different sets of g-values show that the two environments of Fe(III) are different, since the g-values also reflect the nature of the ligands attached to the Fe(III). (In general, for a given geometry of the metal ion, g-factors for low-spin Fe(II) compounds vary according to the coordinated ligands in the order $O > N > S$. (One important point here is that in ferrodoxins, the Fe atoms occur in pairs consisting of high-spin Fe(III), which has five unpaired elecrons, coupled to high-spin Fe(II), which has four unpaired electrons. This pair behaves as a single unit of $S = \frac{1}{2}$.) However, for proteins, where changes in the coordination geometry of the metal ion can occur, this order is indicative rather than diagnostic of a particular ligand coordination. (e) Mn(II), aqueous, has a g-value of 2 and is isotropic—the splitting is from the nuclear spin of ^{55}Mn ($I = \frac{5}{2}$). The ZFS is small (and not resolved), which suggests that symmetry is almost spherical. For Mn(II) in creatine kinase, the ZFS is not averaged out and all transitions except those from $\frac{1}{2}$ to $-\frac{1}{2}$ are inhomogeneously broadened. (Theoretically, the intensity of the $\frac{1}{2} \rightarrow -\frac{1}{2}$ transition is roughly one-fourth of the free signal.) (f) Rubredoxin Fe(III) with rhombic symmetry has very high g-values, which indicate that it is in the high-spin state. (It is distinguished from (b) because the ZFS is slightly smaller here, so "mixing" between the energy levels of the three doubly degenerate spin states occurs (see Figure 7.24).

5. All these contain "effectively" low-spin Fe(III). If the differences in the number of Fe atoms at the active site are considered, the differences in the EPR spectra are rather slight. The method is indicative of a particular structure rather than diagnostic. The linewidths are dominated by the short value of T_1 at ordinary temperatures (which is dominated by spin-orbit interactions). At low temperatures, T_1 increases dramatically and its contribution $(1/T_1)$ to the linewidth decreases.

6. (a) The broadening of the resonance in the ^{57}Fe-substituted spectrum arising from the ^{57}Fe hyperfine structure proves that Fe is involved in the paramagnetic center. It is possible to just make out the 1:2:1 triplet pattern on one of the peaks, which further shows that *two* Fe atoms are involved (for further details, see J. C. M. Tsibris et al., *Proc. Natl. Acad. Sci. U.S.A.,* 59(1968):959–965). (b) The nuclear spin of ^{33}S means that there is the possibility of *superhyperfine* interaction with the electron. Although this is not resolved, it can contribute to a broadening of the linewidth, as is observed here. It also shows that the S is directly coordinated to the Fe.

7. Since the motion is fast and axial, $A_{zz} = A_{\|}$, and A_{xx} and A_{yy} are averaged to give $A_{\perp}$. To plot the polarity profile, calculate a_0 each time from the equation $a_o = \frac{1}{3}(A_{\|} + 2A_{\perp})$.

	4PCSL	6PCSL	8PCSL	10PCSL	12PCSL	14PCSL
a_0 (mT):	1.54	1.53	1.47	1.46	1.43	1.43

Note that a_0 is constant up to about the 6-label, and then decreases as one proceeds toward the center of the bilayer. This indicates that the polarity decreases, presumably because water penetration is decreased, so the lipid environment toward the center of the bilayer may be quite dehydrated.

8. The A_{zz} values of haptens (I) and (II) are much larger than those of haptens (III) and (IV). Whereas the former indicate some motional restrictions, the latter are typical of near-isotropic motion of the label. From $a_0 = \frac{1}{3}(A_{zz} + A_{xx} + A_{yy})$ we can calculate that a_0 should be about 1.6 mT, as is observed for (III) and (IV)). This shows the label is out of the site and suggests that both the antibody sites have a depth less than 1.2 nm. (Note, however, that the *lateral* dimensions of the site may well be different, since haptens (I) and (II) are more immobilized in XRPC 25 than in MOPC 460. (for further details, see K. J. Willan, et al., *Biochem. J.*, 165(1977):199–206.

9. Mark out lines for the two g-values, noting that the shift difference between them corresponds to 1 mT (since $g_\perp > g_\parallel$, $g_\parallel$ must be at higher field). Each of these lines is then split into three by the ^{14}N nuclear spin ($I = 1$). This splitting results in three pairs of lines ($a_\parallel - a_\perp$) with separation of 14, 10, and 34 mT, as shown in Figure 7.32. As the molecule rotates, this anisotropy is averaged out. However, the different values of ($a_\parallel - a_\perp$) provide slightly different time scales over which the averaging has to occur. Thus, the central pair will start to collapse at rates around $2.10^8\ s^{-1}$, while the rate has to be greater than $6.10^8\ s^{-1}$ for the high-field pair. This explains why the lines of the averaged spectrum broaden differentially as the rate of rotation becomes slower. (In fact, for the linewidths, the averaging is really proportional to the square of the anisotropy, making the differential effect quite dramatic.)

Chapter 8

1. The consequences of the $\sin^2\phi$ dependence are that the scattering along the z-direction ($\sin 0° = 0$) is zero. The scattering along y is finite ($\sin 90° = 1$). The oscillating dipole in the z-direction will produce the same effect along x as along y. A similar result is obtained if the driving wave is polarized in the xy-plane, but the dependence is now $\cos^2\phi$.

For unpolarized incident radiation, the driving wave may be considered to consist of two equal components at right angles, one polarized in the zy-plane the other in the xy-plane. It therefore follows that the angular dependence in the zy-plane is $(\sin^2 90 + \cos^2\theta)/2 = (1 + \cos^2\theta)/2$.

2. The intercept on the vertical axis of a Zimm plot is $1/\overline{M}_w$; thus, this will decrease by a factor of two as the molecule dimerizes. The value of R_G will increase; thus, the slope will increase, as shown in Figure 8.23.

3. A ray diagram is shown in Figure 8.24. The rays remain parallel; thus, $OR/PQ = c_2/c_1$. But $\sin \theta_1 = PQ/OQ$, and $\sin \theta_2 = OR/OQ$. Also, $n_1 = c_0/c_1$ and $n_2 = c_0/c_2$, by definition.

Thus, $n_1/n_2 = c_2/c_1 = \sin \theta_2/\sin \theta_1$. If $\sin \theta_2 = n_1/n_2$, $\theta_1 = 90°$ ($\sin \theta_1 = 1$). This angle θ_2 is called the **critical angle**, since if it increases further, no light emerges in the medium with the lower refractive index (n_1). It is possible to devise instruments that measure critical angle and thus refractive index very accurately. This is important in some forms of scattering where it is necessary to know dn/dc.

4. $\overline{M}_w = \Sigma NM_i^2/\Sigma NM = [0.001 \times 10^6 + (40)^2]/[.001 \times 10^3 + 40] =$
$[10^3 + 1600]/41 = 63.4$ k dalton.

5. For the circle, the circumference is $2\pi r = 400$ nm; therefore, $R_G = 400/2\pi =$
63.67 nm. For the rod, $R_G = 200/\sqrt{12} = 57.73$ nm.
The equation for $P(\theta)$ is of the approximate form $1 - Q^2R_G^2/3$.
   (a) When $\theta \simeq 20°$, $Q = 4\pi \sin \theta/2/\lambda = 4.36 \times 10^{-3}\,\mathrm{nm}^{-1}$; therefore,
   $Q^2/3 = 6.3 \times 10^{-6}\,\mathrm{nm}^{-2}$.
   For the rod, $P(\theta) \simeq 0.9788$ and for the circle $P(\theta), \simeq 0.9745$.
   Thus the ratio is 1.004.
   (b) When $\theta = 90°$, $Q = 1.78 \times 10^{-2}\mathrm{nm}^{-1}$; therefore, for the rod, $P(\theta) = 0.649$
   and for the circle, $P(\theta) = 0.574$. Thus, the ratio is 1.13.

   Note the increased sensitivity at $\theta = 90°$.

6. Using the crude formula given, we estimate a rotational diffusion coefficient of
$10^{13}/5 \times 15 \times 10^6$. The relaxation time constant will thus be 7.5 μs. Therefore,
the 10-μs time constant probably corresponds to rotational diffusion of the
DNA. The slower relaxation times probably correspond to overall deformations
of the DNA coil in solution (see J. M. Schurr, and V. Bloomfield, *Crit. Rev. Biochem.*, 4(1977):371–431).

7. At 70% D_2O, the observed scattering is essentially all from protein, while at 41%
D_2O, it is from RNA. The scattering density is proportional to the volume of the
scatterer and only to percent D_2O if the exchange properties are the same
throughout. Thus, the ratio of protein volume to RNA volume is 25.9 : 3.1 =
8.35:1.

8. The ratio of molecular weights of Cl to hemoglobin is $820,000/62,000 \simeq 13.2$.
The ratio of R_G values is $12.8/2.5 = 5.12$. For spherical particles, we expect the
molecular weight to be proportional to R_G^3, thus the expected ratio for the R_G
values is $3\sqrt{13.2} = 2.36$. The value of R_G for Cl is thus about twice as large as
expected and the molecule is probably elongated.

Chapter 9

1. The spectra show that as the rat aged (life expectancy is approximately 35 mo),
most of the thiol groups were converted into disulfide bonds. It would be necessary to show that the proteins in the rat lens nucleus remained the same in amino
acid composition and secondary structure (for further details see E. J. East, R.
C. C. Chang, and N-T. Yu, *J. Biol. Chem.*, 253(1978):1436–1441).

2. Resonance Raman involves excitation of an electronic transition of a chromophore. However, absorption into the excited energy level is also probable and
emission, i.e., fluorescence can then occur (see Chapter 5). The intensity of the
emitted light will interfere with the resonance Raman spectrum. Various techniques have been used to remove the fluorescence. These include quenching of
the fluorescence by addition of I^- and making use of the much shorter lifetime
($\sim 10^{-12}$ s) of scattering compared with fluorescence ($\sim 10^{-9}$ s). If very short
laser pulses are used with synchronized photon detection, Raman photons can be
observed before most fluorescence photons reach the detector.

3. (a) The change in frequency is given by the square root of the ratio of the reduced masses. Thus $\nu_{^{18}O_2}/\nu_{^{16}O_2} = \sqrt{\mu_{^{16}O_2}/\mu_{^{18}O_2}} = \sqrt{8/9} = 0.94$.

   Therefore, $\nu_{^{18}O_2} = 0.94 \times 844 = 796$ cm^{-1}.

   (b) In (I) and (II), the symmetrical mode of O_2 binding would result in only one vibrational mode. In structures (III) and (IV), there are two possible modes of attachment of the unsymmetrical ligand, but we cannot choose between them.

4. Irradiation of the absorption band will produce the resonance Raman spectrum associated with the chromophore. If these frequencies can be assigned to tyrosine or histidine residues or to both, then these assignments will give information on the coordinated ligands.

5. The results suggest that the sulfonamide becomes ionized on binding to the protein. The extra features may indicate that a change in geometry in the inhibitor has occurred on binding (for further details, see the review by P. R. Carey, *Quart. Rev. of Biophys.*, 11(1978):309–310 and references therein).

6. The interactions in the combining sites that give rise to the shifts must be very similar in all three antibodies. If these interactions involve several residues, then it suggests that the differences in amino acid sequence do not alter the combining-site interactions with the hapten probed here.

7. The oxygen of the tyrosine is chelated to the zinc, and the azo group is in the trans conformation (for further details, see R. K. Scheule et al., *Proc. Natl. Acad. Sci. U.S.A.* 74(1977):3273–3277).

8. The resonance Raman vibrations for the Fe(III) complex are presumably associated with the coordinated tyrosine–phenolate ion. This suggests that the resonance Raman modes in the Mn(III) enzyme do arise from a coordinated tyrosine. The similarity in the vibrational frequencies probably results because the chemistries of Mn(III) and Fe(III) are very similar.

9. The Soret band is much more intense than the α-β band, so the modes that are in resonance with the Soret band will therefore be more enhanced. Another difference is that when the α-β band is stimulated, there are essentially two excited electronic states rather than one. The electronic excitations distort the molecule,

and vibrations that promote this distortion will be enhanced. If two electronic states are considered rather than one, the symmetry of the vibrations that can interact may differ. If so, the spectra will be different.

Chapter 10

1. (a) The coincidence suggests that the ellipticity changes arise largely from the chromophore itself and not from the changes in environment of residues in the protein.

   (b) In trypsin, the acyl group can probably bind in the specificity pocket, which causes the electronic perturbations that might account for the observed CD spectrum in GB trypsin. In trypsinogen, the binding site may be "underdeveloped", so that mode of binding is not possible (for further details, see M. A. Kerr, K. A. Walsh, and H. Neurath, *Biochemistry,* 14(1975): 5088–5094).

2. (a) Distinct CD bands can arise from the binding of an optically inactive ligand if the binding site is asymmetric. This results in extrinsic Cotton effects in the absorption region of the chromophore—in this case Dnp.

   (b) The difference CD spectra are similar in sign, shape, and magnitude, and demonstrate that the native or gross features of the Dnp–lysine binding site are preserved in both the Fab and Fv fragments. This also demonstrates that the antibody binding site is located in the Fv region (see D. Inbar, D. Givol, and J. Hochman, *Biochemistry*, 12(1973):4541–3).

3. The CD bands result from the Tnp ligand and are induced by electronic interactions between the ligand and the neighboring groups in the binding site. The CD difference spectrum for the intact combining site is almost identical with that on the light chain. This means that the residues responsible for the hapten asymmetry must mainly be on the light chain (see R. M. Freed, J. H. Rockey, and R. C. Davis, *Immunochemistry*, 13(1976):509–515).

4. (a) The hydrophobic peptide is predominantly α-helix. The water-soluble peptides seem to be mainly random coil, although the 1–40 peptide has some β-sheet character.

   (b) The composite spectrum matches that of the intact glycophorin A. This indicates that the consituent peptides do not interact with each other in the intact molecule in a way that affects their conformation (see Schultes and Marchesi, *Biochemistry*, 18(1979):275–279).

5. (a) From a graph of $[\theta]_{270}$ versus R, the end point is reached when $R \simeq 0.4$. This suggests that the gene-5 protein binds to the DNA along its length, at every 2.5 bases, and changes its conformation—possibly converting the DNA to single strands. (The CD spectra at 270 nm are from the DNA only.)

   (b) (i) The reversal of the CD change by Mg^{2+} probably arises because the Mg^{2+} competes for the negatively charged groups on the DNA, thus displacing the protein.

   (ii)–(iv) The effect of chemical modifications does not cause any change in the protein structure backbone, since the protein CD spectrum between 200 and 250 nm does not change. The modified tyrosines and lysines are likely to be surface ones and involved in the DNA binding site on gene 5. A possible mechanism would be interaction with positive lysines and negative DNA with interaction of tyrosines between the bases (for further details, see R. A. Anderson, Y. Nakashima, and J. E. Coleman, *Biochemistry*, 14(1975): 907–917).

6. The calculated $[\phi]$ values from the model compounds and the experimental values of HA_2, HA_3 & HA_4 are as follows:

α_4	β_4	$\beta_3\alpha_1$	$\beta_2\alpha_2$	$\beta_1\alpha_3$	HA_4	$\lambda(nm)$
−2418	+2164	+1019	−126	−1272	+1220	365
−853	+746	+346	−53	−453	+400	589

	β_3	$\beta_2\alpha_1$	$\beta_1\alpha_2$	α_3	HA_3	
	+1610	+465	−681	−1827	+2300	365
	+554	+155	−245	−645	+920	589

	β_2	$\beta_1\alpha_1$	α_2		HA_2	
	+1056	−90	−1236		+958	365
	+363	−37	−437		+342	589

From the data, HA_4 has the $\beta_3 \alpha_1$ composition. HA_2 has β_2 composition. HA_3 has an enhanced positive rotation, which is larger than the theoretical value for the β_3 linkage. The discrepancy is probably due to the formation of another chiral center that has not been allowed for in the calculation, such as hydrogen bonding within the chain. This is a limitation of this type of calculation. Interactions other than those expected can alter the $[\phi]$ value. (An advantage of the procedure is that if the primary structure is known, these interactions can sometimes be deduced; for further details see P. J. Ferrier in *The Carbohydrates: Chemistry and Biochemistry*, ed. W. Pigman, and D. Horton, Academic Press (New York: 1980), pp. 1354–1375 and D. Ashford. et al., *Biochem. J.*, 201(1982):199–208.

Note the use of ORD rather than CD in this example. This is because of the long "wings" of ORD dispersion spread over a wider range of wavelengths than CD curves; thus, even though the chromophores here absorb at about 260 nm, the ORD curves can still be detected at 589 nm.

Chapter 11

1. From Figure 11.23, the angle $\gamma = \alpha + \beta$. If the angles are small, $\gamma = h/OC + h/IC$. When the ray is parallel to the axis, $\gamma \approx h/CF$. Therefore, $1/OC + 1/IC = 1/CF$ or $1/u + 1/v = 1/f$.

2. From $eV = mV^2/2$, $V = (2eV/m)^{1/2}$. From $\lambda = h/mv$, $\lambda = h/(2meV)^{1/2} = (1.22 \times 10^{-9})/m/V^{1/2}$. When $V = 5 \times 10^4$ v, $\lambda = 0.0054$ nm. (This solution neglects relativistic effects.)

3. Shadow casting produces "shadows" of length l from an image of height d (see Figure 11.24). Since the angle of incidence of the metal from the evaporation source is approximately constant, d/l is a constant. Thus, the virus diameter is $80 \times 105/185 = 45$ nm.

4. The assembly comprises eight spheres at the corners of a cube.

5. One would expect to see structures similar to those in Figure 11.25(a) with occasional views like those in Figure 11.25(b) (In fact, the view in (a) has a helical character.)

6. Changes in fluorescence polarization suggest a change in the mobility of the fluorescein molecules in the cytoplasm. This change has, in fact, been used as an empirical test for the "activation" of lymphocytes. Activation is a sequence of biochemical events induced by the binding of certain molecules at the cell surface (see R. C. Nairn and J. M. Rolland, *Endeavour,* 5(1982):167–171).

Chapter 12

1. The volume of the reciprocal lattice that must be sampled will increase by $(0.6/0.2)^3 = 27$. Thus, about $27 \times 400 = 10{,}800$ reflections will need to be analyzed to obtain a resolution of 0.2 nm.

2. The volume of the unit cell is $6.65 \times 8.75 \times 4.82 = 280.5 \times 10^{-27}\,m^3$. The mass of the unit cell is density $\times$ volume $= 1.26 \times 2.805 \times 10^{-25} \times 10^6 = 3.53 \times 10^{-19}$ g. The mass of protein per unit cell is 50% of this value, or 1.77×10^{-19} g. Since the molecular weight is 55,000, the mass of one protein molecule is 55,000 divided by Avogadro's number (6.023×10^{23}), or 0.913×10^{-19} g. Thus, there are $1.77/0.913$, or about 2 molecules per unit cell. If there are eight asymmetric units per unit cell, the protein must have four subunits per molecule.

3. The equation relating wavelength to velocity is $\lambda = h/mv$. Thus,

$$\lambda = \frac{6.62 \times 10^{-34}\ J \cdot s(kg \cdot m^2 \cdot s^{-1})}{1.67 \times 10^{-27}kg \times 2.8 \times 10^3 m \cdot s^{-1}}$$

$$= 0.142\ nm$$

4. The scattering is proportional to the atomic number (C = 6, Hg = 80). There will be three kinds of intensity in the Patterson map: those generated by C–C products, those generated by Hg–C, products and those generated by Hg–Hg products. These will be in the ratio $(6)^2: 80 \times 6 :(80)^2$. The Hg–Hg peaks are thus more than 10 times stronger than the others and will dominate the map. The Patterson function then reduces to only three peaks—one at the origin and two others, as shown in Figure 12.33.

Molecule   Patterson function

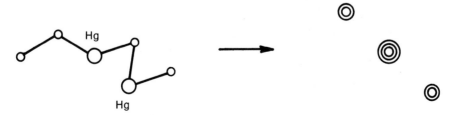

Figure 12.33 The Patterson function of a hypothetical molecule containing five carbon atoms and two mercury atoms.

This means that the Hg positions can be determined readily from the diffraction pattern.

5. If a group takes up several different positions in different molecules in the crystal, or if it moves while the diffraction data are being collected (this procedure may take many hours), then the scattering contribution will be averaged out over the crystal. This means that conformational flexibility *or* crystal packing, which allows different conformations, will lead to undetected regions of the polypeptide chain. This effect is common with lysine side chains and terminal regions of polypeptide chains.

6. Since the enzyme is still active in the crystal, the real substrate is cleaved. It is thus normally only possible to study inhibition and extrapolation must be made to the situation with substrate. Attempts are now being made to carry out X-ray analyses at very low temperatures in the hope that significant amounts of an ES complex may be accumulated to allow this to be observed directly (see, for example, M. W. Makinen and A. L. Fink, *Ann. Rev. Biophys. Bioeng.*, 6 (1977):301).

7. The G-6-P presumably causes conformational changes in the molecule that cannot be accommodated by the crystal lattice.

8. The strong density around 2.2 nm corresponds to the phosphate groups, which have a relatively high scattering cross section (P = 15). The low density around 0 nm corresponds to the terminal CH_3 groups. These have low density (H = 1) and are also probably disordered compared with the glycerol backbone. The effect of disorder can also be seen in the decrease in density in the 1.8–0.5-nm range (see, for example, N. P. Franks and W. R. Lieb, *J. Mol. Biol.*, 133(1979):469).

Index